PERSPECTIVES IN GEOMORPHOLOGY

VOLUME FOUR

By the Same Author

- Climate Change: Bio-Diversity and Green Economy with *S. Padmaja* and *Ganesh Sharma*
- Indian Geomorphology
- Mathematical Modelling in Geographical Information System Global Positioning System and Digital Cartography with *D.R. Ram*, *Rama Prasad* and *P.R. Binda*
- Modeling in Resource Management and Environment Through Geomatics with *P.R. Binda*
- Ranthambhore Sanctuary: Dilemma of Eco-Development
- Ravine Erosion in India
- Sustainable Development: Issues and Case Studies with *Sri K. Chattopadhyay*
- Tropical Geomorphology: A Morphogenetic Study of Rajasthan

About the Editor

Prof. H. S. Sharma (*b.* 1945) holds post-graduate and doctoral degrees in Geography from the University of Saugar, Madhya Pradesh. Prof. Sharma is a former Professor and Head, Department of Geography, University of Rajasthan, Jaipur. He is also a former Director of the Indira Gandhi Centre for Human Ecology, Environment and Population Studies (HEEPS), Academic Staff College and Dean, Faculty of Science, University of Rajasthan.

Prof. Sharma has over 100 research papers in National and International journals like *Nature* and 16 books to his credit. He has supervised more than three dozen Ph.D.s. Prof. Sharma has travelled widely across U.K., U.S.A., Germany, Holland, Canada, Norway, Seychelles and Malaysia to attend and chair technical sessions in International Conferences and Exchange Programmes.

Prof. Sharma has successfully conducted several research projects sponsored by the U.G.C., Ministry of Environment and Forests, DRDO, New Delhi, NORAD, Norway and ISRO (SAC) Ahmedabad.

Prof. Sharma has been President of IIG, IGI and NAGI. He is also a President of the Commonwealth Human Ecology Council (India Chapter). Recently, he has been awarded Senior Fellowship by Indian Council of Social Science Research, New Delhi to work on Climate Change and its Impact on Land and Water Resources of Rajasthan.

Concept's International Series in Geography No. 2

PERSPECTIVES IN GEOMORPHOLOGY

VOLUME FOUR

ESSAYS ON INDIAN GEOMORPHOLOGY

Edited by
H.S. Sharma

CONCEPT PUBLISHING COMPANY PVT. LTD.
NEW DELHI-110059

ISBN-13: 978-81-7022-123-4 (Set)

Edition 2016

Published and Printed by

Concept Publishing Company Pvt. Ltd.
A/15-16, Commercial Block, Mohan Garden
New Delhi-110059 (India)
T: 25351460, 25351794, *F*: 091-11-25357109
E: publishing@conceptpub.com, W: www.conceptpub.com
Editorial Office: H-13, Bali Nagar, New Delhi-110 015, India.

Cataloging in Publication Data--Courtesy: D.K. Agencies (P) Ltd. <docinfo@dkagencies.com>

Perspectives in geomorphology / edited by H.S. Sharma.
volumes cm. -- (Concept's international series in geography ; no. 2)
Reprint. Originally published: 1981-1982.
Includes bibliographical references and index.
Contents: volume 1. Recent trends -- volume 2. Quantitative fluvial geomorphology -- volume 3. Applied geomorphology volume 4. Essays on Indian geomorphology.
ISBN 9788170221234 (set)

1. Geomorphology. 2. Geomorphology--India. I. Sharma, H. S. (Hari Shanker), 1945- editor. II. Series: Concept's international series in geography ; no. 2.

GB401.5.P47 2016 DDC 551.4 23

PUBLISHER'S NOTE

THIS IS the second set of volumes in our International Geography Series. It follows close upon the heels of the five-volume set on *Perspectives in Agricultural Geography* edited by Dr. Noor Mohammad. We take this opportunity at the very outset to acknowledge the co-operation we have received from the Editor Dr. Noor Mohammad and the numerous contributors from all over the world who have made Series One possible. The wide acclaim and the heartening response from our readers—researchers and students in the field of agricultural geography—have encouraged us to plan the publication of this second set entitled *Perspectives in Geomorphology* edited by Dr. Hari Shanker Sharma. Here again we would like to thank the Editor and the contributors from various countries of the world who have given a truly international perspective to these volumes. We are confident that this series will also be equally well-received all over the world.

The Concept Series in Geography is now an on-going one and we are very happy to announce that plans are afoot to make available a third set of volumes in the very contemporary subject of Urban Geography. This series is expected to run into fifty volumes and the publication programme, hopefully, will commence early in 1982. The response to our brochure has been very encouraging and more than 150 articles have already been received including several from overseas contributors.

With the publication of the initial volumes of this new series we hope to announce the subject of the fourth series so that both students and prospective contributors will receive information in advance.

In the next few years we hope in this manner, to provide geographers with a complete compendium of the latest advances in major areas in geography. Here, we might emphasize again, that the benefits that the students in this country as well as in other developing countries will obtain will be immeasurable.

Finally, we crave once again, the indulgence of our readers for any shortcomings in these volumes. We shall learn with experience.

PREFACE

IN THE last three decades there has been considerable progress in all branches of science. The science of geomorphology has also achieved a great deal during this time, and geographers, geologists, hydrologists, pedologists and specialists from other fields who have contributed to the study of geomorphology, have become aware of the comparative suddenness of the change within the context of land form studies. As a matter of fact, it is becoming virtually impossible for a single scholar to command all aspects of geomorphology. However, there has been increasing specialization at all levels on the one hand and a need for synthesising the results of these researches, to benefit the scholars working in the field of geo-science.

The present project which has been sponsored by the Concept Publishing Company, New Delhi is an humble attempt to disseminate the knowledge of concepts, theories, models and recent trends in geomorphology with reference to the present conceptual revolution in this field. With this ambition in view original research papers dealing with modern quantitative techniques were invited from geo-scientists from all over the world. The basic themes along with the tentative titles of the papers were sent to the contributors in various branches of geo-science. The contributors have responded well with original research papers.

Later on, these papers were thoroughly edited and the papers found worthy for publication were classified and arranged into four volumes.

The first volume of the series contains the papers dealing with the modern trends in geomorphology. This volume has been divided into five sections. Section one deals with the stream lines of geomorphic research, climatic and climato-genetic geomorphology, planation surfaces, granitic inselbergs and duri-crusts. Section two discusses Karst processes and land forms. Papers dealing with glacial and peri-glacial land-forms have been included in Section three. Section four contains papers dealing with the coastal processes and land forms in tropical region. Section five contains the papers dealing with remote sensing and geomorphology.

As considerable amount of work has been done in fluvial geomorphology, therefore, volume two has been devoted to the papers dealing with the various aspects of quantitative fluvial geomorphology.

Volume three contains papers on applied geomorphology. Papers dealing with land classification, geomorphological mapping, terrain evaluation, and application of photo-interpretation and remote sensing in geomorphological mapping have been included in this volume.

Geo-scientists in India have also become aware and realised the significance of land form studies in their researches. As a matter of fact considerable amount of work has been done in various aspects of geomorphology. The fourth volume of the series contains papers on Indian geomorphology. It is earnestly hoped that these volumes will serve the purpose of enlightening the readers regarding the kind of research which is being taken up by modern geomorphologists today. As the editor of these volumes and, in addition a contributor, I count myself highly fortunate in my fellow contributors. All those whose papers are included in these volumes have prepared and supplied their material in the face of manifest difficulties—mainly the difficulties of time which pressed hard upon every busy academician. Most of my fellow contributors have patiently dealt with manuscripts and with queries relating to their papers. Those who sent their materials early have been tolerant of the subsequent delay, as late comers have tolerantly borne pressure from myself to complete their papers.

It would be ungenerous to conclude this introduction without acknowledging the guidance, encouragement and help of a number of senior scientists.

I am indebted to Professor Moonis Raza for his suggestions in making the outline of the project during my visit to the Centre for Regional Development, Jawaharlal Nehru University, New Delhi from January to April, 1980. I am highly thankful to Prof. Vaidyanadhan of Waltair University and Prof. Pramila

Kumar of Saugar University for their suggestions and encouragement. I am also thankful to Shri P.S. Dhinwa and Shri Ram Avatar Sharma for their assistance throughout the preparation of these volumes. Staff members of the Geography Department, University of Rajasthan, Jaipur, also deserve thanks for their co-operation and kindness in the preparation of this project. Special thanks are due to the publisher who has calmly suffered the repeated redistribution of authors throughout the world and who maintained effective contacts with myself. In addition I may suitably take this occasion to make a block statement of gratitude, on behalf of all the contributors, to the numerous friends and colleagues who have at various stages read part or all of the manuscripts and have supplied comments, criticisms and suggestions.

Department of Geography
Rajasthan University
Jaipur (India)

H.S. SHARMA

Independence Day
15 August 1981

CONTENTS

LIST OF CONTRIBUTORS

Jag Mohan Agrawal
Principal Civil Hydrographic Officer, Naval Hyrographic Office, Dehra Dun, (Uttar Pradesh)

K.N. Das
Reader and Head, Department of Geography, Bhagalpur University, Bhagalpur, (Bihar)

Pratap Singh Dhinwa
Research Fellow, Department of Geography, University of Rajasthan, Jaipur, (Rajasthan)

Ram Deo Gujar
Department of Geography, Rajasthan University, Jaipur, (Rajasthan)

S.N.P. Gupta
Teacher Fellow (UGC), Post-Graduate, Department of Geography, Ranchi University, Ranchi, (Bihar)

Kalyan Mal Hironi
Lecturer in Geography, D. A. V. College, Ajmer, (Rajasthan)

Anil Kumar
Reader in Geomorphology, P. G. Department of Geography, Bhagalpur University, Bhagalpur, (Bihar)

Pramila Kumar
Reader, Department of Geography, University of Saugar, Sagar, (Madhya Pradesh)

Sudershan Mukerjee
Research Officer, National Atlas and Thematic Mapping Organisation, Calcutta, (West Bengal)

J. G. Mulay
Department of Geology, University of Poona,
Pune, (Maharashtra)

Ram Niwas Pandey
Research Fellow, P. G. Department of Geography,
Bhagalpur University, Bhagalpur, (Bihar)

V.V. Peshwa
Department of Geology, University of Poona,
Pune, (Maharashtra)

Satya Prakash
Principal Civil Hydrographic Officer, Naval Hydrographic Office, Ministry of Defence, Dehra Dun, (Uttar Pradesh)

Nageshwar Prasad
Lecturer, Post-Graduate, Department of Geography,
Bhagalpur University, Bhagalpur, (Bihar)

Miss M. Uttara Prabhu
Research Scholar in Geography, Sri Venkateswara
University, Tirupathi, (Andhra Pradesh)

P.K. Roy
Geologist, Geological Survey of India, Western Region,
Bani Park, Jaipur, (Rajasthan)

P.B. Saxena
Senior Lecturer and Director of Research Project of ICSSR, Department of Geography, D.B.S. College, Dehra Dun, (Uttar Pradesh)

Hari Shanker Sharma
Career Scientist (UGC), Department of Geography,
University of Rajasthan, Jaipur, (Rajasthan)

A. Singh
Assistant Professor, Department of Geography,
P. G. College, Ambah, Jiwaji University, Gwalior,
(Madhya Pradesh)

B.V. Singh
Principal and Head, Department of Geography,
P. G. Collge, Ambah, Jiwaji University, Gwalior,
(Madhya Pradesh)

R.P. Singh
Department of Geography, Magadh University,
Bodh Gaya, (Bihar)

Surendra Singh
Geomorphologist, Central Arid Zone Research Institute,
Jodhpur, (Rajasthan)

O.P. Singh
Lecturer, Department of Geography, H. R. Degree College,
Khalilabad, Basti, (Uttar Pradesh)

Miss G. Soundaravalli
Research Scholar in Geography, Osmania University,
Hyderabad, (Andhra Pradesh)

R.P. Verma
Geologist, Geological Survey of India, Western Region,
Bani Park, Jaipur, (Rajasthan)

LIST OF PLATES

Chapter Two

Chapter Eleven

Chapter Twenty

LIST OF FIGURES

LIST OF TABLES

INTRODUCTION

IT IS paradoxical that the geomorphological studies in India have received the least attention from geographers and geologists. It is only within the last decade that the geo-scientists have realised the importance of this branch of the environmental sciences. It is now well established through experience that all national programmes of regional developments should be conformal with the natural setting and its long term equilibrium conditions. In this context the study of landforms is most vital and it is legitimate to study landforms and landforming processes in order to provide sound understanding of the physical landscape.

In the modern age of resource consciousness and planning various disciplines have not only developed their applied aspects but looked more and more providing scientific, analytical and quantitative basis for developmental projects. Quantitative analysis of a region of different scales is also a field of integrated approach for allied disciplines like Geology, Hydrology, Pedology, Forestry, Mining, Urban Planning, Military Science and Agricultural Planning etc. Quantitative analysis of landforms thus provides fundamental background for multiple developmental planning.

Geomorphological studies in India received great fillip by the works of geologists and geographers. There has been an emphasis on various aspects of Geomorphology from regional to coastal, structural, fluvial, climatic and applied Geomorphology. The introduction of aerial photographs and Landsat Imageries has given an impetus to landform studies. The Geological Survey of India, National Remote Sensing Agency, Space Applications Centre, ISRO, Ahmedabad, Indian Institute of Earth Sciences, Trivandrum, Central Arid Zone Research Institute, Jodhpur are specialised agencies which are dealing with landform studies. The University Departments, Department of Geography, University of Andhra, Waltair; Department of Geography, University of Ranchi; Department of Geography, Magadh University; Centre for Advanced Studies in Geomorphology and Structural Geology, Saugar University, Sagar; Department of Geography, University of Rajasthan, Jaipur are also conducting geomorphological studies. Two symposiums on Geomorphology at Saugar in 1965 and at Delhi in 1978 are indicative of greater concern of geographers and geologists in this newly developed geo-science.

Regional Geomorphology

Most of the early studies in Geomorphology in India has been done in the field of regional geomorphology with an emphasis to establish the denudation chronology and recognition of erosion surfaces. Almost all the geomorphic regions of India particularly Chota Nagpur plateau (Singh, R.P. 1956-79, Chatterjee, S.P. 1940); Madhya Pradesh; (Pramila Kumar 1972, Dixey and Choubey 1971); Deccan plateau, Rajasthan (Heron, 1938, Pramila Kumar 1960, Sen, D. 1972, Sharma 1969-72,76,78); Cuddapah Basin of Andhra Pradesh (Vaidyanadhan 1978); Tamil Nadu and partially the Himalayas (Kaushik 1965, Bose, S.C. 1967); and Indo-Gangetic plain have been covered by geologists and geographers.

Quantitative Basin Studies

Recent emphasis is on quantitative geomorphology of drainage basins by various morphometric techniques and measures to establish the interrelationship of basin parametres. Studies of Gupta, Arijit and Chakraborti, S.C. (1965); Ghosh, Bimal Pandey, Singh and Gheesa Lal (1967); Sharma, Vijay (1968); Singh, R.P. and Kumar, A. (1969); Singh, Labh (1970 & 1971); Kharkwal, S.C. (1970 & 1971), Pal, S.K. (1972), Verma, V.K. and Tandon, S.K. (1971 & 1974); Agrawal, Meera (1972); Sidhu, G.S. and Pandey, I.C. (1974), Anil Kumar (1973-1975), Sharma, H.S. and Padmaja, G. (1977; 1978); Srivastava, Renu (1978), Singh, Savindra (1978 & 1980), Subramanyam, V. (1976); Singh, R.L. (1967) are worth mentioning on basin morphometry.

Ravine Morphology

The study of ravine erosion is one of the significant aspects of geomorphology. In recent years ravine geomorphology has made significant advances particularly in the European countries and in the United States of America. In India a significant contribution in this branch of geomorphology has been made by Gorrie (1957), Tajwani and Ahuja (1956), Ahmad (1968, 1970), Sharma (1968, 1976, 1980), Kaul (1967), Mishra and Bhola (1967), Bali (1972) and Gadkary, and Rama Rao (1955). Ahmad, Enayat (1968) and Sharma, H.S. (1968, 1980), have given a new dimension to the problem of ravine genesis and suggested that ravine erosion in different ecological settings in India is an outcome of lowering of local base level of erosion caused by the cymatogenic warping of the Peninsular shield as the side of the last phase of the Himalayan Orogeny.

Arid Processes

The Central Arid Zone Research Institute, Jodhpur, has been carrying out systematic studies on arid processes and forms. Significant studies by the scientists working here have been made on sand dune cycle, grain size analysis, former drainage lines, palaeo-flood plains and pediments etc.

Coastal Processes

Significant studies have been done by Indian geo-scientists on coastal forms and processes. Ahmad has unfolded for the first time the characteristics, classification and distribution of coasts in India in his book *Coastal Geomorphology*. Systematic studies on the evolution of coastal landforms in Krishna, Godavari and Mahanadi deltas have been made by Vaidyanadhan (1978, 1979) and his team. The comparative study of toposheets, naval hydrographic charts, aerial photographs of different years spread over about half a century of these river delta fronts have brought to light changes in coastal landforms. Of particular importance are the spits which grow into bars enclosing lagoons. Further addition of detritus brought by the distributaries of river Krishna fills in the lagoons thus resulting in the progradation of the delta. The studies of Niyogi (1968), Mahadevan (1958), Ranapathi,

Venkatarama (1958) and Bapu (1975, 1978) are also worth mentioning on coastal geomorphology.

Slopes

Though the considerable work on various aspects of geomorphology have been done but the studies on slopes, glaciated topography and palaeo-climatic studies are still lacking. On glaciated topography studies of Bose, S.C. (1976) and Kaushik (1965) are significant. The problem of slope evolution in India has not been thoroughly studied. Few studies of Pandey (1968), Rai (1971), Subramanyam (1976), Sharma and Padmaja (1977, 1978) and Anil Kumar (1978) are also significant. Studies on palaeo-geomorphology with the help of Landsat Imageries is a new introduction. CAZRI, Jodhpur research scientists and a team of scientists of pre-history and archaeology of Poona are some workers who are concentrating in this branch of geomorphology. The work on UGC project on morphogenetic regions in Rajasthan : A study in climatic Geomorphology is progressing well by the author himself. On completion it would unfold a number of mysteries of climatic geomorphology of Rajasthan.

The present volume contains twentythree original research paper written by experts in the field of geomorphology. The paper of Singh, R.P. deals with the "Progress of Geomorphological Research in India" since very beginning. Singh has stated that significant research has been done in different branches of Indian Geomorphology.

The paper by Sharma, H.S. deals with the "Morphology of Ravines of the Morel Basin, Rajasthan." He has analysed ravines at a micro-level and discussed geomorphic and hydrological factors that are pertinent to ravine and gully erosion. While discussing various stages of gully growth he has attributed gully erosion to the lowering of the local base level of streams caused by recent upliftment.

The paper of Das, K.N. discusses the "Fluvial Processes in Kosi catchment and their impact on morphology of flood plain". The paper is divided into two sections. In Section A he describes general characteristics of the catchment including extent, geology, seismic character, relief conditions, glaciers, rainfall, fluvial erosional and depositional features. In Section B he has described the topographic, fluvial and hydraulic conditions of Kosi catchment in Bihar. On the basis of his study he has concluded that the fluvial character of the Kosi has a direct bearing to the problem of floods in this region.

The paper of Saxena, P.B. and Satya Prakash attempts to determine the stage of landform development and denudation chronology of the Nayar Basin in Garhwal Himalayas through the analysis of morphometric determinants.

The paper by Nageshwar Prasad on "Some Aspects of Meandering Streams of the Barakar Basin and their Sinuosity Indexes" deals with the sinuous nature of streams. He concludes that the streams of Barakar Basin are sinuous of the low SSI, and they in the maturity or in the middle stage between youth and old stage of the cycle of erosion. The entrenched character of the meandering streams leads to the conclusion that the region has experienced rejuvenation which has increased the dominating influence of the topographic sinuosity, and thereby slackened the development of the flood plains reducing the amount of the hydraulic sinuosity index (HSI).

The paper of the same author deals with the geomorphological analysis of the stream line surfaces of the Barakar Basin on the basis of this study he has deduced the evolution of exacting planation of surfaces.

The paper on "Morphometric Evaluation of Landforms in Palamau Upland Region" by Singh, O.P. attempts to evaluate the landforms of this region on the basis of Weaver's method. The area, thus has been divided into landform units and their details have been enumerated in the paper.

The paper by Hironi, Kalyan Mal attempts to trace the evolution of antecedent course of the Banas at Rajmahal. This study supports the views of Heron, A.M.

The paper by Verma and Roy on "Paleo-drainage studies of the Lower Luni Basin" makes an interesting case history to understand the role of aerial photo studies and geomorphology in studying the problems connected with floods in such tracts. The study has also been substantiated by field evidences.

The paper on "Isostatic Readjustment, their Rejuvenation and Fluvial Geomorphology in the Chambal Basin" by Singh, B.V. and Singh, A. is an attempt to analyse the tectonic features and their impact on fluvial morphology of the Chambal System.

The study also upholds the earlier views expressed by Sharma, H.S. (1968-1970) in his study on Geomorphology of the Lower Chambal Valley.

The northwestern part of Indian sub-continent is arid and semi-arid. The area is characterized by a chain of sand dunes and sand-sheets. The paper by Surendra Singh attempts to study "The types and formation of sand dunes in Rajasthan desert". The study is based on aerial photographs and Landsat imageries substantiated by intensive field-study. Singh has distinguished various types of sand dunes like simple, compound and complex. The author has also discussed the origin of various types of sand dunes. The factors responsible for the formation of these dunes are winds of sufficient strength and duration, adequate source of sand supply, bio-physical barriers, wind-deflation and processes of sand transportation. The significant relationship between the wind regime and dune orientation has been thoroughly analysed. The study contradicts the earlier views that the source of dune sands is Runn of Kutch and establishes that it has been locally derived from the existing landforms and of Aeolian origin.

The north eastern part of Indian sub-continent has not been extensively studied from geomorphological point of view. The paper by Mukerjee, Sudarshan on "Structural Geomorphology of Nagaland" is a welcome study which fills the lacuna in Indian Geomorphology. In his study Mukerjee has discussed the structural framework of Nagaland and concludes that besides being time and process, structure is inherently a very important factor in landscape evolution.

The paper by Pramila Kumar on "Diastrophic Forces and their Geomorphic Expression in Ranchi Plateau" is a significant contribution to geomorphology. On the basis of her study she concludes that diastrophic forces working in the geological past have left their imprints on Ranchi Plateau and to some extent they are responsible for the present height of the plateau as well as for the present base level of erosion for the river of the area.

The paper of Singh, O.P. also deals with the intermittent uplifts of Palamau upland in Tertiary period : A part of Chota Nagpur Highland. The author has also tried to examine the applicability of Penck and King's views for interpretation of the various erosion surfaces in the area.

Aerial photographs have proved to be significant tools for geomorphic study. The paper by Mulay and Peshwa indicates that though aerial photographs are of great help in geomorphological studies, they don't eliminate the need of either field work or the use of topographic sheets.

The paper on "Quantitative analysis of relief of Hazaribagh plateau region" by Anil Kumar and Pandey, R.N. attempts to use various morphometric techniques like—area altitude relationship, hypsometric curve, frequency histogram, intensity of relief, relative relief and dissection index, etc.

The next paper by the same authors analyses "The salient characteristics of slope profiles in Hazaribagh Plateau". On the basis of 6 slope profiles and applying Young's method, the evolution of slopes have been analysed.

The paper on "Study of Slopes in the Berach Basin, Rajasthan" by Dhinwa, P.S. deals with hillside slopes, regional slope, Strahler's slope, height curves and histogram and factors and processes that give rise to slope development. On the basis of this study the author has differentiated the development of slope on Pre-Aravalli and Vindhyans. The Vindhyan slopes exhibit all the four elements of slope while scarp face is missing on the Pre-Aravallis.

In India very little effort has been made to study landslides. The paper by Saxena, P.B. on the study of landslides in Alakananda Valley is a new study which attempts to analyse various causes of landslides and other mass movement processes. The paper also discusses the zoning of landslides and their preventive measures.

The paper on "Inselberg topography" by Uttara Prabhu and Soundaravalli attempts to examine the spatial distribution of inselbergs in Tamil Nadu in the light of existing morphogenic environment. The study reveals that although lithological control is in evidence locally in the inselberg landscape of Tamil Nadu, it is rather difficult to differentiate zoning of this topography on a map at regional level.

As has been referred earlier that coastal landforms have attracted attention of geo-scientists. The paper on "morphology of the submarine terrain of the North Indian ocean and its geonomic potentiality" by Satya Prakash and Saxena, P.B. is indicative of the increasing importance of the Indian ocean. The paper discusses the complexity of the Indian Ocean. Of the three major world oceans the Indian Ocean have now become the arena of power rivalry amongst the super world powers due to its geo-strategic and geonomic potentiality. The study further analyses the morphology of sub-marine terrain which is creditable to perform the marine cycle of resource transformation. The authors have also outlined briefly the conservational measures of the marine resources for the economic planning of the maritime state.

The Arabian Sea is the northwestern part of Indian Ocean and is pivot of interest of various countries of Middle East and Indian Sub-continent. This part of the Indian Ocean has become an arena of power rivalry among the super world powers due to its expected petroleum potentials on the Bombay High and Kori High, proximity to the Gulf countries, its link with Atlantic Ocean through Suez Canal and the growing geopolitical conflicts in the vicinity. The paper by Satya Prakash and Agrawal discusses the "Morphometry of North East Arabian Sea from Kosi High to Bombay High". The study is based on the bathymetric data collected during the recent ocean expeditions particularly "Oceanovax Cruises" by Indian Naval Ship *Darshak* in 1973-74.

The paper by Ram Dev Gujar discusses some aspects of geomorphology of Tonk district (Rajasthan).

SUGGESTED READINGS

Agrawal, Meera, 1972 : Morphometric analysis of some drainage basins of Jabalpur plateau, *The National Geographical Journal of India*, Volume 18, pp. 140-168.

Ahmad, E., 1968 : Distribution and causes of gully erosion in India, Paper presented in *the 21st International Geographical Congress*, New Delhi, Section IV, Complex Physical Geography.

Ahmad, E., 1970 : *Proceedings of Symposium on Erosion Surfaces, 21st International Geographical Congress India*, 1968, National Committee for Geographers, Calcutta.

Babu, P.V.L.P., 1975 : Morphological evolution of the Krishna Delta, Photonirvachak, *Journal of Indian Sociological Photointer*, Volume 3, No.1.

Babu, P.V.L.P., 1978 : Geomorphic evolution of the Mahanadi delta, *Proceedings of Symposium on Morphology and Evolution of Landforms*, Department of Geology, Delhi University, Delhi, pp. 216-221.

Bali, 1972 : *Ravine Reclamation Needs for India.*

Bose, S.C., 1967 : *Geomorphology of the Glaciated Regions of the Western Himalaya*, Seminar on geomorphological studies in India, Department of Applied Geology, University of Saugar, Sagar, pp. 19-29.

Chatterjee, S P., 1957 : Geomorphic regions of east Rajasthan, Pathar, *Geographical Review of India*, Volume 19, Part 4, pp. 42-47.

Chhiber, H.L., 1954 : The development of landforms in the Himalayas, (Presidential Address of Indian Science Congress of Geological and Geographical Society), *Bulletin of National Geographical Society of India*, No. 20, pp. 17-31.

Choubey, V.D., 1971 : *The Narmada-Sone Line Thrust : The Great Boundary Fault along the Southern Margin of the Vindhyan Basin, Central India*, W.D. West Felicitation Volume, University of Saugar, Sagar, pp. 420-438.

Dixey, F. and West, W.D., 1971 : *The Geomorphology of Madhya Pradesh, India*, Felicitation Volume, University of Saugar, Sagar, pp. 195-224.

Gadkary and Rama Rao, 1955 : Reclamation of Kothar lands, *Journal of Soil and Water Conservation in India*, Volume 3, pp. 150-160.

Ghose, B., Pandey and Singh, 1967 : Quantitative geomorphology of the drainage Basins in Central Luni Basin in Western Rajasthan, *Annals of Geomorphology*, New Folge Band 11, Heft 2, Sciten, 146-160.

Gorrie, 1957 : Torrent types for recent corrections, *Journal of Soil and Water Conservation in India*, Volume 5, pp. 133-138.

Gupta, A. and Chakraborty, S.C., 1965 : Quantitative analysis of the landforms, *Proceedings of the Seminar on Geomorphic Studies in India*, University of Saugar (M.P.), pp. 190-199.

Heron, A.M., 1938 : The Physiography of Rajputana, Presidential address, *Proceedings of the Indian Science Congress*, 25th Session, Calcutta.

Kaul, 1967 : Management of Chambal ravines in Rajasthan, *Indian Forester*, Volume 88, pp. 725-730.

Kaushik, 1965 : Rejuvenation of the fluvial geomorphology of the Central Himalayas, *Proceedings of the Seminar on Geomorphological Studies in India*, pp. 92-96.

Kharkwal, S.C., 1968 : Morphometric study of Himalayan Basin : A sample study, *National Geographical Journal of India*, Volume 16, Part 1, pp. 47-60.

Kharkwal, S.C., 1971 : Slope studies on Himalayan terrain, *The National Geographical Journal of India*, Volume 15, Part 1, pp. 1-15.

Kumar, A, 1973-75 : Pair-wise relationship of Basin area and stream length : A case study of the Upper Burha Basin, *Geographical Outlook*, Volume 10, pp. 49-58.

Kumar, A., 1978 : Some observations on slope profiles in the Pat region, Ranchi, *Proceedings of the Symposium on Morphology and Evolution of Landforms*, Department of Geology, pp. 41-54.

Kumar, Pramila, 1972 : Has Deccan foreland rejuvenated? *The Geographer*, Volume 19, pp. 34-46.

Misra, R. and Bhola, 1967 : *Economics of Ravine Afforestation with Special Reference to Chambal Ravines*, Kota.

Niyogi, D., 1968 : Morphology and evolution of Subarnarekha delta, India, *Saertryk of Geografisk Tidsskrift*, Volume 67, pp. 230-241.

Pal, S K., 1972 : A Classification of morphometric methods of analysis, *Geographical Review of India*, Volume 34, Part I, pp. 61-85.

Pandey, S., 1968 : Evolution of hillslopes around Jodhpur, *National Geographical Journal of India*, Volume 14, Part 1, pp. 32-40.

Rai, R.K., 1980 : *Geomorphology of the Sonar Bearma Basin*, Concept Publishing Company, New Delhi.

Sen, D., 1972 : Geomorphology of the Aravalli Range, Rajasthan and a reinterpretation of residual erosion surfaces, *Geographical Review of India*, Volume 34, No. 3, pp. 265-286.

Sharma, H.S., 1968 : Genesis and pattern of ravines of the Lower Chambal Valley, *Geographical Review of India*, Volume 30, No. 4, pp. 14-24.

Sharma, H.S., 1970 : Micro-Geomorphic regions of Lower Chambal Valley, *Indian Geographical Journal*, Volume 45, pp. 1-14.

Sharma, H S., 1970 : Geomorphology of the Ramgarh Dome, Rajasthan, *Geographical Review of India*, Volume 32, No. 3, pp. 186-194.

Sharma, H.S., 1972 : Recognition of erosion surfaces in the Lower Chambal Valley (India), *International Geography*, Volume 1, Montreal, Canada.

Sharma, H.S., 1973 : Geomorphic evidence of rejuvenation in Lower Chambal Valley, Paper presented before 6th All Rajasthan Geographers Conference at Jaipur.

Sharma, H.S., 1976 : The problem of ravine erosion and its effects on land-use pattern in the Lower Chambal Valley, *Essays in Applied Geography*, Ali Memorial Volume, Saugar University, Sagar, pp. 155-182.

Sharma, H.S., 1976 : Morphology of the ravines of the Moral Basin, Rajasthan, *International Geography*, Moscow, USSR, Volume 1.

Sharma, H.S. and Padmaja, G., 1977 : Quantitative geomorphic characteristics of streams of the moral Basin, Rajasthan, *Geographical Review of India*, Volume 39, No. 4, pp. 353-366.

Sharma, H.S. and Padmaja Rao, G., 1978 : Evolution of hillside slopes of the Moral Basin (Rajasthan), *Proceedings of the Symposium on Morphology and Evolution of Landforms*, Delhi University, Delhi, pp. 177-188.

Sharma, H.S., 1979 : *Physiography of the Lower Chambal Valley and its Agricultural Development*, Concept Publishing Company, New Delhi.

Sharma, H.S., 1980 : *Ravine Erosion in India*, Concept Publishing Company, New Delhi.

Sharma, H.S. and Dhinwa, P.S., 1981 : Some aspects of geomorphology of the Menali Basin, Rajasthan, Paper presented before II annual conference of NAGI, held at Tirupati, in January 1981 (sent for publication in Professor R.P. Singh's felicitation volume).

Sharma, V., 1977 : Preliminary investigation of pleistocene valley glaciation in part of Kangra District, Himachal Pradesh, *Journal of Geological Society of India*, Volume 18, No 1, pp. 23-30.

Sindhu, G.S. and Pandey, I.C., 1974 : Parametric analysis of Patelgad and drainage basin east of Mussourie, *The National Geographical Journal of India*, Volume 20, Part 4, pp. 223-235.

Singh, Labh, tr., 1970 : Basin circularity ratio as a technique (terrain type element) A case study of Hoshiarpur-Kangra Tract, The *Deccan Geographer*, Volume 8, pp. 119-129.

Singh, O.P., 1976 : Slope studies of Palamau uplands : A quantitative approach, *Indian Geographical Research Bulletin*, No. 6, pp. 15-23.

Singh, R.L., 1967 : Morphometric analysis of terrain, *National Geographical Society of India Bulletin*, No. 22, pp. 1-24.

Singh, R.P., 1958 : Structure drainage and morphology of the Chota Nagpur Highlands, *Geographical Outlook*, Ranchi, Volume 2, No. 3, pp. 23-32.

Singh, R.P. and Kumar, A., 1969 : Geomorphological evolution of stream orders of Topa and Silphi basin, Ranchi, *The National Geographical Journal of India*, Volume 15, Part I, pp. 38-44.

Singh, R.P., 1960 : Geomorphological evolution of highland of Rajmahal, *National Geographical Journal of India*, Volume 6, Part 1, pp. 1-13.

Srivastava, Renu, 1978 : Morphometric study of the open and closed links of the Belan Basin, *Proceedings of the Symposium on Morphology and Evolution of Landforms*, Department of Geology, Delhi University, Delhi, pp. 75-82.

Singh, S. and Srivastava, Renu, 1977 : A statistical analysis of the average slopes of Belan Basin, *The Deccan Geographer*, Volume 15, No. 2, pp. 307-316.

Singh, S. and Srivastava, Renu, 1975 : A quantitative study of the relief of the upper reaches of the Belan Basin, *National Geographer*, Volume 10, pp. 27-28.

Singh, S., 1978 : Physiographic regions, landforms and erosion surfaces of the Ranchi Plateau, *National Geographer*, Volume 30, No. 1, pp. 43-65.

Subramanyan, V., 1976 : Hill slope analysis of the Deccan traps around Sagar, Madhya Pradesh, *Geological Survey of India Symposium on Deccan Traps and Bauxites*, Pune.

Tandon, S.K., 1971 : A study of the stratigraphic frequency distribution of various III order basin parameters in a part of Kumaon Himalayas, *National Geographical Journal of India*, Volume 17, Parts 2 and 3, pp. 127-133.

Vaidyanadhan, R. and Rama Rao, K.L.V. 1977 : Recognition, correlation and age of erosion surfaces in the iron ore ranges of Bihar and Orissa in *Proceedings of Symposium on Geomorphology of Indian Shield*, 21st International Geographical Congress, pp. 33-34.

Vaidyanadhan, R. and K. Nageshwar Rao, K.L.V., 1978 : Geomorphic features of the Krishna delta and its evolution, *Proceedings of the Symposium on Morphology and Evolution of Landforms*, Department of Geology, Delhi University, Delhi, pp. 120-130.

Verma, V.K. and Tandon, S.K., 1971 : Stream [fortuosity : A new method of visual estimation *Z. Geomorph. N.F. Suppl, Bd.*, Berlinstuttgart, Volume 12. pp. 165-171.

West, W.D. and Choubey, V.D , 1964 : The Geomorphology of the country around Saugar and Katanji, M.P. : An example of superimposed drainage, *Journal of Geological Society of India*, Volume 5, pp. 41-55.

ONE

R. P Singh

PROGRESS OF GEOMORPHOLOGY IN INDIA

THE STUDY of geomorphology has received scant attention in India. No detailed study of the geomorphology of India or its parts was made till 1956 although the country possesses significant chunks for making advance studies of geomorphology.

Growth and development of geomorphology in India is rather late. It was initiated by administrators, explorers and geologists as registration of physical description in course of their field studies. In fact, scientific explanation of the origin of landforms has commenced after 1940 with works of eminent scholars belonging to the field of geology and geography. Geologists have to be credited for shaping geomorphology in India not unlike other countries. Amongst geologists whose contributions are immensely valuable, mention may be made of Heron, Wadia, Dunn, West, S.C. Chatterjee, Auden, Arogyaswamy, Radha Krishnan, Dey and others. At an early stage initiative was also taken up by geographers like S.C. Chatterjee (1945 and 1948), H.L. Chibber (1948) and R.P. Singh (1950 and 1954) on 'Some aspects of the geomorphology of the Ranchi Plateau, 'Physiographic evolution of Chota Nagpur'; 'Classification and Origin of Plateaus with special reference to India'; 'Hydrography of Bihar'; and 'Landscape of Rajroppa' respectively. Also, a pioneering, study of landforms has been made by R.P. Singh.

Geomorphological studies in India received great fillip by the works of geologists and geographers since 1956. There is an emphasis of various aspects of geomorphology, from regional to coastal, fluvial, structural, climatic and applied geomorphology. Equipped with adequate geological memoirs, records and maps, topographical maps and aerial photographs, a significant research has been conducted in different branches of Indian geomorphology.

As the present work is in itself a work of regional geomorphology, a brief review of researches done in the field is essential particularly of those related to the Peninsular India.

Regional geomorphology integrates both qualitative and quantitative approaches of geomorphological studies. Various methods have to be applied to construct the stages of landform or surface evolution on the basis of quality maps. Such studies again incorporate the analysis of process, variables of landscape and their interrelationship. They greatly depend upon field studies to supplement the conclusions drawn from maps. Much depends on reasoning and evaluation of data at each step of investigation. The quantitative approach emphasises measurements and collection of numerical data.

Most of the work in our country has been done in the field of regional geomorphology with an emphasis to establish the denudation chronology of the country and recognition of erosion surfaces. R.P. Singh (1956) made a significant contribution in the field of regional geomorphology through his pioneer study of the geomorphology of the Chota Nagpur Highlands. He has been since then continuously engaged in geomorphological studies of the Peninsular India, Rajmahal Highlands, Shillong Plateau and other areas. Mention can also be made of the contributions made by E. Ahmad (1958), Bagchi and Sengupta (1958).

Almost all physiographic regions of the Peninsular India particularly Chota Nagpur Highlands, Madhya Pradesh, Andhra Pradesh including Vindhyans and the Aravallis, Deccan Plateau, Rajasthan, Southern Plateaus and Peninsular India have been wholly or partially covered. Recent emphasis is on quantitative geomorphology of drainage basins by various morphometric techniques and measures to establish the interrelation of basin parameters. Few recent studies related to channel characteristics, rills and gullies, meanders, floods, slope profiles and characteristics in the fields of fluvial geomorphology and slope analysis are encouraging. Application of morphometric techniques in classification of landforms into morphological units of various order has been initiated by R.L. Singh (1967). The use of comparable morphometric variables in classification of morpho-units of terrain has encouraged a host of young scholars to apply morphometric symbols in their respective areas of regional study. A review of geomorphological studies reveals that the range of its studies in India is now becoming wide and extensive. Recent researches in regional geomorphology incorporates the study of the processes modifying landforms.

Contributions made by the geomorphologists in India have been mentioned earlier but an attempt is being made to focus the contributions made by some of the outstanding geomorphologists.

Singh, R.P.'s Contribution to Geomorphology in India

R.P. Singh (1957) precisely and distinctly traced the stages of structural growth of Chota Nagpur. Singh (1957) analysed landscape evolution of southeast Chota Nagpur. The work is based on intensive field study and mapping of the area. Singh (1958) reviewed structure drainage and morphology of Chota Nagpur Highlands with a comparative note on King's cycles established in South Africa. Singh (1960) advanced geomorphic research by making an analysis of Geomorphological Evolution of Highlands of Rajmahal. Singh (1961) applied the knowledge of geomorphology in treating Bokaro—the Future Steel Town. Singh (1966) depicted changing landscape of Chota Nagpur throwing light on the effect of physical landscape on cultural landscape.

Singh (1966) made a critical review of the various methods of geomorphological mapping. Singh (1966) also presented the concept of relict feature as interpreted primarily in geomorphology. Singh (1967) presented a lively review of denudation chronology of the Himalayas, Singh (1967) turned his attention towards Peninsular geomorphology by outlining stages of structural growth of the Peninsular India. Singh (1967) established for the first time the Landscape Cycles of the Peninsular India. Singh (1968) treated successfully geomorphology of Shillong Plateau of Assam. Singh (1968) presented a learned Presidential address, Indian Council of Geographers, Varanasi on Morphology of the Indian Peninsula. He states that the story of the morphology of the Indian Peninsula consists of several cycles of denudation, sedimentation and igneous activities in harmony with orogeny, epeirogeny, and cymatogeny, effusion, metamorphism of deep seated rocks, tearing eustatism and widespread resurrection. He concludes by stating, "Some of its specific chapters have been outlined, but many are to be sorted out from its mutilated, amputated limbs and accentuated, buried and covered configuration to fill up hyphens and blanks of its chapters". Singh 1968 traces structure processes and stages of geomorphology of the Indian Peninsula. He concludes in the following words : "Topographical, erosion surfaces, unconformities, discordance, nicks, escarps, corrugations, faults, rifts, berms, uplifted peneplains, tilted landscape, warped countryside and many other bewitching topographic expressions like waterfalls, gorges, rapids punctuating the senile or matured landscape await scientific and systematic study".

Singh and Kumar (1969) analysed the geomorphological evolution of stream orders of the Tobpa and Silphi basins in Ranchi.

Singh (1969) outlined concept of tilting giving examples from India. Singh (1970) made a unique treatment of topography and towns in the Chota Nagpur Highlands by tracing the significance of rocks and surfaces in the distribution of urban settlements. Singh (1971) reviewed the concept of periglacial landscape illustrating it by Block diagrams. Singh (1972) distinctly added a new chapter of denudation chronology in Indian Peninsula. In fact this paper was presented before Geomorphology section of International Geographical Union, New Delhi, 1968. Singh (1974) traced effusion and landscape in India from Pre-Cambrian to the Tertiary. Each chapter of landscape has been carved out in relation to evolving drainage plains.

Singh (1975) made a significant analysis of drainage and landscape resulting from warping. Singh (1975-76) presented the geomorphological evolution of the ranges and basins of Cuddapah crescent.

Singh (1976) analysed structure, drainage, and morphology of Pat country in Bihar. Singh (1976) also traced glaciated landscape in the Indian Peninsula. Singh (1977) prepared a critical survey of geomorphology and its recent trends. Prof. Singh's other research works which have not been published can be consulted in the abstracts of Geology and Geography Section of All India Science Congress. Singh (1979) delivered a lecture on applied geomorphology with special reference to the Chota Nagpur Highlands. He threw light on the role of geomorphology in the spheres of economic geology, hydrology, communication settlements, etc. He also delivered next lecture in the Summer School, Jaipur on the Surface history of the Peninsular India. He systematically carved out various chapters to give a precise knowledge of the peninsular surface history. Both the lectures are being published in the Proceedings of Summer School held in July-August 1979 in Geography Deptt. of Jaipur University.

Review of Contributions of other Workers in Geomorphology in India

Ahmad, Enayat has devoted himself in carrying research also in the field of geomorphology. Ahmad (1958) reviewed some of the earlier works and presented his own views in his published paper Geomorphic Outline of Chota Nagpur. Ahmad (1958) has treated the geomorphology of the Peninsular India by tracing the landscape characteristics of different segments of the Indian Peninsula. Ahmad & Devi P. (1965) traced the origin of Chota Nagpur scarps. Ahmad (1970) reviewed the erosion surfaces in the Proceedings of Symposium on Erosion Surfaces 21st I.G.C. India, 1968. Ahmad (1975-76) traced geology of outer Himalayan, Himachal and in 1977 presented Geology, structure and geomorphology of Trans Himalayan, Nepal. Ahmad has unfolded for the first time the characteristics, classification and distribution of coasts in India in his book *Coastal Geomorphology*. Ahmad (1979) presented two lucid papers in the Summer School held in Geography Department Jaipur University in the field of applied geomorphology. He distinctly pointed out the role of geomorphology in the spheres of communication. His papers are being published in the Proceedings of the Summer School of the above institution.

Ahmad, F. (1966) lucidly documented post-Gondwana faulting in Peninsular India and its bearing on the tectonics of the sub-continent.

Ahmad, S.A. (1975) described the formation of central plain of Manipur.

Agrawal, M. (1972) presented morphometric analysis of some drainage basins of Jabalpur Plateau. Anatapadmonasam, N. (1967) traced the shifting of divides and stream captures in the Western Ghats. Arunachalam, B. (1972) treated some salient geomorphic features of South Ratnagiri.

The study of landform elements in relation to settlement has been attempted in India by Asthana, V.K. (1968); Singh, K.N. (1968); Kharakwal (1969); Agrawal, M. (1971); Singh, O.P. (1976); Gupta, B.P. (1977) and Singh, S.B. (1977); Yadava, S.R.S. (1979), Bagchi, K. and Sengupta, A.K. (1958) divided Ranchi Plateau into physiographic regions. Bagchi, K. (1960) made notable contribution on the drainage patterns in Ganga Delta. Bandyopadhyay, M.K. and Bandyopadhyay, G. (1971) presented a paper on Geomorphological Peculiarities of the Kusi River Basin in Meghalaya in the symposium on Humid Tropics. Basu, S. (1976) has outlined some of the problems in the formation of meander in the Bhagirathi. Betal, H.R. (1970) has identified slope categories in Damodar valley while in 1973 he mapped Frazerganj coast. Bedi, N. (1978) has thrown light on the geomorphology of the Mahi River Basin. Bhattacharya, R. and Das, P. (1973) wrote a geomorphological account of Kaliajhari River Basin.

Bose, S.C. (1967) presented a paper on Geomorphology of the Glaciated Regions of the Western Himalayas in the Seminar on Geomorphological studies in India at Sagar and in 1972 he traced the denudation chronology of the East and West Coasts of the Deccan Plateau. Budel, J. traced some of the relief types in Madras.

Chatterjee, S.C. (1945) threw a light on some important aspects of the geomorphology of the Ranchi Plateau. In 1946 he traced physiographic evolution of Chota Nagpur and in 1966 he brought significant aspects of the Deccan Traps.

Chatterjee, S.P. (1940) threw a light on the gneissic topography of the Ranchi Plateau. Chatterjee, S.P. and Bagchi, K. (1957) described the characteristics of geomorphic regions of East Rajasthan Pathar.

Chhibber, H.L. (1953) traced the landforms, life and radial drainage of Mount Parasnath. He in 1954 in his presidential address, Geology and Geography Section lucidly expressed the development of landform in the Himalayas.

Choubey, V.D. (1967) presented an interesting study of the erosion surfaces in the Sagar, Damoh, Jabalpur and Narsinghpur district, Madhya Pradesh in the seminar on geomorphological studies in India at Sagar. Choudhary, L.N. (1973) outlined the laterites of West Bengal. Das, P. (1975) gave an appraisal of flood incidence of the Kaliaghai river basin. De Souza, N.G. (1972) enumerated geomorphic features of the laterites of Goa. Dhabriya, S.S. (1967) presented a geomorphological study of the Udaipur Valley. Dhillon, S.S. (1977) outlined the impact of landform on agricultural land use in Mahasu district. Dikshit, K.R. (1970) distinctly described polycyclic landscape and the surfaces of erosion in the Deccan Trap country with special reference to upland Maharashtra. Dikshit, K.R. and others (1972) noted some geomorphological observations in South Konkan. Devi, P. (1968) describes scarps of Chota Nagpur. Dube, R.S. (1968) made a study of erosion surfaces in the Rewa Plateau. Dube, R.S. (1970) advanced the study of erosion surfaces on the Rewa Plateau. Dutt, G.K. (1970) traced the evolution of landforms in the lower Son valley. Ghori, G.K. (1967) treated the geomorphology of the Mysore Plateau. Ghose, B. Pandey and S. Lalji (1966) described geomorphology of the Central Loni Basin. Ghose B. and Pandey, S. (1967) also treated geomorphology of Ahor Block. Gupta, S.C. (1972) notes topography of Mahabaleswar Plateau.

Kar, A. Bandhyopadhya (1974) traced mechanism of rills. Kayearkar, M.V. and Wadhawan, S.K. (1970) presented a geomorphic classification of terrain. Khan, M.N. (1967) described physiographic divisions of the Lower Indus Valley. Kharkwal, S.C. (1968) classified Kumaon Himalaya into morpho-units noting their characteristics. Kharkwal, S.C. (1970) made a morphometric study of a Himalayan Basin : A sample study. Kharkwal, S.C. (1971) also presented slope studies in the Himalayan Terrain. Krishnan, M.S. (1953) gave a brilliant account of the structural and tectonic history of India. Kulkarni, D.G. and Sinha, B.N. (1972) traced the morphological evolution of landform in Lower Malaprabha—Krishna Doab.

Kumar, A. (1969-1971) made a geomorphic interpretation of drainage and relief of Eastern Ranchi Plateau. Kumar, A. (1972) treated with grace the rock, slope and denudation chronology of Eastern Ranchi Plateau. Kumar, A. (1972) made a morphological analysis of soil in S.W. Ranchi Plateau. Kumar, A. (1974) also classified landform regions of S.W. Ranchi Plateau. Kumar, A. (1973-75) made pair-wise relationship of basin area and stream length of the upper Burha Basin. Kumar, A. (1978) made some observations on slope profiles in the Pat Region, Ranchi. Kumar, A. and Pandey, R.N. (1975) noted morphometric properties of the Polygenetic Drainage Basin. Kumar, A. and Pandey, R.N. (1977) made a systematic numerical analysis of small drainage basins of Hazaribagh Plateau Region. Kumar, A. and Singh, N.P. (1978) presented a quantitative classification of landform regions of the Western Higher Plateau of Bihar.

Kumar, P. and Rai, R.K. (1972) has made observations regarding rejuvenation of Deccan foreland. Mithal, R.S. (1968) traced the physical and structural evolution of the Himalayas.

Mukhopadhyay, S.C. (1968) traced the erosion surfaces in the Sanjay River Valley. Mukhopadhyay, S.C. (1968) described some aspects of surface water hydrology of the lower Subarnarekha basin with special reference to its problems of flood control. Mukhopadhyay, S.C. (1969) also described some aspects of geomorphology of the Subarnarekha basin around Mahali Murup, Bihar. He in 1970 gave an account of the geomorphology of some residual hills, Burus Dungris of upper Subarnarekha basin. Mukhopadhyay, S.C. (1973) made an analysis of river terraces of the Subarnarekha basin. Mukhopadhyay, S.C. (1977) described geomorphology of a part of Lower Kharkai Basin.

Munsi, S.K. and Das, R.K. (1967) traced the evolution of the topography in Saranda, south Singhbhum. Niyogi, D. (1970) enumerated the morphology of the terraces of the Subarnarekha River, India. Padmaja, G. (1975) dealt some aspects of quantitative drainage characteristics of the Dhund Basin. Pal, S.K. (1973) analysed quantitative geomorphology of drainage basins in the Himalayas. Pal, S.K. and Bagchi, K. (1975) described recurrence of floods in Brahmaputra and Kosi Basins. It reflects a study of climatic geomorphology. Pandey, J.N. (1978) describes geomorphology of Kaimur Plateau.

Pandey, R.N. (1979) made a morphometric analysis of the geomorphology of the Damodar Valley Region. Pandey, S. (1966) describes geomorphology of granite landforms in India's arid zone. Pandey, S. (1968) makes analysis of the evolution of hill slopes around Jodhpur. Pandey, S. (1971) deals some aspects of geomorphology of Indian arid zone. Pandey, S.N. (1978) emphasized the significance of drainage patterns and anomalies in the geomorphic evolution of the northern Jabalpur District. Pofali, R.M.; Hirekenur, L.R. and Bhattacharjee, J.C. made geomorphological studies in the vicinity of Nagpur. Pathak, M. (1967) made a study of the landforms in the Jethian Regions, Gaya District. Pathak, M. (1970) analysed the geomorphology of Chhechhari Basin. Pradhan, P.N. and Sinha, B.N. (1973) made a morphometric analysis of the Rapti Valley. Prakash, A. (1976-77) gave evidences of the origin and evolution of the Tertiary scarps of the Pat region with special reference to the Jaspur Pat. Prasad, N. (1973) made a study of the geomorphic subregions of the Barakar Basin. He also presented a geomorphic treatment of Barakar Basin and was awarded D.Lit., degree from the Ranchi University in 1979.

Radhakrishna, B.P. (1969) made a deep study of the Western Ghats of the Indian Peninsula. Rai, R.K. (1969) presented a *Geomorphology of the Sonar-Bearma Basin*, (M.P.). Ram, L.N. (1968) determined uplift through profile study in south-east Chota Nagpur. Ramamurthi, K. (1972) described Maharashtra Plateau Region of the Deccan Traps. Ramana Rao, K.L.V. and Vaidyanadhan, R. (1974) noted the significance of laterites in the recognition of Planar surfaces in the Keonjhar Region. Rao, G.V. (1967) described geology and geomorphology of the Ṣon Valley area, M.P. Rode, K.P. (1964) traced the geomorphology and evolution of the Rajasthan Desert.

Saran, A. (1958) traced the evolution of drainage system in the Damodar Basin. Satpathi, D.P.P. (1970) drew a broad outline of geomorphology of Singhbhum. Satpathi, D.P.P. (1972-73) made a quantitative analysis of landform illustrating study from the Deo River Basin, Singhbhum. Satpathi, D.P.P. (1975-76) identified the landscape cycles of Singhbhum. Satpathi, D.P.P. (1976-77) made a systematic study of meander machanism, geometry of meanders and sinuosity indexes of the major rivers of Singhbhum. Sen, D. (1972) not only described geomorphology of the Aravalli Range, Rajasthan but also made reinterpretation of residual erosion surfaces. Sen, R.N. (1977) treated recent aspects of the geomorphic evolution of Bandi River Basin, Eastern Rajasthan. Sen, S. (1956) noted changes in drainage conditions in the lower reach of the Damodar. Sengupta, A.K. (1957) described physiography of the Pat countries of the Western Ranchi Plateau. Sengupta, A.K. (1970) discussed denudation in Central Ranchi Plateau, Bihar. Sharma, H.S. (1967) made morphometric analysis of the Jaipur Region, Rajasthan. Sharma, H.S. (1968) traced the evolution of the Chambal scarps. Sharma, H.S. and Padmaja, G. (1977) noted quantitative geomorphic characteristics of streams of the Morel Basin, Rajasthan. Sharma, H.S. and Rao, G.P. (1978) analysed the evolution of Hill side slopes of the Morel Basin, Rajasthan. Sharma, H.S. (1970) explained geomorphology of the Ramgarh Dome, Rajashan. Sharma, S.N. (1968) discussed the problem in drainage development in western Chota Nagpur. Sharma, V. (1973) described post-glacial terraces of the Weed and Willow Creeks of Central Alberta, Canada. Sharma V. (1977) made a preliminary investigation of Pleistocene Valley Glaciation in part of Kangra District, Himachal Pradesh. Shrestha, R.K. (1976) showed impact of altitude and slope on land use in Khandabari, Nepal. Singh, G.N. made a systematic study of Rajmahal Highlands. Singh, N.P. (1975) made a critical study of the geomorphology of the Western Higher Plateau of Bihar for which he was awarded Ph. D. degree. Singh, R.L. (1967) made a morphometric analysis of Terrain. Singh, S. and Srivastava, R. (1974) made a morphometric study of the tributary basin of upper reaches of the Belan River. Singh, S. (1977) traced the evolution of Granito-gneissic tors and cut-off spur mounds at Ranchi Plateau. Singh, S. (1976) prepared a note on the quantitative parameters of the computation of drainage density. Texture and frequency of a part of Ranchi Plateau. Singh, S. (1978) described physiographic regions, landforms and erosion surfaces of the Ranchi Plateau. Singh, K.N. made a study of landforms and settlement distribution in Southern Uplands of Eastern Uttar Pradesh. Singh, O.P. (1976) has done work on *Palamau Uplands : A Study in Landforms and Settlement*. Singh, S.B. (1977) made a study of landforms and settlements of Simla and its environs. Sharma, K., has made critical study of the geomorphological evolution of Kaimur Plateau.

Sinha, N.K.P. (1968) made a commendable study of the geomorphic evolution of northern Repununi, British Guiana, Sircar, P.K. (1953) described river basins and geomorphological regions. Subba Rao, S. (1970) noted the landforms of the Deccan Traps. Shukla, M.K. and Verma, V.K. (1973) accounted for the occurrence of waterfalls around Pachmarhi, Hoshangabad District, Madhya Pradesh. Subramanyam, V. (1967) edited the proceedings of the Seminar on geomorphological studies in India. Tamaskar, B.G. (1959) treated geomorphology and human settlement of the Jabra Basin. Jewari, A.K. (1967) made a study on some aspects of geomorphology of Jaunsar Himalayas.

Vaidyanadhan, R. (1967) gave an outline of the geomorphological history of India south of N. Latitude 18°. Vaidyanadhan, R. and Ramana Rao, K.L.V. (1972) made recognition, correction and age of erosion surfaces in the iron ore ranges of Bihar and Orissa. Vaidyanadhan, R. (1977) traced recent advances in geomorphic studies. Verma, P. (1958) presented a systematic study of Ranchi Plateau, its geomorphology and human settlements. Verma, P. (1960) made a study of the geomorphology of Udaipur Region. Verma, P. (1961) described diastrophic forces and their geomorphic expressions in Ranchi Plateau. West, W.D. and Choubey, V.D. (1964) made a neat study of the geomorphology of the country around Sagar and Katangi, M.P. They also provided an example and discussion of superimposed drainage in the said country.

Yadava, S.R.S. (1979) has studied geomorphology of *Hazaribagh Plateau—a Study in Landforms and Settlement*.

SUGGESTED READINGS

Acharya, S., 1967 : Structure and stratigraphy control in the geomorphic evolution of daiteri range, Orissa, *Seminar on geomorphological studies in India*. Dept. of Applied Geology.

Agrawal, M., 1972 : Morphometric analysis of some drainage basins of Jabalpur plateau, *Nat. Geog. Journal India*, Vol. XVIII, part 3 & 4, pp, 148-168.

Agrawal, Meera, 1971 : Morphometric evolution of landforms and settlements in Jabalpur and its environs, Ph.D. thesis (Unpub.) B.H.U. Varanasi.

Ahmad, Enayat, 1958 : Geomorphic outline of Chota Nagpur, *Geog. Outlook*, Vol. II, No. III, pp. 16-22.

Ahmad, E. and Devi, P., 1965 : Origin of Chota Nagpur scarps, *Geog. Outlook*, Vol. IV, pp. 81-85.

Ahmad, E., 1970 : *Proceedings of Symposium on Erosion Surfaces*, 21st. Int. Geog. Cong. India, 1968, Nat. Comm. for Geog., Calcutta.

Ahmad, E., 1975-76 : Geology of outer Himalayan, *Himachal Geog. Outlook*, Ranchi.

Ahmad, E., 1977 : Geology, structure and geomorphology trans, *Himalayan Nepal, Geogr. Outlook.*

Ahmad, E., 1966 : Post Gondwana faulting in Peninsular India and its bearing on the tectonics of the sub-continent. *Annals of the Geology Dept.* Aligarh Muslim University, Vol. II, pp. 1-64.

Anatapadmonavham, N., 1967 : Shifting of divides and stream captures in the Western ghats, Bombay, *Geog. Mag.* Vol. XV, No. 1, pp. 53-63.

Anjorin, E.A., 1972 : Geomorphological evolution of the Cavery Kabani doab, in *Proc. of Sym. on Geomo. of Indian Shield*, 21st Int. Geog. Con. Calcutta.

Arunachalam, B., 1972 : Some salient geomorphic features of South Ratnagiri Region. *Proc. of Symp. Geomo. of Indian Shield*, 21st. I.G.C., Nat. Com. Geog. Calcutta, pp. 27-32.

Asthana, V.K., 1967 : Morphometric evolution of landforms in Almora and its environs., *Nat. Geog. Ju. India*, Vol. XIII, No. 1, pp. 37-54.

Bagchi, K. and Sengupta, A.K., 1958 : Physiographic regions of the Ranchi plateau, Bihar, *Geog. Rev. of India*, Vol. 20, pp. 127-131.

Bagchi, K., 1960 : Drainage pattern in Ganga Delta, *Geog. Rev. of India*, Vol. 22, No. 4, pp. 51-53.

Bandyopadhyay, M.K. & Bandyopadhyay, G., 1971 : Geomorphological peculiarities of the Kulsi River Basin in Meghalaya, India, *Proc. Sym. on Humid Tropics*, 21st. I.G.C., 1968, Nat. Com. Geog., Calcutta, pp. 8-11.

Basu, S., 1976 : Some fundamental problems of meander formation with special reference to the Bhagirathi River, *Geog. Rev. of India*, Vol. 38, 1, pp. 54-67.

Baulig, H., 1928 : *Le Plateau Central de la France*, Colin, Paris.

Betal, H.R., 1970 : Identification of slope categories in Damodar Valley. India, *Selected Papers*, Vol. 1, (Physical Geography) 21st. I.G.C. 1968. Nat. Comm. Geog., pp 4-6.

Betal, H.R., 1973 : Geomorphological mapping Frazerganj Coast., *Geog. Rev. of India*, Vol. 35, No. 2, pp. 19-115,

Bedi, N., 1978 : Geomorphology of Mahi river basin in Gujarat, India, *Symp. on Morphology and Evolution of Landforms*, Deptt. of Geology, University of Delhi.

Bhattacharya, R. & Das, P., 1973 : A Geomorphological account of Kaliajhu River Basin., *Geog. Rev. of India*, Vol, 35, No. 4. pp. 365-75,

Bose, S.C., 1967 : Geomorphology of the glaciated regions of the western Himalayas, *Seminar on Geomorphological Studies in India*, Deptt. of Applied Geology., Univ. of Saugar, Sagar, pp. 19-29.

Bose, S.C., 1972 : Denudation chronology of the east and west coasts of the Deccan plateau, *Proc. of Symp. on Geomo. Indian Shield*, 21st. Int. Geog. Cong. 1968., Nat. Comm. Geog. Calcutta, pp. 7-10.

Budel, J., 1965 : Die Relief typen der flachenspulzone sub-Indians am ostabfall dekans geoger Madras, *Colloquium Geographicum*, Vol. 8.

Chatterjee, S.C., 1945 : Some aspects of the geomorphology of the Ranchi plateau, *Cal. Geog, Rev*. vii, pp. 31-35.

Chatterjee, S.C., 1946 : Physiographic evolution of Chota Nagpur, *Cal. Geog. Rev*. vol. VIII, No. 3-4, pp. 39-42.

Chatterjee, S.C., 1966 : The Deccan traps, *Bhu-vidya*, vol. 27.

Chatterjee, S.P., 1940 : Gneissic Topography of the Ranchi plateau, *Cal. Geog. Rev.*, vol. III, pp. 45-48.

Chatterjee, S.P. & Bagchi, K., 1957 : Geomorphic regions of east Rajasthan pathar, *Geog. Rev. Ind.*, vol. 19, part 4.

Chhibber, H.L., 1953 : The landforms, life and radial drainage of Mount Parasnath, Hazaribagh Dist., Bihar, *Bul. No. 18 of Nat. Geo. Soc. of India*, pp. 11-51.

Chhibber, H.L., 1954 : The Development of landforms in the Himalayas, (Presidential Address of I.S. Con. Geol. and Geog. Sec.), *Bull. No. 20. of Nat. Geog. Soc. of India*, pp. 17-31.

Choubey, V.P., 1967 : A study of the erosion surfaces in the Sagar, Damoh, Jabalpur and Narsinghpur District, Madhya Pradesh, *Seminar on Geomorphological Studies in India,* Deptt. of Applied Geology. University of Saugar, Sagar (M.A.), pp. 164-173.

Choudhary, R.N., 1973 : Laterites of West Bengal, *Geog. Rev. of India*, vol. 35, No. 1, pp. 61-72.

Das, P., 1975 : Appraisal of flood incidence of the Kaliaghai river basin, *Geog. Rev. of India*, vol. 37, No. 1., pp. 73-80.

De Souza, N.G., 1972 : Geomorphic features of the laterites of Goa in *Proc. of Symp. on Geom. of Indian Shield*, 21st Int. Geog. Con., Calcutta., pp. 22-26.

Dhabriya, S.S., 1967 : A Geomorphological study of the Udaipur valley, *Seminar on Geomorphological Studies in India,*, Dept. of Applied Geology., Univ. of Saugar, Sagar, pp. 42-48.

Dhillon, S.S., 1977 : The impact of landform on agricultural landuse in Mahasu district., *Geog. Rev. of India*, vol. 39, No. 2, pp. 183-92.

Dikshit, K.R., 1970 : Polycyclic landscape and the surfaces of Erosion in the Deccan trap country—with special reference to upland Maharashtra, *Nat. Geog. Ju. of India*, vol. XVI, part 3 & 4, pp. 236-52.

Dikshit, K.R.; Raja Guru, S.N.; Ghate, N.S. & Jog. S.R., 1972 : Some geomorphological observations in South Konkan, *Nat. Geog. Ju. Ind,*. vol. XVIII, p. 1-4.

Devi, P., 1968 : Scarps of Chota Nagpur, *Geog. Outlook.*, vol. V, pp. 81-93.

Dube, R.S., 1968 : The Study of erosion surfaces in the Rewa Plateau, Presented at the Int. Geog. Cong. New Delhi.

Dube, R.S., 1970 : Erosion surfaces on the Rewa plateau, M.P. India. *Selected Papers*, vol. 1, (Physical Geography) 21st I.G.C., 1968, Nat. Com. Geog. Calcutta, pp. 35-42.

Evans, P., 1970 : *The Tertiary Plateau of Bihar* : *Studies in Earth Sciences, West Commemoration*, Today, Tomorrow, Pub. New Delhi, pp. 195-224

Ghori, G.K., 1967 : Geomorphology of the Mysore Plateau, *On Seminar on Geomorphological Studies in India*, Univ. of Sagar, pp. 15-18.

Ghose, B., Pandey, S., 1966 : Geomorphology of the central Luni Basin, western Rajasthan, *Annals Arid zone*, vol. V., No. 1.

Ghose, B. & Pandey, S., 1967 : Geomorphology of Ahor Block, *Seminar on Geomorphological Studies in India*, University of Sagar, pp. 71-77.

Gupta, S.C., 1972 : Notes on topography of Mahabaleswar plateau, *Proc. of Symp. on Geom. of Indian Shield*, 21st I.G.C., Nat. Comm. Geog., Calcutta, pp. 61-63.

Kar, A., Bandhyopadhya, 1974 : Mechanism of rills : An investigation in micro-geomorphology, *Geog. Rev. of India*, vol. 36, No. 3, pp. 204-215.

Kayear, M.V. & Wadhawan, S.K., 1970 : A Geomorphic classification of Terrain, *Selected Papers*, Vol. 1 (Phy. Geog.) 21st I.G.C. 1968, Nat. Comm. Geog. Calcutta, pp. 68-71.

Khan, M.N., 1967 : Physiographic Divisions of the Lower Basin, *Seminar on Geomorphological Studies in India*, Dept. of Applied Geology, Univ. of Sagar.

Kharkwal, S.C., 1968 : Classification of Kumaon Himalayas into morph-units., *Nat. Geog. Ju. of India*, vol. XIII, Pt. 4, pp. 223-38.

Kharkwal, S.C., 1969 : *Landform and Settlements in Nainital and its Environs*, Ph.D. thesis, B.H.U. (Unpublished).

Kharkwal, S.C., 1970 : Morphometric study of a Himalayan Basin—A sample study, *Nat. Geog. Ju. of India.*, vol. XVI, p. 1 pp. 47-60.

Kharkwal, S.C., 1971 : Slope studies in the Himalayan Terrain, *Nat. Geog. Ju. of India*, vol. XVII., Pt., 3 p. 1.

Krishnan, M.S., 1953 : *The Structural and Tectonic History of India*, vol. 81, pp. 1-109.

Kulkarni, D.G. & Sinha, B.N., 1972 : Morphological evolution of land-form in lower Malaprabha—Krishna Doab, *Proc. of Symp. on Geomorphology*, Indian shield, 21st I.G.C., Nat. Comm. Geog, Calcutta, pp. 41-44.

Kumar, A., 1969 : *Geomorphology of Simdega and its adjoining area*, Bihar, Ph.D. thesis (in press), Bhagalpur Univ.

Kumar, A., 1971 : Geomorphic interpretation of drainage and relief of eastern Ranchi plateau, Uttar Bharat. *Bhoogol Patrika*, vol. VII, No. 1, pp. 18-33.

Kumar, A., 1972 : Lock, slope and denudation chronology of eastern

Ranchi plateau—Uttar Bharat, *Bhoogol Patrika*, vol. VIII, No. 2, pp. 91-100.

Kumar, A., 1972 : Morphological analysis of soil in S.W. Ranchi Plateau, *Geog. Prog. Bull.*, No. 1, pp. 73-80.

Kumar, A.,, 1974 : Morphological classification of landform regions of S.W. Ranchi Plateau, *Ind. Geog. Res. Bull.*, No. 2, pp. 11-16.

Kumar, A., 1973-75 : Pair-wise Relationship of Basin Area and stream length—A case study of the Upper Burha Basin, *Geog. Outlook*, vol. X, pp. 49-58.

Kumar, A., 1978 : Some observations on slope profiles in the Pat—Region, Ranchi, *Symp. on Morphology and Evolution of Landforms*, Dept. of Geology, University of Delhi.

Kumar, A. & Pandey, R.N., 1975 : Morphometric properties of the polygenetic drainage basin, *Geog. Rev. of India*, vol. 37; No. 4, pp. 329-338.

Kumar, A., & Pandey, R.N., 1977 : Numerical analysis of small drainage Basins of Hazaribagh plateau region, *Geog. Bull. Ind.*, vol. 2, pp. 1-13.

Kumar, A. & Singh, N.P., 1978 : Quantitative classification of landform regions of the western higher plateau, Bihar, *Geog. Rev. of India*, vol. 40, No. 4, pp. 378-385.

Kumar, P. & Rai, R.K., 1972 : Has Deccan foreland rejuvenated, *The Geographer*, vol. XIX, pp. 34-46.

Mithal, R.S., 1968 : The physiographical and structural evolution of the Himalayas, the mountains and rivers in India, *Nat. Comm. for Geog.*, Calcutta.

Mukhopadhyay, S.C., 1968 : Erosion surfaces in the Sanjay River Valley (The Subarnarekha system), Bihar, 21st session. *Int. Geog. Congr.* Proc., 5-25 Ranchi University, India, pp. 56-60.

Mukhopadhyay, S.C., 1969 ; Some aspects of geomorphology of part of the Subarnarekha basin around Mahali Murup, Bihar, *Geog. Rev. of India*, vol. 31, No. 2, pp. 33-40.

Mukhopadhyay, S.C., 1970 : Geomorphology of some residual hills; Burus Dungris of the Upper Subarnarekha basin, Bihar, *Orbit*, Calcutta, pp. 7-12.

Mukhopadhyay, S.C., 1970 : Erosion surfaces in Sanjay Valley in Bihar, India, *Proc. Symp. on Erosion Surfaces*, 21st I.G.C., 1968, Nat. Comm. Geog. Calcutta, pp. 56-61.

Mukhopadhyay, S.C., 1973 : River terraces of the Subarnarekha Basin. *Geog. Rev. of India*, Vol. 35, No. 2, pp. 155-168.

Mukhopadhyay, S.C., 1974 : *Geomorphology of the Subarnarekha basin with a special reference to its cycles of erosion*, Ph.D (Thesis) unpublished.

Mukhopadhyay, S.C., 1977 : Geomorphology of a part of Lower Kharkai Basin, *Geographical Review of India*, *Geogr. Soc. of India*, Calcutta, Vol. 39, No. 3, Sept. 1977, pp. 248-258.

Munsi, S.K. & Das, R.K., 1967 : Evolution of the topography in Saranda,

South Singhbhum, *Seminar on Geomorphological studies in India*, Deptt. of applied Geology, University of Saugar, pp. 158-163.

Niyogi, D., 1970 : Morphology of the terraces of the subarnarekha river, India, *Selected Papers*, vol. I (Phy. geog.) Nat. Com. Geog. Calcutta pp. 84-88.

Padmaja, G,, 1975 : Some aspects of quantitative drainage characteristics of the Dhund Basin, *Geog. Rev. of India*, vol. 38. No. 2, p. 158.

Pal, S.K., 1973 : Quantitative geomorphology of drainage basins in the Himalaya, *Geog. Rev. of India*, vol. 35., No. 1., pp. 81-101.

Pal, S.K., & Bagchi, K., 1975 : Recurrence of floods in Brahmaputra and Kosi Basins : A study in Climatic Geomorphology, *Geog. Rev. of India*, vol. 37, No. 3, pp. 242-248.

Pandey, J.N., 1978 : Geomorphology of Kaimoor plateau, Ph.D. thesis (unpub.) Magadh Univ., Bodh Gaya.

Pandey, R.N., 1979 : Geomorphology of the Damodar Valley Region, Ph.D. thesis (unpublished) Bhagalpur University.

Pandey, S., 1966 : Geomorphology of Granite Landforms in India's Arid Zone, *Ind. Jr. Geog.* Vol 1, No. 1.

Pandey, S., 1966 : Evolution of Hillslopes Around Jodhpur, Nat. *Geog. Jr. India*. vol. XIV, p. 1, pp. 32-40.

Pandey, S., 1971 : Some aspects of geomorphology of Indian arid zone, *Proc. of Symp. on Arid Zone*, 21st. Int. Geog. Cong. 1968, Nat. Com. Geog. Calcutta, pp. 28-32.

Pandey, S.N., 1978 : Significance of drainage patterns and anomalies in the geomorphic evolution of the northern Jabalpur Dist. *Symp. on Morphology* and *Evolution of Landforms*, Univ. of Delhi.

Pofali, R.M., Hirekenur, L.R. & Bhattacharjee, J.C., 1973 : Geomorphological studies in the vicinity of Nagpur, *Geog. Rev. of India*, vol. 35, No. 3, pp. 289-98.

Pathak, M., 1967 : A study of the landforms in the Tethian Regions, Gaya district, Bihar, in *Seminar on Geomorphological Studies in India*, University of Saugar, pp. 49-55.

Pathak, M., 1970 : Geomorphology of Chhechhari Basin, India, *Proc. Symp. on Erosion Surfaces*, 21st. I.G.C. 1968, Nat. Comm. Geog., Calcutta, pp. 49-51.

Pradhan, P.N. & Sinha, B.N., 1973 : Morphometric analysis of the Rapti-valley, *Nat. Geog. Jr. Ind.*, vol. XIX, p. 3 & 4. pp. 204-211.

Prakash, A., 1976-77 : The Pat Regions of Chota Nagpur, evidences of the origin and evolution of the tertiary scarps with particular reference to the Jaspur Pat, *Geog. Outlook*, Vol. XII, pp. 111-120.

Prasad, N., 1973 : Geomorphic Sub-Regions of Barakar Basin, *Ind. Geog. Stud. Research Bull.* No. 1, pp. 50-57.

Radha Krishna, B.P., 1967 : The western ghats of the Indian Peninsula, *Seminar on Geomorphological Studies in India*, Dept. of Applied Geology. Univ. of Sagar, pp. 4-14.

Rai, R.K., 1969 : Geomorphology of the Sonar-Bearma Basin (M.P.), Concept Publishing Company, New Delhi.

Ram, L.N., 1968 : Determination of uplift through profile study in South East Chota Nagpur, *Transaction*, Vol. 5, pp. 117-126.

Ramamurthi, K., 1972 : Maharashtra plateau region of the Deccan Traps, *Proc. of Symp. on Geom. of Indian Shield*, 21st. Int. Geog,, Nat. Comm. Geog., Calcutta, pp. 55-60.

Ramana Rao, K.L.V. & Vaidyanadhan, R., 1974 : Significance of laterites in the recognition of Plamar surfaces in the Keonjhar Region, Eastern India, *Geog, Rev. of India*, Vol. 32, No. 2, pp. 93-101.

Rao, G.V., 1967 : Geology Geomorphology of the Son Valley area, M.P., *Seminar on Geomorphological Studies in India*. Dept. of applied Geology, Univ. of Sagar, pp. 56-63.

Rode, K.P., 1964 : Geomorphology and evolution of the Rajasthan Desert, *Proc. Jen. Symp. Probl. Ind.* Arid zone Jodhpur.

Saran, A., 1958 : Evolution of drainage system in the Damodar Basin, *Nat. Geog. Ju. of India*, Vol. IV, Part 3, pp. 130-36.

Satpathi, D.P.P., 1970 : A broad outline of geomorphology of Singhbhum, India, in E. Ahmad (Ed.) *Symposium on Erosion Surfaces*, 21st. Int. Geog. Cong., pp. 68-70.

Satpathi, D.P.P., 1972-73 : Quantitative analysis of landform : A case study in the Deo River Basin, Singhbhum, *Geog. Outlook*, Vol. IX, pp. 57-66.

Satpathi, D.P.P., 1973 : Geomorphology of Singhbhum, D. Litt. Thesis (unpublished), Ranchi University.

Satpathi, D.P.P., 1975-76 : Landscape cycles of Singhbhum, *Geog. Outlook*, Vol. XI, pp. 27-35.

Satpathi, D.P.P., 1976-77 : Meander mechanism, geometry of meanders and sinuosity indexes of the major rivers of Singhbhum, India; *Geog. Outlook*, Vol. XII, pp. 11-22.

Sen, D., 1972 : Geomorphology of the Aravalli Range, Rajasthan and a reinterpretation of residual erosion surfaces, *Geog. Rev. of India*, Vol. 34, No. 3.

Sen, R.N., 1977 : Recent aspects of the geomorphic evolution of Bandi River Basin, E. Rajasthan; *Geog. Rev. of India*, Vol. 39, part 1, pp. 12-18.

Sen, S., 1956 : Changes in drainage conditions in the lower reach of the Damodar, *Geog. Rev. of India*, Vol. 38, No. 4, pp. 24-38.

Sengupta, A.K., 1957 : Physiography of the Pat Countries of the Western Ranchi Plateau; *Observer*, Vol. 2.

Sengupta, A.K., 1970 : Denudation in Central Ranchi Plateau, Bihar, India, *Selected Papers*, Vol. 1 (Phy. Geog.) 21st. I.G.C. 1968. Nat. Com. Geog. Calcutta, pp. 111-113.

Sharma, H.S., 1967 : Morphometric Analysis of the Jaipur Region (Rajasthan), *Geog. Thought.*, Gorakhpur, Vol. III, No. 2, pp. 102-110.

Sharma, H.S., 1968 : Evolution of the Chambal Scarps, *Uttar Bharat Bhoogol Patrika*, Vol. IV, No. 2, pp. 99-103.

Sharma, H.S. & Padmaja, G., 1977 : Quantitative Geomorphic Characteristics of streams of the Morel Basin, Rajasthan, *Geog. Rev. of India*, Vol. 39, No. 4, pp. 353-366.

Sharma, H.S. & Rao, G.P., 1978 : Evolution of hill side slopes of the Morel Basin (Rajasthan) *Symp. on Morphology and Evolution of Landforms*, Deptt. of Geology, Univ. of Delhi.

Sharma, H.S., 1970 : Geomorphology of the Ramgarh Dome-Rajasthan, *Geog. Rev. of India*, Vol. 32, part 3, pp. 186-94.

Sharma, K.D., 1979, Geomorphological evolution of the Maikal Plateau M.P., India, Ph.D. Thesis Magadh, University, Bodh Gaya.

Sharma, S.N., 1968 : A problem in drainage development in western Chota Nagpur; *Geog. Outlook*, Vol. V.

Sharma, S.N., 1972-73 : Origin of the Singhbhum Plain, Geog. *Outlook*, Vol. IX, pp. 37-56.

Sharma, V., 1973 : Post-Glacial terraces of the weed and willow creeks, *Central Alberta, Canada Quaternaria*, vol. XVII.

Sharma, V., 1977 : Preliminary Investigation of Pleistocene Valley Glaciation in Part of Kangra Dist., Himachal Pradesh, *Journal of the Geological Society of India*, Vol. 18, No. 1, Jan. 1977, pp. 23-30.

Sidhu, G.S. & Pandey, I.C., 1974 : Parametric analysis of Patel Gad drainage Basin (East of Masuri) *Nat. Geog. Jr. India*, Vol. XX, part 4, pp. 223-235.

Singh, G.N., 1977 : Geomorphology of Rajmahal Highlands, Ph.D. Thesis (unpubl.) Magadh University, Bodh Gaya.

Singh, N.P., 1975 : Geomorphology of the western higher plateau of Bihar, Ph.D. Thesis (unpubl.) Magadh Univ. Bodh Gaya.

Singh, O.P., 1976 : A quantitative approach to the classification of morpho-units : A case study of Palamau upland, *Nat. Geog.* Vol. XI, No. 1, pp. 79-87.

Singh, O.P., 1976 : Slope studies of Palamau uplands : A quantitative approach, *Ind. Geog. Res. Bul.* No. 6, pp. 15-23.

Singh, O.P., 1976 : Palamau upland : A study in landform and settlements (Unpublished Thesis), B.H.U. Varanasi.

Singh, R.L., 1967 : Morphometric analysis of Terrain, *N. G. So. of India, Bull.* No. 22., pp. 1-24.

Singh, K.N., 1968 : Southern uplands of Eastern Uttar Pradesh : A study in landforms and settlement distribution. Ph.D. Thesis (unpublished), B.H.U. Varanasi.

Singh, R.P., 1956 : *Geomorphological Evolution of the Highlands of Chota Nagpur and the Adjoining Districts of Bihar*; Ph.D. Thesis, London, Micro Methods Ltd., East Ardsley, Wakefield, Yorkshire.

Singh, R.P., 1957 : Stages of structural growth of Chota Nagpur, *Jr. of Univ. of Bihar*, Vol. III, No. 1, pp. 82-92.

Singh, R.P., 1957 : Landscape of southeast Chota Nagpur and its evolution, *N.G.J. of India*, Vol. III, pt. 3 & 4, pp. 137-168.

Singh, R.P., 1958 : Structure, drainage and morphology of the Chota Nagpur highlands, *Geog. Outlook*, Vol. II, No.3, Ranchi, pp. 23-32.

Singh, R.P., 1960 : Geomorphological evolution of highlands of

Rajmahal, *National Geographical Journal of India*, Vol. VI, Pt. I March, pp. 1-13.

Singh, R.P., 1961 : Bokaro—The future steel town, *Nat. Geog. Jr. of India*, Vol. II, Pt. 4 Sept., pp. 245-256.

Singh, R.P., 1966 : Changing Landscape of Chota Nagpur, *Patna University Journal*, Vol. 21, No. 1, Jan. 1966, pp. 70-93.

Singh, R.P., 1966 : Geomorphological Mapping, Applied Geography, *Proceedings of the Summer School*, Dept. of Geography, B.H.U. (Nov. 1-10, 1966), pp. 211-269.

Singh, R.P., 1967 : Denudation chronology of the Himalayas, *Trans. of the Indian Council of Geographers*, Vol. 4., pp. 16-22.

Singh, R.P., 1967 : Stages of structural growth of the peninsular India, *Patna College Magazine*, May, 1967, pp. 34-42.

Singh, R.P. 1967 : Landscape cycles of the Peninsular India, *The Seminar on Geomorphological studies in India*, 1965, Proceedings 1967, pp. 144-157, Centre of Advanced Studies in Geology, University of Saugar.

Singh, R.P., 1968 : Geomorphology of the Shillong Plateau of Assam. *Proceedings of the Pre-Congress Symposium and Field-Study Meeting on the Physical Geography of Eastern Himalayas and the Meghalaya Plateau* (S6 E. 6 of I.G.U.); Gauhati Nov. 68, Dept. of Geography, Gauhati University, Gauhati, pp. 1-9.

Singh, R.P., 1968 : Geomorphology of the Indian Peninsula, *Essays in Geography*, B.H.U. Bhagalpur, pp. 45-66.

Singh, R.P.. 1969 : Concept of Tilting, *Journal of the North-East Indian Geographical Society*, Vol. I, No. 12, Jan. Jy, pp. 1-5.

Singh, R.P., 1970 : Topography and towns in the Chota Nagpur Highlands., *The Magadh University Journal, Bodh Gaya*, Vol. III, No. 1, pp. 93-103.

Singh, R.P., 1971 : Concept of periglacial landscape, *National Geographer*, Vol. VI, Allahabad, pp. 1-8.

Singh, R.P., 1972 : Denudation Chronology in Indian Peninsula, *Journal of the North East India*, Geographical Society, Vol. III, No. 2. Vol. IV, No. 1 & 2 July, 71: Jan. & July 1972, pp. 1-18.

Singh, R.P., 1974 : Effusions and landscape in India, *The Indian Geographer*, Vol. 14, Nos. 1 & 2. Dec. 1974, pp. 45-63.

Singh, R.P., 1975 : Concept of Warping, Indian Geographical Studies, *Research Bulletin*, No. 4, March 1975. Geographical Research Centre, Patna 5, pp. 1-4.

Singh, R.P., 1975-76 : Geomorphological evolution of the range and basins of Cuddapah Crescent, *Geog. Outlook*, Vol. XI, pp. 7-14.

Singh, R.P., 1976 : Structure, drainage and morphology of Pat Country in Bihar, Geomorophology Section Moscow, *International Geography*, 76. Vol. I.

Singh, R.P., 1976 : Glaciated Landscape in the Indian Peninsula. *International Geography*, 76, Vol. I. Moscow.

Singh, R.P., 1977 : Geomorphology and its recent trends; *Nat., Geogr.*, Vol. XII, No. 1. pp. 34-45.

Singh, R.P., 1978 : Unconfirmities in the Highlands of Chota Nagpur. *Paper presented in Geol. & Geog Section*, Bhubaneshwar Abstract Geology and Geography Section.

Singh, R.P., 1979 : Applied geomorphology with special reference to Chota Nagpur Highlands. *Proceedings of Summer School*, Geography Deptt. Jaipur University, Jaipur.

Singh, R.P., 1979 : Surface history of the Peninsular India, *Proceedings of Summer School*, Geography Deptt. Jaipur University, Jaipur.

Singh, R.P. & Kumar, A., 1969 : Geomorphological evolution of stream order of the Topa and Silphi Basins in Ranchi, *N.G.I. of India*, Vol. XV, pp. 38-44.

Singh, S. & Srivastava, R., 1974 : A morphometric study of the tributary Basin of upper reaches of Belan River, *Nat. Geog.*, Vol. IX, pp. 31-44.

Singh, S., 1972 : Altimetric analysis : A morphometric technique of landform study. *National Geographer*, Vol. VII, Allahabad.

Singh, S., 1975 : Methods and approaches in the study of landforms. *Indian Geographical Studies*, *Research Bull.* No. 4. March '75, Geographical Research Centre, Patna-5, pp. 5-12.

Singh, S., 1977 : Evolution of granites—gneissic tors and cut-off spur mounds at Ranchi Plateau, *Nat. Geog*. Vol. XII, No, 1., pp. 93-97.

Singh, S., 1976 : On the quantitative parameters of the computation of drainage density, texture and frequency : A case study of a part of Ranchi plateau, *Nat. Geog.* Vol. XI, No. 1, pp. 21-31

Singh, S., 1978 : Physiographic Regions, landforms and erosion surfaces of the Ranchi plateau, *Nat. Geogr.*, Vol. XIII., No. 1., pp. 43-65.

Sinha, N.K.P., 1968 : Geomorphic Evolution of northern Rupununi, British Guiana, *Tech. Report* No. 12 (Ph.D. Thesis) Geography Branchis Office of Naval Research, Washington.

Sircar, P.K., 1953 : River Basins and geomorphological regions, *Geog. Rev. of India*, Vol. XV, No. 4,

Subba Rao, S., 1970 : Landforms of the Deccan traps., *Selected Papers*, (Phys. Geog.) Vol., 21st. I.G.C. 1968, Nat. Comm. Geog. Calcutta, pp. 127-132.

Subramanyam, V., (Ed.) 1967 : Seminar on geomorphological studies in India, *Proceedings*, Deptt. of App. Geol. Univ. of Saugar.

Sukla, M.K. & Verma, V.K., 1973 : Occurrence of waterfalls around Pachmarhi, Hosangabad Distt, Madhya Pradesh, *Nat. Geog. Jr. Ind*, Vol. XIX, part 3 & 4, pp. 200-203.

Tamskar, B.G., 1959 : The Jabra Basin : Its geomorphology and human settlement. *National Geographer*, Vol. II, Allahabad.

Tewari, A.K., 1967 : Some aspects of the geomorphology of Jaunsar Himalayas, *Seminar on Geomorphological Studies in India*, Deptt. of Applied Geology, Univ. of Saugar, pp. 30-41,

Vaidyanadhan, R., 1967 : An outline of the geomorphic history of India South of N, latitude 18°., *Seminar on Geomorphological studies in India*, Deptt. of Applied Geology, University of Saugar, pp. 121-130.

Vaidyanadhan, R. and Ramana Rao, K.L.V., 1972 : Recognition, correlation and age of erosion surfaces in the iron ore ranges of Bihar and Orissa, in *Proc. of symp. on Geom. Indian Shield.* 21st Int. Geog. Cong. pp., 33-34.

Vaidyanadhan, R., 1977 : Recent advances in geomorphic studies, *A Review, Ind., J. Earth Sc.* (S. Roy Volume), pp. 13-35.

Verma, P., 1958 : Ranchi Plateau, its geomorphology and human settlements, Ph. D. Thesis (Unpub.) Allahabad University.

Verma, P., 1960 : Udaipur Region—A Study in geomorphology, *Ind. Geog.*, Vol. 5, No 122.

Verma, P., 1961 : Diastrophic forces and their geomorphic expressions in Ranchi plateau, *Ind. Geog. Jr.* Vol. XXXVI, No. 4.

West, W.D. & Choubey, V.D., 1964 : The Geomorphology of the country around Sagar and Katangi, M.P. An example of Superimposed Drainage, *Jr. Geol* Soc. Ind., Vol. 5.

Yadava, S.R.S., 1979 : Hazaribagh Plateau : A Study in landforms and settlements Ph. D. (Thesis) unpublished B.H.U.

TWO

Hari Shanker Sharma

MORPHOLOGY OF RAVINES OF THE MOREL BASIN RAJASTHAN (INDIA)

THE LITERATURE on the problem of ravine and gully erosion in the arid and semi-arid climates of the world has been extensive and has been quite thoroughly investigated. In India, the references on the morphological study of ravines and gullies are rather scant.

The principal purpose of the present study is to analyse the morphological features of ravines of the Morel Basin and to establish the model of ravine genesis.

The Morel Basin lies between 25°5′—27°10′ N and 75°45′—77°0′ E (Fig. 2.1). This basin lies east of the Aravalli range and is a part of the East Rajasthan Plain. The basin slope decreases from northwest to southeast. The maximum elevation is about 630 metres above sea level in the north-western Jaipur Hills while it decreases to 200 metres in the southeastern plain. The average height is about 415 metres (Fig. 2.1). The problem of ravine erosion is severe in the area between 200 metres to 360 metres.

Material and Methods

The present study is based on the fieldwork of five years from 1970-1975. Various sites have been chosen for field measurement of ravines (Fig. 2.2.). Measurements of the gully length, width and depth have been made with the tape and the slope of head and side scarp have been measured with Abney level. The Survey of India quarter inch topographic maps have been used as base maps. Seven soil samples were taken from different

Note : The short form of this paper was published in *International Geography*, Vol. I, 1976 (23rd International Geographical Congress, Moscow).

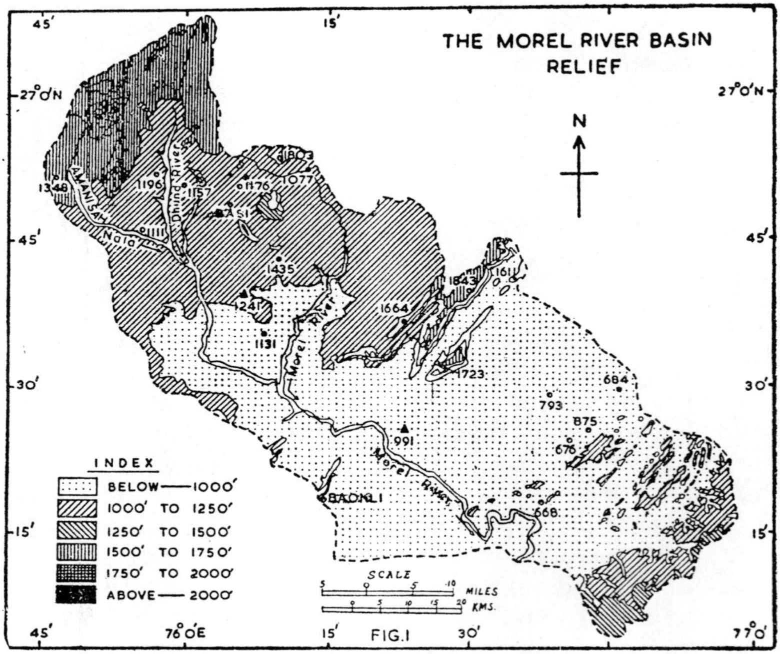

FIG. 2.1 : *The Morel River Basin Relief*

parts of the basin in order to analyse the mechanical properties of the various soil types. Soil samples from the ravine top and bottom were taken to show the variation in the soil character of these two sites.

Distribution of Ravines

The ravine cut in recent and sub-recent alluvium is a common and most striking geomorphic feature of the basin. Generally, the banks of the Morel and its tributaries have been dissected into ravines. Broadly, three compact ravine zones may be observed in the basin (Fig 2.2).

These are :

1. along the Dhund river between heights from 300 and 360 metres,

1. Alluvial Knoll with a Zone of Kankar Pan—Dhund River

2. Alluvial Knoll with a Zone of Kankar Pan in the Sub-soil

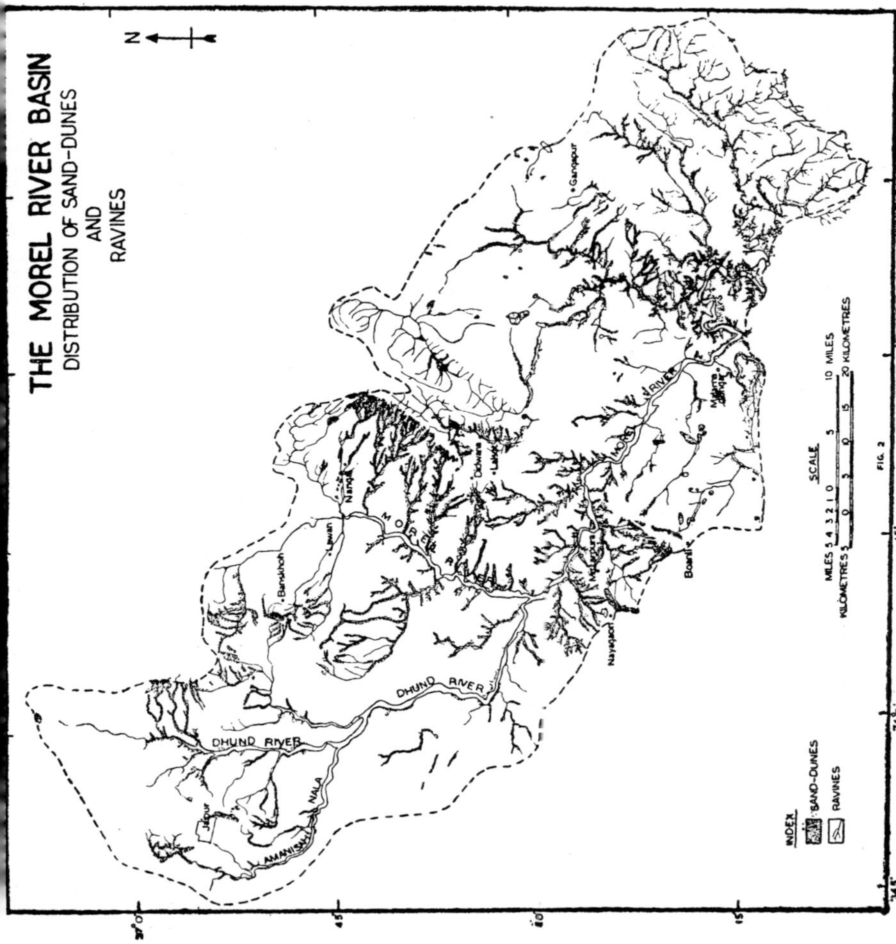

FIG. 2.2 : *The Morel River Basin, Distribution of Sand-dunes and Ravines*

2. along the Morel and its tributaries between the Bassi-Jatwara hills and Boanli-Lalsot hills at a height from 300 to 360 metres, and
3. along the lower reaches of the Morel river below 180-190 metres.

Ravines of the Dhund

The Dhund is a tributary of the Morel. It originates from Achroal hills in the northwest. An intricate network of ravines is developing along the Dhund between Achroal hills and its confluence with Amanisah Nala. The sand sheets of the western foothills of Jaipur have also been dissected into ravines which are wide V-shaped. The ravines of the Dhund are characteristically wide V-shaped with steep head and side scarps. Ravine in ravine is also a feature at many places. The peripheral area above the gully head is remarkably flat and then one finds an abrupt drop with the slope of about 85° to 90°. The depth is generally 5 to 10 metres.

Ravines of the Morel between Bassi-Jatwara and Boanli-Lalsot hills

Another compact zone of ravine erosion appears along the Morel and its small tributaries. Ravine erosion is common in the upper reaches of the Morel. This type of erosion is most active along the streams that descend from the Lalsot hills. The shape of the ravines of this section is wide V-shaped. The head scarp is vertical while the side scarp is comparatively gentle. The bed slope is concave. The depth is generally 10 to 15 metres.

Ravines of the Lower Morel and its Tributaries

Again a compact belt of ravines is perceptible along the lower reaches of the Morel and its tributaries which descend from the Vindhyans scarp. The ravines of this zone are much more wide and deep. The depth is more than 20 metres. Multistoried ravines have been observed. Side and head scarps are steep but there are breaks in the slopes of side scarps which are due to different layers of soils. The courses of the Morel and its tributaries are much more meandering and incised.

Geomorphic and Hydrological Factors

The formation of ravine topography in the basin appears to be due to several geomorphic and hydrological factors. Ravine erosion is a function of soil type, geology, slope, character of streams, rainfall and upliftment. The complicated nature of erosion is expressed as an equation.

$$\text{Ravine Erosion} = T \times G \times S \times C \times R \times U$$

where

T=soil type, G=geological structure,
S=slope, C=character of streams,
R=rainfall, U=upliftment.

In the following lines various factors are being discussed :

Geological Structure

The basin may be regarded as consisting of two types of areas so far as geology is concerned. Firstly, there are areas where solid geology appears on the surface; and secondly, are those tracts where a layer of soil and sediment occur between the surface and solid geology. The hills of the basin are composed of Archaean and Purana rocks. The plateau in the south-eastern part of the basin comprises the Vindhyans mainly of sandstone, shales and limestone (Fig. 2.3). The area in which ravines are developing is mainly *free from solid geology* and is composed of recent and sub-recent alluvium.

The northern and southeastern parts of the basin are characterized by the presence of faults. These faults have a significant bearing on the formation of ravines in this area. One fault may be traced west of the Morel in Bassi-Banskho hills. The fault is mostly concealed under alluvium. Two other faults are in the southeastern part. These are western extensions of the Great Boundary Fault of Rajasthan. These faults divide the older Archaean rock from the Younger Vindhyans. Heron (1938) suggested the period of faulting as Cretaceous.

Character of Soils

Broadly, the soils of the basin may be classified as sandy loam to loamy sand and clay loam to clay. It has been studied that loamy sand to sandy loam are comparatively much more liable to erosion than clay loam to loamy clay. The soils of the northern

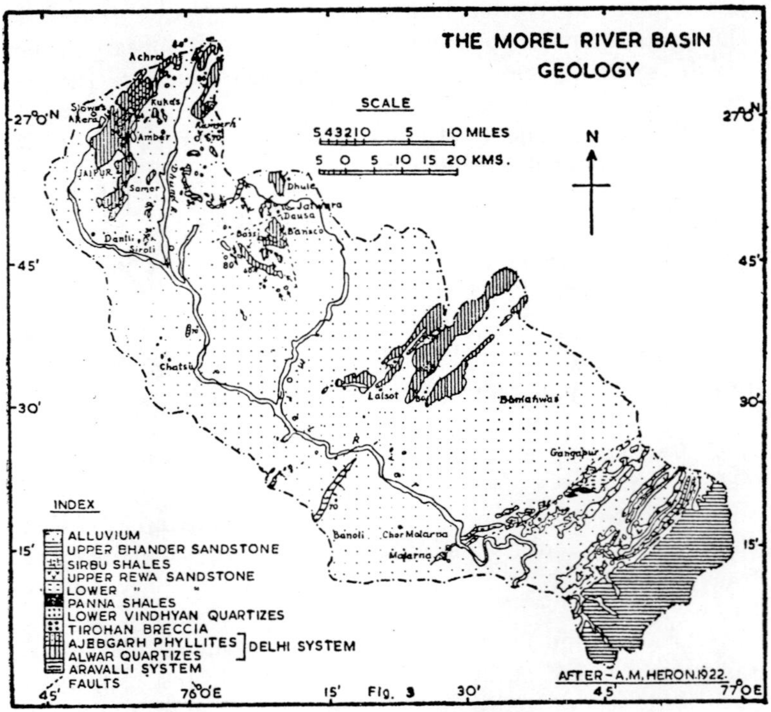

FIG. 2.3 : *The Morel River Basin Geology*

part of the basin are sandy loam with some patches of clayey loam. The southern and southeastern portions are covered with clay loam, which become sticky during wet season.

The mechanical properties of the soils are tabulated as follows :

TABLE 2.1 : **Mechanical Properties**

Sample Number	Site	Sand Percentage	Silt Percentage	Clay Percentage
1	Lawan Village	65.35	11.55	18.16
2	Lawan Village	65.45	15.06	15.15
3	Malwas-Chor Nala	67.85	9.15	20.52
4	Jhir Road Side	65.35	13.51	18.16
5A	Dhund River Ravine Bed	63.65	18.57	15.09
5B	Dhund River Peripheral Land	55.15	35.65	14.65
6	Malarana	64.75	13.55	11.63

It is evident from the table that the sand percentage is more in all the profiles. Sand percentage ranges from 55.15 to 67.85. Silt percentage in profile 5B is high. Clay percentage ranges from 11.63 to 20.52.

The significant feature of the soils of the basin is the presence of *Kankar Pan*. This feature plays a very important role in the growth of ravine through swallow holes and tunnelling. The calcium carbonate pan is a very common feature in all the profiles. Generally, it is concealed under alluvium but at places the calcium carbonate pan is exposed at the surface where the soil is removed by erosion. The pan of this basin is very thick, hard and compact, and when dry it is whitish-grey to yellowish in colour (Plate 1).

Slope

Slopes are also important in ravine erosion. The general lay of the land comprises flat tablelands with less than 3 per cent slopes. The ravine heads suddenly drop to 10-15 metres with a slope 85° to 90°. The bed slopes of ravines are generally concave.

Character of Rainfall

The semi-arid nature of the area is characterized by a wide range of temperature and an irregularly distributed rainfall. The annual rainfall generally varies from 60 to 67 cm. Most of the rain is received in July, August and September. The average intensity of rain in the basin is 2.38 cm. The average monsoonal intensity is 2.03 cm. The average intensity of dry months from November to June is 0.79 cm.

Character of the Morel and its Tributaries

The Morel is a tributary of the Banas and the Banas in turn is a tributary of the Chambal. In this respect the Morel is extreme northern river which aids water to the Chambal. However, this basin is a part of the Chambal drainage system and therefore this part may not be free from the phenomenon occurring in the Chambal Valley. The Chambal is a rejuvenated stream (Sharma, H.S., 1968). The rejuvenation along the lower Morel is also evident by the presence of incised meanders along the lower course and the breaks in the long profile of the Morel (Fig. 2.2). The courses of the other tributary streams of this section are also incised and meandering. It may be assumed that this is a weak zone of the basin in the sense that it is characterized by faults, and therefore it is more susceptible to any change occurring along the Chambal. Ravine is an expression of such change. The depth of the thalweg of the Morel and ravines is comparatively more than the upper sections.

Shape of Ravines

The study of ravines in the field shows that the size and areal pattern and rate of encroachment are mostly related to soil character and its thickness. In the basin where the soil is underlain by fine grained clay, which is resistant to rapid erosion, ravines are V-shaped. Such types of ravines have been observed along the Dhund near Ramgarh Road site. U- shaped ravines developed where the soils and sub-soils are commonly friable and easily cut by flowing water. This type of ravines are common along the Morel and its tributary streams west of the Lalsot hills.

Rate of Headward Growth of Ravines

In the present study the rate of headward extension of ravines has been assessed with the field measurements on few ravines. The following table shows the measurements and results.

TABLE 2.2 : **Rate of Growth**

Number of Ravine	Site	Soil type	Length of ravine in Dec. 1970 (in metres)	Length in July 1975 (in metres)	Distance increased (in metres)	Rate per year (in cm.)
1	Dhund	Clay loam	15.5	19.5	4.0	80
2	Ramgarh Road	Clay loam				
3	Lalsot Road Side	Sandy loam	10.00	13.00	3.0	60
	Rohuwas Nala		15.00	20.00	5.0	100
4	Near Morel	Clay loam	12.30	16.50	4.20	82
5	Near Morel	Clay loam	13.00	16.00	3.00	60
6	Near Morel	Clay loam	8.00	12.35	4.35	85
7	Near Confluence of the Morel with Dhund	Sandy	7.00	12.00	5.00	100

The 5-6 years period (1970-1975) shows ravines to be encroaching at the rate of 60 cm. to 100 cm per year depending upon the soil morphology and size of the ravine. The ravine extension has been expressed as an equation (Sharma, H.S., 1968) :

$$R_e = \frac{D_i}{T_e}$$

where

R_e=is rate of extension.
T_e=is time of erosion, and
D_i=distance increased.

Stages of Headward Growth of Ravines

The ravines extend on the catchment area generally through waterfall erosion and plunge pool action. In the area under investigation, ravines mainly undergo four stages of development, namely swallow holes, tunnelling, collapsing of tunnel roof and recession of the side scarp by slumping and soil creep (Fig. 2.4 and 2.5). The description of these stages as given under :

Swallow Hole Stage

In the initial stage parallel swallow holes develop on the flat ground along the river by the mutual action of rain water and the soil particles (Plate 2). The development of swallow holes are more evident on the soils containing calcium carbonate pan. The Kanker nodules provide more space to water to penetrate into sub-soil.

Tunnelling Stage

The second stage is the formation of tunnels like holes. The development of pipes is an important element in the headward extension of many vertical cuts. Near ravine wall these tunnels often extend into unravined alluvium for distance of several feet. Even at such distances swallow holes can often be seen. Tunnel is presumably often initiated through rodent burrows. However, the existence of tunnels is so prevalent in the basin that the action of rodents could hardly be considered a principal cause.

Tunnelling undoubtedly represents zones where water has

FIG. 2.4 : *Showing Mechanism of Ravine Extension through Swallow Hole and Tunnelling Stages*

passed from the unravine floor of the swallow holes down to the base of the vertical ravine wall along which the water removed not only salts in solution but also fine grained particles of soils.

Collapsing Stage

The movement of water through the *Kankar Pan* actually transported fine particles and salts away from their original place of deposition and left void. These voids then concentrated water accelerating the process of soil removal and resulting eventually in the formation of collapsed tunnels from which fullfledged ravine develops.

3. Swallow Hole

4. Swallow Hole, 1 metre deep

5. Showing Swallow Holes and subsequent Development of Tunnel (Dhund River)

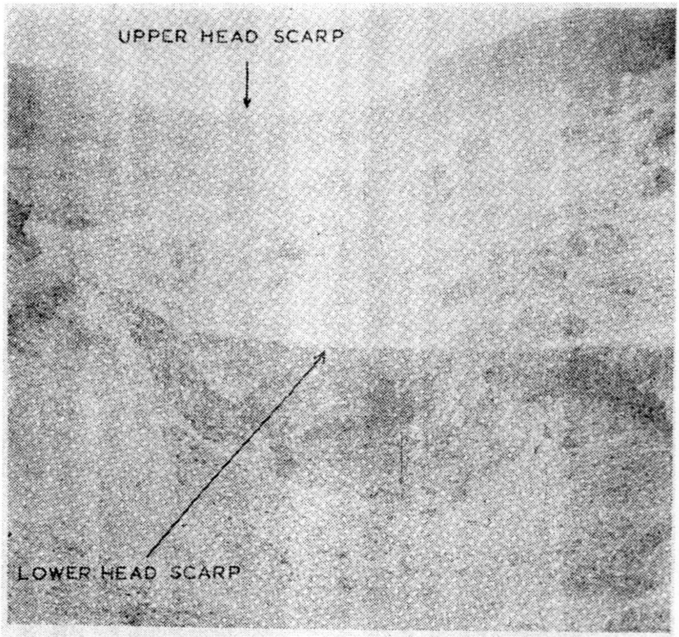

6. Showing Ravine in Ravine (Dhund River)

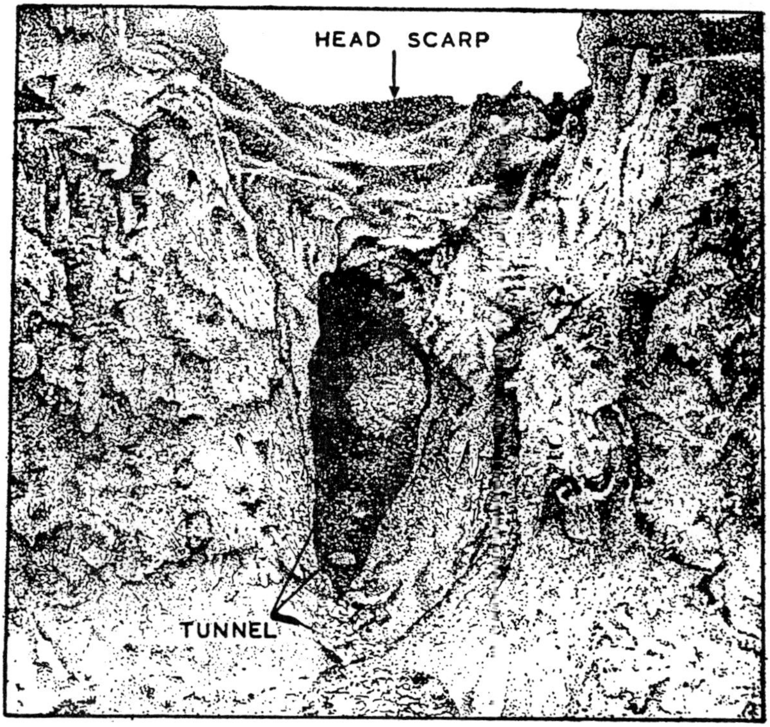

FIG. 2.5 : *Showing Mechanism of Ravine Extension through Swallow Hole and Tunnelling Stages*

Recession Stage

The collapsed material of the tunnel is subsequently transported by the concentrated flows. The side scarps of the ravines then recede continuously by the process of soil creep. One can observe talus cones at the base of the ravine scarps. With the continuous recession of the sides alluvial bluffs are disconnected or isolated. In this stage ravines enlarge their width and size (Plate 3).

Genesis of Ravines

There are three conflicting models of ravine genesis. These are—overgrazing model, climatic model and rejuvenation model.

Overgrazing Model

Bennett (1955) and Brice (1960) are exponents of this model. They associate gully and ravine erosion exclusively to land use or misuse, the chief cause amongst is overgrazing.

Climatic Model

The followers of this model consider that gullies are result of climatic changes wherein degradation occurs in dry periods and aggradation in the wet. The discovery of pre-historic gullies in the American southwest has led to the formation of two different and conflicting theories even in the climatic model.

The Bryan-Antev Model : Associates gully cutting with drought and a poor vegetative cover and gully filling with higher rainfall and an improved vegetative cover.

Martin-Schoenwetter Model : Associates gullying with increased summer high intensity rainfall, when there may actually be a greater annual rainfall and an increased vegetative cover. Tuen (1966) presented evidence to support the Martin-Schoenwetter Model.

Rejuvenation Model

Associates gullying to the lowering of the local base level of erosion caused by the recent upliftment. Sharma (1968) and Ahmed (1968) are exponents of this model.

The basic cause of marked gully and ravine erosion in this basin is probably tectonic uplift. The study shows that the intensity of ravine erosion occurs apparently indifferent to geology, elevation, climate and vegetative cover. It may be assumed that the rigid Peninsular block is passing against the Himalayas and as a result of this, certain areas, particularly where the margin of the shield occurs, are being uplifted. The uplift is the maximum in the zone where the distance between the Himalayas and Peninsular upland is the shortest e.g. the area between the Aravallis on the west and the Raj Mahals on the east. The Morel Basin lies east of the Aravallis, therefore this area has experienced upliftment as side effects of the Himalayan orogeny right from the Miocene period up to the recent.

CONCLUSION

The foregoing study on the morphology of the ravines of the Morel Basin upholds the rejuvenation model. Generally, geomorphologists consider only river terraces, incised meanders, rapids and gorges and not gullies and ravines as results of rejuvenation. But under present geographical conditions found in the Morel Basin, it seems that ravines are one of the features developing due to rejuvenation and not exclusively due to defective land use and changes of climate. Thinking in terms of cycle of erosion ravines may be taken as a feature of early maturity. The first stage for the ravine is of course entrenching the master stream itself. Therefore, due to entrenching the stream the banks are cut and dissected in the form of ravine instead of well defined river terraces.

SUGGESTED READINGS

Ahmed, E., 1969 : *Distribution and Causes of Gully Erosion in India,* Selected Papers, Vol. 1, 21st I.G.U., pp. 1-3.

Antev, E.V., 1952 : Arroyo-cutting and Filling, *Journal of Geology,* Vol. 60, pp. 375-385.

Bennett, H.H., 1955 : *Element of Soil Conservation,* McGraw-Hill Book Company.

Brice, J.C., 1966 : Erosion and deposition in the loess-mantled great plain Medicine Creek drainage basin, Nebraska., *U.S.G.S. Professional Paper* No. 352-H, pp. 255-335.

Bryan, K., 1941 : Pre-Columbian agriculture in southwest, as conditioned by periods of alluviation., *Annals of Association of American Geographers*, Vol. 31, pp. 219-242.

Heron, A.M., 1938 : The Physiography of Rajputana, *Proceeding 25th Indian Science Congress*, Presidential Address, Vol. 25. Part 2, pp. 119-132.

Martin, P.S., 1963 : *The last 10,000 years, A Fossil Pollen Record of the American South-West* (Tucson), Arizona, University of Arizona Press, pp. 61-64.

Scheonwetter, J., 1962 : The pollen analysis of eighteen archaeological Sites in Arizona and New Mexico, Chapter 8 in chapters in the

Pre-history of Eastern Arizona Fieldiana. *Anthropology*, Vol. 53, Chicago National History Museum.

Sharma, H.S., 1968 : *Genesis of Ravines of the Lower Chambal Valley India*, Selected Papers, Vol. 1, 21st I.G.O., pp. 114-118.

Tuen, Yi Fu., 1966 : New Mexican Gullies : A critical review and some recent observation, *Annals of the Association of American Geograpers*, Vol. 56, No. 4, pp. 573-97.

THREE

K. N. Das

FLUVIAL PROCESSES IN THE KOSI CATCHMENT BASIN AND THEIR IMPACT ON MORPHOLOGY OF THE FLOODPLAIN

INTRODUCTION

THE KOSI a mightly Himalayan stream of the Nepal Himalayas is, in fact, a combination of seven streams. The united stream is known in Nepal as the Sapt Kosi. It assumes its popular name of Kosi on entering the plain at Chatra further downstream.

From physical and drainage points of view, the whole Kosi basin has two conspicuous divisions: (i) northern, and (ii) the southern. The behaviour of the Kosi in the lower basin, lying for the most part in Bihar, is dependent on the upper catchment basin. The river apparently has almost no middle course. Its upper courses in the mountainous and hilly Himalayas end near Chatra where the 'Sapt' Kosi leaves the gorge and enters into an almost level plain of Nepal terai. Right from this point downstream, the Kosi engage itself in building up an alluvial cone by forming braided channels to accomplish the task of building up land of the Kosi region in Bihar, south of Hanuman Nagar on the Bihar-Nepal border.

In so far as the physical framework of the region has a direct bearing on population and land use, a brief examination of some problems of Kosi, dealt with in the following pages, will enable us to understand the issues in a clearer perspective.

Much of the cultural personality and economic sequence of the Kosi floodplain are the result of the behaviour of the Kosi.

The Kausiki (Kosi)

The Kosi, known for its antiquity, is mentioned as Kausiki in epics such as the *Skandha Puran*, Ramayana and the *Mahabharata*. In ancient literature, the river is said to be a large and powerful stream. According to *Skandha Puran*, this is said to be the daughter of Kausik Raja, a celebrated king of Gadhi. This nymph (Kausiki) has a brother named Vishwamitra, a worshipper of Para-Brahma or the Supreme God. His sister Kausiki, although daughter of a Kshatriya, was wedded to a Brahmin sage, Richik by name, who had attained perfection by virtue of worshipping gods. The saint later got furious with his wife who bore a son, who unlike his father, preferred fighting to worshipping gods. Richik cursed his wife and prayed to God to change his wife (Kausiki) into a river.[1]

The Kausiki, according to another mythical writing, was a mermaid goddess, a sea woman having the head and body of a lovely lady and her waist ending in the tail of a fish. She was worshipped by the people inhabiting the *Matsyadesha*, the land of the fishes. The kingdom at that time was delimited between the old beds of the Brahmaputra and the Karatoya to which the Kosi was connected by a waterway.[2] In the great Hindu epic Ramayana too, the Kausiki is associated with the sage Vishwamitra. Satyavati, the elder sister of the sage, was married to a saint Rehika. She followed her husband to the heaven from where she was transformed into a big river, later to be known as the Kausiki, and it was on its banks that Vishwamitra used to live in his *ashram* (hermitage).

In the Mahabharat, Vishwamitra is known to have created a sacred stream-like lake for the purpose of holy bath as this was regarded to be as good as the performance of an *Ashwamedha Yajna* (a holy celebration by kings in ancient times on their attainment of great power and perfection). During that time, the pilgrims used to go to Tamrauna, and Tribeni (latitudes 26° 55′N. long. 87° 10′ E), former probably the Tamur river and latter the junction of the Tamur and the Sun with the Arun. References of Tamu (Ancient Tamrauna), Arun and Tribeni exist even today.[3]

Realm of the Catchment Basin

With a geographical area of 23,000 sq. miles, the catchment basin[4] of the Kosi is the third largest in area in the whole of India, ranking next only to those of the Indus (1,18,400 sq. miles) and the Brahmaputra (99,200 sq. miles). It is hemmed in between latitudes 26°28′ and 29°5′ N and longitudes 85°27′ and 89°10′ E (Fig. 3.1).

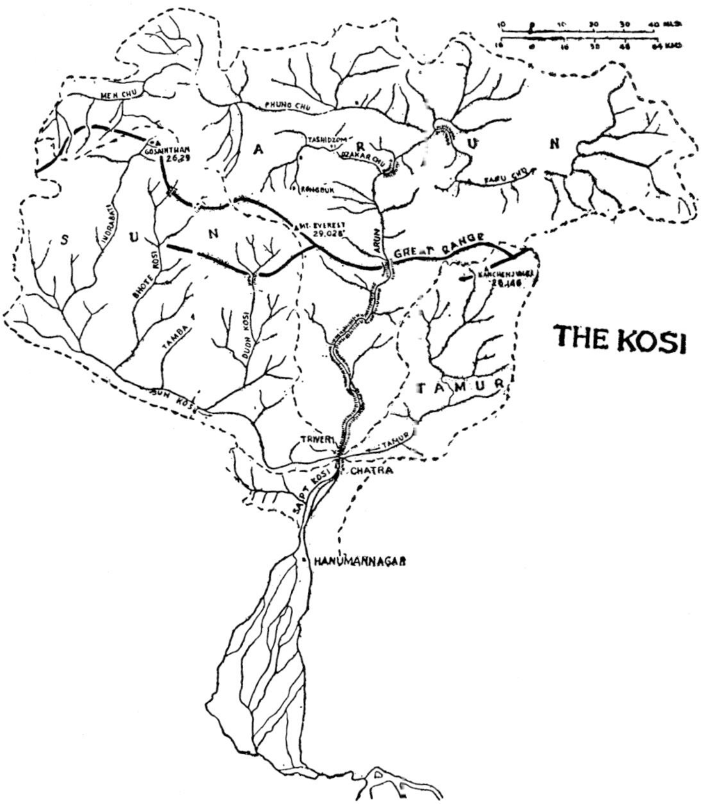

FIG. 3.1 : *Upstream Section of the Kosi*

The geology, topography, precipitation, etc., of the catchment area have important bearing on the hydrological characteristics of the Kosi river. The catchment basin, which includes a vast trough of land between Gosainthan peak in the west and

Kanchenjunga in the east, extends far into the Tibetan plateau north of the Nepal-Tibet watershed while it stretches as far south as Chatra in Nepal terai bordering the northeastern Bihar plain.

Geology

Geologically, the catchment basin represents varied facies of rocks ranging from Pre-Cambrian to present day alluvium. Much of this region is, however, geologically terra incognito and awaits survey and investigation.

Stratigraphically, the rocks fall into three broad zones. From north to south these are :

(i) the Tibetan Zone,
(ii) the Central Zone,
(iii) the Siwalik Zone.

The Tibetan zone lies to the north of the Nepal-Tibet watershed and is composed of fossiliferous and marine sedimentary rocks belonging to Cambrian and later geological periods—all of which have been highly metamorphosed.[5] There also occur igneous rocks with small patches of muscovite and biotite. Between the Great Range and the sedimentary beds of the Tibetan plateau lie thick beds of limestone which underlie Jurassic shales, and the unfossiliferous. They are intruded by granite, and consequently the parent rocks have changed into crystalline limestones and calc-schists due to contact metamorphism.[6]

Lying to the south of the above, the central zone is met with along the northsouth of the main Himalayan axis. The geological structure is very much complicated here and the bulk of rocks belong to a group of unclassified crystalline rocks generally composed of hard strata of granite, gneisses, etc. In addition to these, unfossiliferous sedimentary rocks are found, chief among which are slates, quartzites, limestones, conglomerates. Some of them overlie the metamorphic beds.

The geological facts about most of the peaks of high Himalaya are but little known and are based on empirical inferences rather than on scientific observations. The Everest group of rocks, in particular, is a pile of altered sedimentary rocks (originally shales and limestones) metamorphosed into banded hornfels, finely foliated calc-schists and crystalline limestones. They exibit a northerly dip where they pass into folded Jurassic Spiti shales of Tibet. Downwards they pass beneath banded biotite gneiss which has been deeply penetrated by sills and dykes of muscovite-tourmaline granite.[7]

Again, calcareous beds along with crystalline rocks of schists, quartzites and phyllites and a host of others, viz., saccharoidal marbles, calciphyres, tremolite, epidote and garnet-biotite-schists occur here and there alternating with each other in a zone traversed by the Arun and its tributaries.[8] Auden's observations have revealed that further downstream along eastwest of Barahkshetra, boulder, state, carbonaceous coal and dolomite of Gondwana age are bounded by the thrust plane against the Nahan series. The age of the Blaini boulder bed of glacial origin is taken to be upper carboniferous (Talchir). Probably the Gondwana ice-sheet extended over this section of the Himalayas for a comparatively long duration owing to which the coal and banded carbonaceous shales do not have any imprint of marine fossils.[9]

The boundary of the Siwalik zone is set by the southern limit of the Lesser Himalayas and includes Siwalik ranges of low hills, forming the Mahabharat range and related hills above the Chatra gorge. The region was uplifted fairly late in the Himalayan orogeny during the Tertiary era. Both in structure as well as lithological character, this presents rather simple geological features as contrasted to the above mentioned complex units in the Central and Trans-Himalayas (Fig. 3.2).

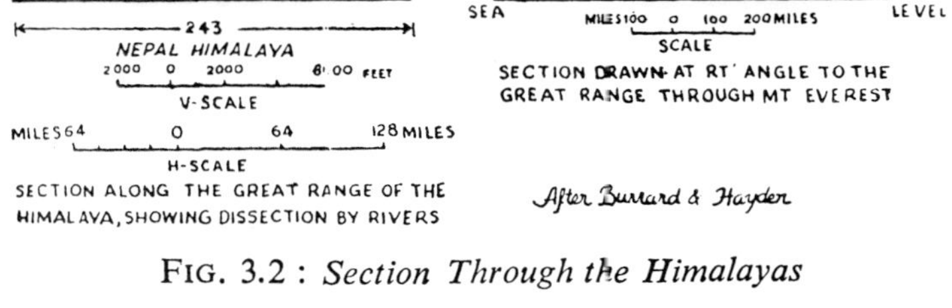

FIG. 3.2 : *Section Through the Himalayas*

The rocks of the Siwalik system (mid-Miocene—lower Pleistocene) are generally composed of fossiliferous limestones, boulder conglomerates, pebbly-grit, shales, clay, sand-rock and even river alluvia of silt and sediments. The boulder conglomerates are fairly coarse-grained while the rest are truly uncompact and soft. They easily get disintegrated and break up in sand particles when pressed even slightly.[10] As will be seen later, a high degree of erodibility of the Siwalik strata is, in large measure, responsible for the high amount of sand and silt which the Kosi drainage system carries and deposits in the north Bihar plain further downstream.

Seismic Character of the Kosi Catchment

A large section of the catchment basin lies in an active seismic belt as there are weak zones and thrust planes. The occurrence of periodical earthquakes of great intensity, major shocks and minor earth tremors of low intensities cannot be ruled out. General geological instability and consequently seismic disturbance along the mountains of a comparatively recent origin lead to minor tremors (Fig.3.3). Records show that the catchment basin normally experiences three minor tremors in a day while seven feeble shocks are not quite uncommon.

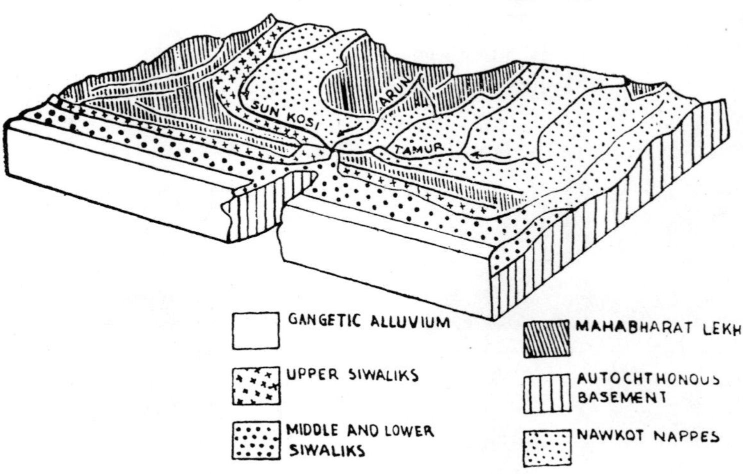

FIG. 3.3 : *Tectonic Block Diagram around Tribeni*

The epicentre of the great Bihar-Nepal earthquake (15th Jan. 1934) lay close to the Kosi region. Minor shocks evidently occur in a large part of the basin in its upper regions. It was partly because of the seismic nature of the Kosi catchment, particularly in the headwaters of the Sapt Kosi, and partly because of difficult terrain conditions as well as doubtful efficacy of an ambitious project (original Kosi Project Plan) that proposals of Belka dam near Barahkshetra (CWINC'S 1946 Dam Project) had to be abandoned. As to its seismic character, it was, however, suggested in the report[11] that in California as well as many parts of the United States which fall in the epicentres of high seismic intensity, dams have been constructed. What is, therefore, needed in the Kosi catchment basin is that "most stringent precautions should be taken in the design of any dam constructed on the Kosi to allow for earthquakes of miximum intensity". West does not attach much importance to the seismic factor in the design of structures in this region.

Relief Conditions

The topographic expression of the upper catchment has a profound influence on the characteristics and behaviour of the Kosi (Fig. 3.1). The features of relief are as varied and complex as the structural differences.

From physiographic point of view, the upper catchment basin can be broadly divided into : (i) the northern plateau region of Tibet, (ii) the central zone of rugged mountains, and (iii) the region of southern low hills.

The northern plateau region lying to the north of the Great Range is generally above 15,000 ft in elevation. It is here that the tributaries of the Arun, viz., the Men Chu, Phung Chu and Yaru Chu take their rise. The country north of the watershed is undulating. The hills, plateau and mountains of trans-Himalayas have an east-west extent and are truncated by river valleys. For some distance their courses run east-west, parallel to the grain of the mountains and then take a sudden southward bend. This is particularly typified by the Phung Chu,* Yaru Chu, etc. (Fig. 3.4) which have formed

* Chu is a Tibetan name for a stream.

impressive gorges. The Yaru near Jikyop and Rongme while the Arun on its junction with the Dzakar Chu southeast of Tashidzom have transverse defiles along their courses.

The central region is a rugged and high mountainous area of varied relief with numerous snowclad peaks above 20,000 ft. Some towering peaks, unrivalled in the world for their great heights and lying on one alignment, i.e., on the main Himalayan axis, are associated with large glaciers,[12] feeders of the headwaters of the Kosi's major tributaries—the Arun, Sun and Tamur. They are snowy white in appearance. Some of them viz. Mt. Everest—29,028′ (27°59′16″, 86°55′44″), Kanchenjunga—28,146′ (27°42′9″, 88°9′0″), Gosainthan.—26,291′ (28°21′7″, 85°46′55″), Makalu—27,790′ (27°53′23″, 87°5′29″), Gourishankar—23,440′ (27°57′52″, 86°20′16″), Cho Oyu—26,750′ (28°5′32″, 86°39′51″) and a host of others like Ngojumba Kang—25,730′, Chomo Lonzo—25,640′, Chamlang—24,012′ etc.,[13] lie within short distance of the Himalayan crest. The tributaries of the Sun Kosi such as the Indravati and the Tamba rise on the southern face of the Great Range while the headwaters of the Dudh and the Bhote lie beyond the range. The Arun has formed magnificent gorges from across the crest of the Himalayas and pursues southward and provides an example of antecedent drainage.

The southernmost section of the basin is generally less than 15,000′ and dwindles further south into insignificance with a low country of hills around 5,000 ft or even less. The lesser ranges merge imperceptibly into the Siwalik hills of only 1,000 ft in elevation. The southern extension of these low but uneven hills ends at Chatra whence forward the river (Sapt Kosi) debouches onto the plain. The basins of the Sun Kosi and Gandak traverse through an outer parallel Mahabharat range (traced east-west in Nepal from lat. 27°15′, long. 86°0′ through 27°45′—85°0′ and 28°0′, 83°0′ to lat. 29°15′, long. 80°30′). The peaks of this range varying between 6,000 and 8,000 ft dwindle to 5,000 ft near the left bank of the Kosi. This is serrated throughout like the edge of a saw and has, in fact, a controlling effect on the drainage courses as both the western (Sun) and eastern (Tamur) branches of the Kosi are obliged to change their directions to unite at Dangkera (Tribeni) with the southflowing Arun.[14]

Transverse Gorges of the Kosi

The transverse gorges of the Arun and its tributaries have evoked great geomorphic interest, and much light appears to have been thrown since 1937 when Wager[15] first proved it to be a case of antecedent drainage. This has, of late, been corroborated by Sparks for whom antecedence as a hypothesis "should be the last resort of a geomorphologist seeking to explain an inconsequent pattern of drainage, as it is, except in ideal cases, undemonstrable".[16] He agrees with the field observations of Wager as antecedent course of the Arun would better explain the phenomena than either river capture or glacial diversion.

The Arun and its tributaries lie at an altitude of 16,000 ft in Tibet and flow through a rugged country to the south between 19,000 ft (north) and 28,000′ (main crest of the mountain to the south). For a great distance, the structural grains that are met with in the south and central section are hard and resistant coupled with an increased height. The Arun nevertheless persistently pursues its southerly march by piercing through a high and hard ridge of the Himalayas. The conditions that obtain in the area would not allow a river, whatever erosive power it might exhibit, to cut into a thick and hard strata of the Great Range, and Wager was impressed by his observations to attribute the southerly flow to antecedence.

Wager's findings may be summarized in the following lines. (i) In the Yo Ri Gorge (Fig. 3.4) terraces have been developed in largely spread gravel beds of post Quarternary glacial age near the confluence of the Dzakar Chu with the Arun, (ii) in the region of Nyonno Ri range along the Yaru Chu at Rongme, thickness of the terraces varies between 400 and 500 ft above the bed of the river, (iii) uplift of the Nyonno Ri has caused local upwarping and apparently terraces lie upward in the Rongme and Jikyop gorge sections, (iv) evidences of glacial or morainic deposits supporting a widespread extent of ice sheet are almost lacking, (v) downstream, about the junction of the Kharta Chu with Arun upto Kimathangka the descent of the latter is 3,300 ft in its eighteen mile flow without any sign of a waterfall. The village Lungto rests on a terrace bench 1200 ft above the riverbed, (vi) narrow and deep defiles are flanked by precipitous walls veneered with thick strata of terrace deposits of fluviatile origin, and lastly, (vii) the longest gorge section of lower Arun, 40 miles below Kimathangka, bears testimony to the erosion going along with uplift of the Himalayas.[17]

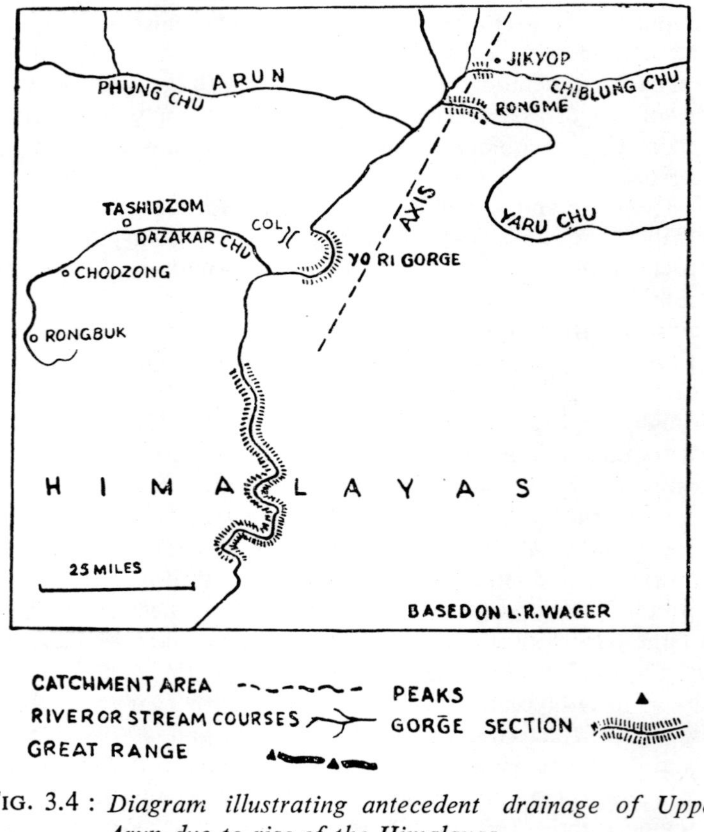

FIG. 3.4 : *Diagram illustrating antecedent drainage of Upper Arun due to rise of the Himalayas.*

From the above account, one might very well conclude that the Arun and its tributaries in the upper reach provide fine examples of an antecedent pattern. They are antecedent to the Himalayas because the uplift of the mountains and stream erosion went on pari passu, leaving an impression that the Arun is of an age anterior to the rise of the Himalayas.[18]

Glaciers and Rainfall

The catchment area, which is in the form of a segment of a circle with radius about 130 miles and with centre at Tribeni, is fed by large glaciers and a comparatively heavy rainfall. That a river system of relatively small length of less than 600 miles, as compared with the Indus, Brahmaputra, Sutlej and Tista should drain the third largest area among the Himalayan rivers and should have second largest runoff is due mainly to large glaciers and an adequate rainfall. Many of the flooding and other problems that are aggravated in the lower reach of the Kosi are chiefly the effects of glacial and precipitation characteristics. Figures for maximum recorded floods as mentioned in the accompanying table also indicate that the Kosi's discharge is second highest among the Indian rivers.

TABLE 3.1 : **Hydrological Data of Some Himalayan Catchment**

River	Catchment at debouchment in Plains (Sq. Mls.)	Area under glaciers (Sq. Mls)	Max. recorded floods (cusecs)	Mean annual rainfall (inches)	Runoff (acre ft.) (average)
Indus	1,18,400	14,415	10,00,000	17.14	87,355,000
Brahmaputra	99,200	9,400	8,00,000	—	15,000,000
Sutlej	23,000	2,468	4,00,000	19.71	13,938,000
Tista	3,200	300	—	80.08	1,150,000
Kosi	23,000	2,228	9,25,000* (Oct. 5, 1968)	60.00	40,600,000

Source : Kayastha, S.L. (1955) *NGJI* (Quoted from R.M. Gorrie) Vol. I, Sept., p. 19.

* The recorded flood of the Kosi is modified with reference to 1968 floods.

Glaciers

The entire catchment basin lying between the peaks of Gosainthan and Kanchenjunga along the main axis is drained by glaciers extending over an area of 2,228 sq. miles (about 10 per cent of area). Throughout its length and breadth, the catchment area is transversed by large Himalayan glaciers, viz., Rongbuk, Kanchenjunga, Yalung, Kyebrak, Kanshung, Zemu etc., flowing down the snow-clad mountain slopes. The snow-melt water approximately contribute 13 per cent to total discharge. The snow line of the Himalayas facing the plains is above 14,000 ft in altitude which increases to about 17,000 ft on the Tibetan side by reason of the desiccation of this region as well as the absence of moisture in rainbearing winds.[19] In this area (Nepal Himalayas) a good number of snowy peaks are situated. Each group of peaks is called Himal, meaning the abode of snow, from which the mountain system of the Himalayas has derived its name. Thus Maha Langur Himal dominatess the peak. Mt. Everest, Rolwaling Himal the peak Gourishankar, and Singalila Himal the peak Kanchenjunga.

Among the glaciers, the Rongbuk (14 miles) with snout at an elevation of over 16,000 ft drains the Everest, the Kyebrak (11 miles) lying between the lofty ranges of Choll Yo and Chorapsang in the Rongshar region are typical examples of transverse glaciated valleys. Similarly, the Kangshung (12 miles) drains the northern slope of the Makalu group of peaks.[20] The notable ones in the eastern section are : Kanchenjunga (10 miles), Yalung (13 miles), and Zemu (16 miles) feeders of the Tamur which flow down the southern slopes of the Kanchenjunga. The above noted glaciers apart from increasing the flow of rivers also cause erosion especially in the southern section.

Rainfall Characteristics

The collection of rainfall data by the Meteorological Department, Pune actually started from 1948, and are generally tentative as the difficult terrain conditions present difficulties in setting up raingauge stations. For example, trans-Himalayas (upper Arun), a country of high elevation covering an area of 10,880 sq. miles has no raingauge station while its lower counterpart having an area of 2,450 sq. miles has only seven—all of which are located within an area of 1000 sq. miles. Even in the Sun and Tamur sub-catchments they are very few in number and cannot be taken to be fully representative of actual rainfall conditions in the area.[21] Available data showing mean annual precipitation indicates that it increases from 10″ in the northeast to about 140″ in the vicinity of Gumthang between the Indravati and the Bhote as also towards southeast of Kanchenjunga in the Tamur valley. The decrease in amount north of the Great range may be attributed to rainshadow conditions. The 80-inch isohyet encircles Tribeni. Rainfall contributes 85 per cent of the water flow in the Kosi, although it is significantly aided by the melting of snow. Nearly 80 to 85 per cent of the rain falls during the monsoon months (May-Oct.). It may safely be concluded that the discharge depends mainly on precipitation, i.e., rainfall over a large part of the year as well as snowfall during the winter months. The intensity of rainfall has a great bearing on the discharge of rivers though total runoff remains a direct function of precipitation as well as the melting of snow.[22]

Characteristics of Drainage (Upper Catchment Basin)

The Kosi has seven main streams, viz., the Arun, Sun, Tamur, Indravati, Bhote, Tamba and Dudh (Fig. 3.1). It is for this reason that the united stream assumes its popular Nepali name, Sapt Kosi at Tribeni. Among these, however, the Arun, Sun and Tamur are major tributaries that drain a vast trough between the Gosainthan and Kanchenjunga. The drainage area roughly stretches between 26° and 30° N latitudes, and 85°30′ and 88° E longitudes. The areal extent of the sub-catchment areas may be had from Table 3.2.

TABLE 3.2 : **Major Tributaries and Sub-catchment Areas**

Major tributaries of the Kosi	Sub-catchment areas (sq. miles)	Per cent
Arun	13,330	57.9
Sun	7,330	31.8
Tamur	2,228	9.8
Sapt Kosi (below Tribeni)	112	0.5
Total Catchment Area	23,000	100.00

Source : Mookerjea, D. (1961) : 'The Kosi : A Challenge in River Control', *Jour. of the Institution of Engineers* (India), vol. XLII, No. 3, p. 117.

The Arun from the north, the Sun from the west and the Tamur from the east flow to meet at Dangkera to form a wide depression known as Tribeni. Other branches of the Sapt Kosi flow southward to meet their present stream the Sun Kosi, a longitudinal river, which runs in an easterly direction, opposite the Tamur, to empty into Tribeni. In the headwater portion, main feeders of the Arun are the Men Chu, Phung Chu, Yaru Chu, Chiblung Chu etc. The northern section of the Arun sub-catchment lies north of the Nepal-Tibet watershed at an altitude of over 16,000 ft. This area is characterised by low rainfall but the flow of streams is maintained by snow-melt waters. Probably no tributary of any Himalayan stream except the Indus and Brahmaputra drains so vast an area as the Arun Kosi.

The basins of the Sun and Tamur, which together are smaller than Arun's, lie respectively west and east of Tribeni. The tributaries of the Sun, for a large part, are transverse to the parent stream. The Dudh Kosi, for example, cuts across the mountain range on the southern face of Mt. Everest. The Tamur forms the easternmost tributary issuing forth from near Kanchenjunga, and besides glaciers, is favoured by a high rainfall. Likewise, the sub-catchment area of the Sun receives high rainfall over and above large glaciers on the southern face of the mountains. The Mahabharat range, a low range of hills lying west of the Singalila ridge, has had a profound influence in diverting the courses of the Arun, Sun and Tamur into the stream-like lake at Dangkera.[23]

Erosional and Depositional Features

The various streams, tributaries and *kholas* (small water courses) carry an unusually enormous amount of sand silt. The sedimentary rocks of the Lesser and Siwalik ranges consisting chiefly of uncompact and unconsolidated limestone, shale sandstone, boulders, micaceous sand and other loose materials are easily eroded away by fast running streams. Besides, due to heavy precipitation and a high incidence of landslides in the area, the eroded boulders, sand and silt of various sizes are hurtled down the hill slopes and eventually picked up by numerous hill torrents and kholas. In a study, Bhaskaran[24] has highlighted the problem of landslides which as precursors of active erosion are caused by such active factors as seismicity, heavy precipitation, high permeability leading to slumping action and earth flow, leavering action of tall hefty tree-roots causing landslips; and such passive factors as geological and lithological characters of rocks.

As a result of an accelerated erosion the silt-yield in the Kosi waters is abnormally high. This is particularly so when hill torrents swell to their brim. In a single flood, a deposition of three to five feet deep layer is a common feature. On the basis of available silt-data collected in some accessible parts near Barahkshetra and its neighbourhood, where reconnaissance surveys have been conducted by CWPC and the Kosi Project authorities, it has been found that the rate of silting in the Kosi is high among the Indian rivers.[25] According to a recent report from 1948 to 1959, the Kosi has carried down from the Himalayas a total sediment load of 10,48,000 acre feet which might cover a blanket, about 10 feet thick, over an area of 1 lakh acres. The estimated annual load of silt is 87,000 acre feet of which the coarse and medium portions would be the order of 14,000 and 25,000 acre feet respectively.

TABLE 3.3 : **Discharge and Silt Load in Tributaries of the Kosi.**

Tributaries with average elevation above M.S.L.		Catchment area (sq. miles)	Discharge 1948-49 ('000 Cft)	Silt Load 1948-49			% Sediment Load 1948-49		
				C 2mm	M .075	F .075	C	M	F
Arun : Upper	11,660'	13,330	15.7	7	8	13	35.6	30.0	27.0
Arun : Lower	7,290'								
Sun	7,500'	7,330	18.8	7	13	24	37.6	46.8	49.0
Tamur	3,220'	2,228	8.3	5	7	12	26.8	23.2	24.0
Sapt Kosi (Less than)	1,000'	112	42.2	14	18	—	—	—	—

Source : Ahuja, P.R. *JSWCI*, vol. 3, p. 110. The Silting problem is associated with discharge in the various streams. The data for runoff, on an average, varies between 40 and 48 million acre feet.

The silting problem is associated with discharge in the various streams. The data for runoff, on an average, varies between 40 and 48 million acre feet and the discharge at

Barahkshetra during peak floods has been 7,00,000 (1927), 8,55,000 (1954) and 9,25,000 (1968) cusecs. Similarly, single day fluctuations in discharge have been estimated to be of the order of 2,00,000, 100,000 and 25,000 cusecs for the Sapt Kosi (below Tribeni), Sun and Tamur respectively. Though a small tributary, the sediment yield in the Tamur is higher than in the Arun and Sun which are larger streams. It is annually estimated to be 38 tons per acre in Tamur followed by 33.2 and 11.96 tons per acre in the Sun and Arun.

SECTION B

THE PHYSICAL FRAMEWORK--THE KOSI REGION, BIHAR

Topography of the Kosi Region, Bihar

The area where the river regime of the Kosi, its numerous left-out channels and tributaries extends, is a flat and monotonous aggradational plain without any marked topographical features. The area, geographically speaking, forms the eastern section of the Middle Ganga Plain in North Bihar. The Bihar portion of the downstream course of the Kosi beginning from near Hanumannagar on the Bihar-Nepal border and comprising a uniform relief is known as the Kosi region, the area of investigation of which we are concerned to know the morphology of the Kosi floodplain.

The generalised contours, based on the interpolation of one-mile topographical sheets and modified in the west with reference to 1950 course of the Kosi, have been drawn at 10 feet interval.

They vary from 250 ft in the north at Bhimnagar to 100 ft in the south on approaching the Ganga *diara*, and have a general westeast trend. Their smoothness gets disturbed when they lie athwart a river course. On the whole, the country slope is from north to south as well as slightly from west and northwest to east and southeast.

There is thus little lithological complexity. The plain is composed of river alluvium, e.g., clay, loam, silt and sand in varying proportions. Oldham estimates the thickness of Gangetic alluvium to be 15,000 ft. Dunn pertinently remarks : "The Ganga plains consist of alluvial deposits which extend to a variable but unknown maximum depth. Their greatest depth is thought to be at least 6000 ft but other estimates have multiplied this more than five times. The deepest part appears to form a trough extending WNW-ESE through Motihari".[26] Basing his estimate on geophysical measurements made in Bihar after the 1934 earthquake, Krishnan puts the thickness of alluvium at 6000 ft keeping an upper limit reasonably at 10,000 ft.[27]

Over this flat surface, there has been an interplay of rivers, mostly south-flowing affluents of the Kosi and other Himalayan streams. In the past, some of them were active channels of the Kosi which at some earlier date formed the bed of the river. The region rises slightly to the west and northwest as is exemplified by the courses of the Bhutahi-Balan, Kamla-Balan, Jiwachh, Baghmati-Kareh, and a host of minor streams in the west. So imperceptible and unrecognisable is the relief that it leaves little impression about any topographic variations in the minds of observers. The most noticeable expressions of relief are the sandy wastes and little mounds created by shifting courses, marshy *chaurs* (shallow depressions filled by water during the rains), luxuriant shrubs of *Kans*, *pater* and water-reeds, river bluffs separated here and there by creeks and minor trickles.

Drainage of the Kosi Region in Bihar

Flowing down the high mountains and low Siwalik hills of the Himalayas, the Kosi enters into the plain of Nepal *terai* at Chatra. Thereafter, it persists through its two branches for some miles until uniting again upstream of Hanumannagar. By now the course becomes about a mile wide and then it flows (right from the gorge end) through numerous bifurcating branches and interlacing channels. A branch of the river turned south-east from Chatra and flowed past Purnea town. Rennell referred to this course as the Cossy or Cosah,[28] which by about 1736 was established firmly in the district of Purnea.

Through its westward shift, it traversed laterally over the flat surface of Purnea and Saharsa districts. The drainage conditions of the Kosi, as depicted by J. Inglis (1892), may be seen in their correct perspective even today. "When swollen by the melting of the snows or by the annual rains, the river overflows its banks, and at such times presents the appearance of a broad swiftly flowing sea, for its breadth from bank to bank is often ten and in some places nearly 20 miles across. In the dry season, the waters always of the same milky hue are confined to innumerable channels, some so shallow that the stilted power can wade across, and others running deep and strong with a ceaseless gurgling swish that would sweep the stateliest elephant off its feet, and carry its ponderous bulk far down the stream. . . The face of the country is split up into an infinitude of islands reticulated everywhere by a network of dry channels and shifting sandbanks, and overall, where there is an inch of soil, the stately elephant grass spreads its feathery mantle".[29]

Over a period of nearly two centuries and a half since 1736, the river has persistently moved westward, usurping the streams that came in its path and leaving abandoned channels dry during its march. Some of the old courses occupied by the then flowing Kosi are now, from east to west, known as the old or *mara dhars* (dead streams), viz., Kali, Livari, Dhamdaha, Sursar, Hiran, Daus, Loran, Dhusan, Parwane Tilabeh, Gohi, etc. By 1949, it had already taken river Tiljuga in its sway. These dead and defunct *dhars* (streams) are less powerful today, and normally go dry except during the rains. The present course of the Kosi (Fig. 3.5) has shifted from near Purnea town (1736) to Nirmali (1950) on the Saharsa-

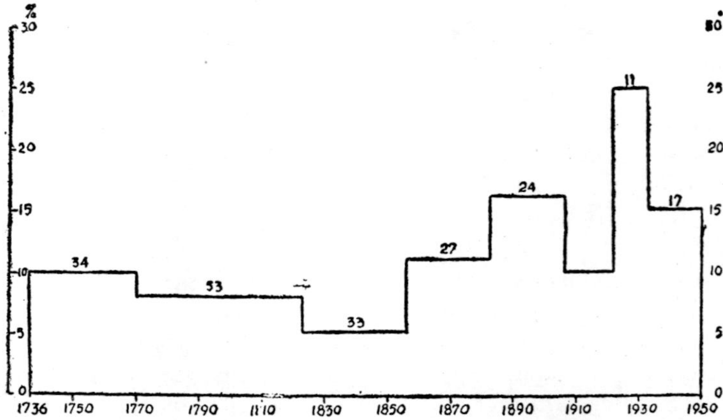

FIG. 3.5 : *Histogram showing rate of movement of the Kosi since 1736.*

Darbhanga border. It is remarkable that there has been little deviation in the line passing between the gorge point (Chatra) and its confluence with the Ganga (Karagola). Around 1800, the point of confluence lay in the neighbourhood of Manihari opposite Sahebganj which later shifted a little to the west near Karagola, 5 miles southeast from Kursela rail bridge, in 1870. Although the westward shift has been remarkable in other parts, this junction with the Ganga has remained intact. About 1941, the length of the Kosi from Chatra to Karagola was 160 miles as compared to about 100 miles in 1870.[30]

The Kosi flowing from near Bhimnagar in the north of Purnea-Saharsa border, forms an arc of a circle and flows through a circuitous route along the Darbhanga-Saharsa district border and then flowing further south again turns to southeast and east until it meets the Ghugri Kosi near Mansi, northeast Monghyr From here onward, it runs almost parallel to the Ganga in a well defined channel as far east as its confluence with the latter. According to Buchanan, the 'Geographical nomenclature of the streams is to the last degree confused'[31] for the river is known by the same name, e.g., the Kamla, Tiljuga, etc., over distances situated miles apart.

Fluvial Morphology of the Kosi Cone

The plain described above lies between the Gandak cone in the west and the Mahananda-Tista cone in the east. The flood-plain of the Kosi is essentially an inland delta formed by the river and its evershifting channels with apex at Chatra, base along the Ganga and arms extending as far east as the Mahananda and as far west as the Bagmati river. According to Geddes, the Kosi cone exhibits typically a depositional plain fanning out from the gorge end at Chatra. Further downstream, the river has pairs of flanking courses which are perennial, meandering and lightly incised. Some minor relief patterns of the cone are accomplished by the processes, firstly, of alluviation, bed scour or incision and then by lateral erosion and shifting. "The depositional pattern consists of the radiating distributary streams of which now one, now another, will dominate in turn so long as deposition still causes a change of direction, as in the Kosi". Associated with this are the spill banks, spill slopes and spill hollows created in the wake of annual flooding and westerly translation.[32]

Soon after its debouch onto the plain, the river begins to flow through many anastomosing channels of braided pattern (Fig. 3.6). They bifurcate through interlacing courses at one place and reunite at the other. For a distance of nearly 130 miles from about the gorge end to as far south as Sonbarsa the river is wholly braided. The bed is nearly a mile wide with islands and bars, embayments and depressions alternating each other at a number of places throughout its longitudinal section. The bed is often incised into a flat country and does hardly

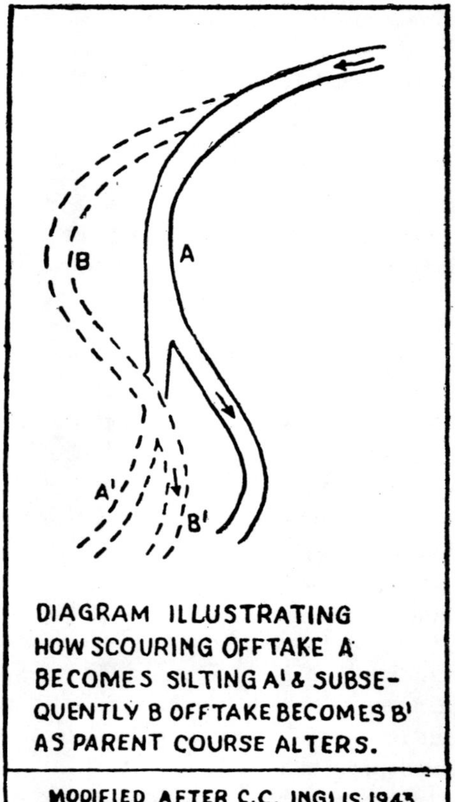

FIG. 3.6 : *Braiding streams*

show raised banks or natural levees. Further southward below Koparia and Dhamaraghat, the meandering courses may be traced around Mansi where the depth of the river increases considerably.

Hydraulic Conditions

A little to the south of the apex of the flat cone, the gradient decreases from .0009 at Chatra to .0002 (feet per foot) near Dhamaraghat[33], corresponding to 4.75 and 1.06 feet per mile respectively. This further dwindles approximately to 1.5, 1.8 and 2 feet per mile along longitudes 86°30′, 87° and 88°E along the Ganga *diara*. Luna B. Leopold, et al. (1964) have mentioned eight interrelated variables, viz., width, depth, velocity, slope, sediment load, size of debris, hydraulic roughness (rugosity) and discharge which are involved in downstream changes in slope and form of the channel.[34] Precisely, the floodplain portion of its long profile exhibits the features both of braided and meandering patterns in order to attain quasi-equilibrium for the discharge of the Kosi's sluggish water (Fig. 3.7).

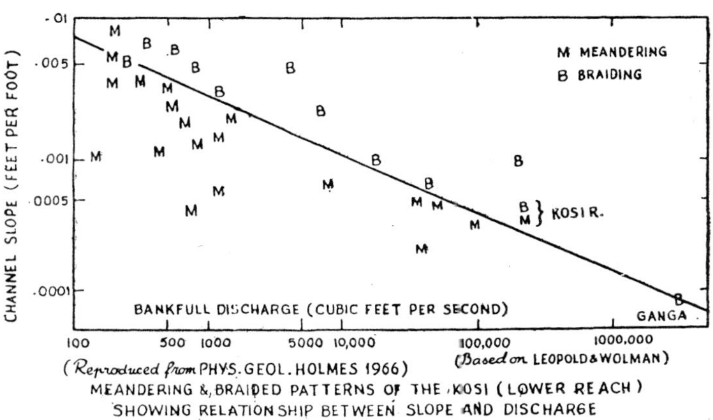

FIG. 3.7 : *Meandering and Braided patterns of the Kosi (lower reach) showing relationship between slope and discharge.*

Scour and Silting

On entering the plain, the Kosi engages itself in eroding its bed and bank for the passage of its runoff. This is done through the processes of scouring on the one side of the channel and making fresh deposits of sand and silt on the other. These processes go on concomitantly in the reach, and where conditions become congenial braided and meandering channels become discernible. The following table reveals an imbalance of silting over scour in four reaches between Chatra and Jhamta, a distance of 85 miles in its downstream course.

TABLE 3.4 : Relationship between Scouring and Silting in the Kosi Basin (Based on observations from 1955-56 to 1962-63)

(Data in '000 acre feet)

Years	Chatra to Bhimnagar (28 miles)		Bhimnagar to Bhapatiahi (14 miles)		Bhapatiahi to Karhara (28 miles)		Karhara to Jhamta (15 miles)		Total Scouring	Total Silting
	Scg.	Stg.	Scg.	Stg.	Scg.	Stg.	Scg.	Stg.		
1955-56	1067	—	—	—	—	711	—	261	1067	972
1956-57	—	533	—	57	—	833	—	42	—	1465
1957-58	405	—	55	—	—	283	—	—	460	283
1958-59	612	—	431	—	—	477	—	—	1043	477
1959-60	—	—	—	149	—	247	—	2917	—	3313
1960-61	494	—	—	390	997	—	2173	—	3664	390
1961-62	-	—	—	307	—	46	250	—	250	353
1962-63	—	364	139	—	—	1888	1024	—	1163	2252
	2578	897	625	903	997	4485	3447	3220	7647	9505
Net result	161	—	—	278	—	3488	227	—	—	1858

Total Scouring (Scg.) 7647 and Silting (Stg.) 9505 acre feet; net result of silting over scouring during the period 1955-56 to 1962-63 comes to about 1,85,800 acre feet in absolute terms.

Source: Based on R.V.P.D., Kosi Wing, Research and Investigation Section, Birpur (Saharsa).

It will thus be seen that the difference of total silting 9,50,500 a. ft and total scouring 7,64,700 a. ft in all the four reaches comes to about 1,85,800 acre feet. This highlights that silting outweigh scour. Hydraulic conditions coupled with scour and fill initiate much of the troubles in the region.[35] The magni-

and bank for the passage of its runoff. This is done through the processes of scouring on the one side of the channel and making fresh deposits of sand and silt on the other. These processes go on concomitantly in the reach, and where conditions become congenial braided and meandering channels become discernible. The following table reveals an imbalance of silting over scour in four reaches between Chatra and Jhamta, a distance of 85 miles in its downstream course.

TABLE 3.4 : **Relationship between Scouring and Silting in the Kosi Basin (Based on observations from 1955-56 to 1962-63)**

(Data in '000 acre feet)

Years	Chatra to Bhimnagar (28 miles)		Bhimnagar to Bhapatiahi (14 miles)		Bhapatiahi to Karhara (28 miles)		Karhara to Jhamta (15 miles)		Total Scouring	Total Silting
	Scg.	Stg.	Scg.	Stg.	Scg.	Stg.	Scg.	Stg.		
1955-56	1067	—	—	—	—	711	—	261	1067	972
1956-57	—	533	—	57	—	833	—	42	—	1465
1957-58	405	—	55	—	—	283	—	—	460	283
1958-59	612	—	431	—	—	477	—	—	1043	477
1959-60	—	—	—	149	—	247	—	2917	—	3313
1960-61	494	—	—	390	997	—	2173	—	3664	390
1961-62	–	—	—	307	—	46	250	—	250	353
1962-63	—	364	139	—	—	1888	1024	—	1163	2252
	2578	897	625	903	997	4485	3447	3220	7647	9505
Net result	161	—	—	278	—	3488	227	—	—	1858

Total Scouring (Scg.) 7647 and Silting (Stg.) 9505 acre feet ; net result of silting over scouring during the period 1955-56 to 1962-63 comes to about 1,85,800 acre feet in absolute terms.

Source : Based on R.V.P.D., Kosi Wing, Research and Investigation Section, Birpur (Saharsa).

It will thus be seen that the difference of total silting 9,50,500 a. ft and total scouring 7,64,700 a. ft in all the four reaches comes to about 1,85,800 acre feet. This highlights that silting outweigh scour. Hydraulic conditions coupled with scour and fill initiate much of the troubles in the region.[35] The magni-

The above table 3.5 shows the rate of movement of the river measured in a line passing through Purnea and Bhelani. According to a report by Claude Cavendish Inglis (1941-42), the river has moved transversely westward at an average rate of 2 miles per year up to 1856 and then at a translation rate of over a mile per year. This further emphasizes that the lateral movement may be considered to have taken place in a series of rapid movements with relatively quiescent periods in between rather than steady and continuous one.[37]

The approximate positions occupied by various dead and defunct *dhars* have been shown by P.C. Ghosh. Available records indicate that the Kosi after issuing forth from Chatra turned east and southeast and met at one time into the Mahananda (E. Purnea). According to Ferguson, the combined waters of the Kosi and the Mahananda then probably flowed further east to join an old course of the Brahmaputra west of Dacca (now in Bangladesh)[38] but this view has little historical support. There is evidence to show that the Kosi was a separate stream west of the Mahananda valley and met the Ganga near Manihari and later in the neighbourhood of Karagola (1856).[39] Subsequent rate of movement of courses of this river, viz., Kalikosi (1736), Livari (1770), Dhamdaha (1807-11), Sursar (1823-39), Hiran (1840), Daus (1873-92), Loran (1893-1921), Dhusan (1922), Parwane (1928), Dhemra-Tilawe-Gohi (1928-33), Purain (1938), Kosi (1950) etc. have been illustrated in Fig. 3.5. At times there has also been slight movement to the east to usurp a course it earlier missed during its westerly translation.

Mechanism of Movement

An excessive sand and silt charge is mainly responsible for the instability of the course of the Kosi. The slope required by the channel to carry its suspended load increases both with increase of grade and load, and when the slope is insufficient for the passage of its full load, it throws down excess sand on one flank and moves on to the other. Where the flank resists scour (due to bank friction), as on the right bank, currents show a strong tendency to cling to the bank. On shallower part of the braided channel where feeble currents percolate into sand bars and islands, as on the left bank, the fill replaces scour[40] (Fig 3.6).

The eastern side of the channel (after the recession of floods and deposition in their wake) presents congenial conditions for silting rather than scouring. As such the western side goes on improving while the eastern one goes on deteriorating. At each bankful stage and after, the repetition of the same hydraulic processes leads to the formation of new cut-off channels to the west. It has been shown elsewhere[41] that the translation is caused owing to flat topographic conditions, low bed slope, extended length through braiding and meandering channels, relative decrease in hydraulic mean depth, excess silt-charge, congestion and stagnation of channels, recurrent flooding etc. To this may enter the 'coriolis force' engendered by the convective mechanism of earth's rotation which, like the planetary wind circulation, also acts on the northsouth course of the Kosi in slightly deflecting its channels to the west.[42]

A rapid change of course is symptomatic of the drainage basin such as of the Kosi which is yet in a formative phase of development. The composition of the cone along its western arm is not complete as yet. The land between the Kosi and the Kamla-Balan, right bank tributary of the Kosi, east of the Sakri-Jaynagar section of the N.E. Rly. is a trough of lowland which, according to Claude Inglis, awaits in-filling. In its normal building up activity the Kosi would have done the job well by making further in-roads into the heart of Darbhanga, but the picture has changed completely since 1955 when the river has been confined within the embankments tied up through the Hanumannagar barrage on Bihar-Nepal border. The westward tendency is yet observed during peak discharges.[43]

Floods and Menace

For centuries the Kosi, by bringing catastrophic floods and ravages, has devastated large areas of Purnea, Saharsa and parts of Darbhanga and Monghyr. Flooding which is associated with discharge both of water and of harmful sand and silt is caused during the rains (June-Oct.). With a mean annual runoff of over 40 million acre feet, floods of very serious proportions have become almost an annual feature in the region. In some portions the floodplain, after the recession of floods, has been known to rise by a foot or more.

Strewn with excess coarse micaceous sand, the flood waters during their westerly translation have turned the rich rice, jute, *bhadai* and *rabi* tracts of the region into a wilderness of sandy and infertile wastes. Conditions that obtain are more or less the same as those in the Hwang Ho river basin of North China Plain. This wayward and obstreperous Kosi largely on account of its heavy sand-silt charge, floods of damaging proportion and calamities wrought in their wake, and above all, its serious impact on the land, people and economy of the region has earned the epithet of the "Sorrow of Bihar".

The average discharge in normal years is about 1,75,000 cfs. but it reaches new heights in extremely abnormal ones, e.g., 7,05,000 (1927), 4,78,442 (1948), 8,55,000 (1954) and 9,25,000 (1968) cfs. Highest ever on record, the 1968 floods (5th October) were caused by sudden cloud bursts over a large part during the departing *Hathia* rains. They created an unprecedented havoc when waters spread through the western earthen embankment, and the newly constructed Hanumannagar barrage was threatened to collapse. Over a million people, 300 villages within the embankments, large number of cattle were affected, and ghastly scenes were seen.

When floods visit the region, there is mass devastation in area as not only do human and cattle populations suffer but there is also a good bye to the standing crops over large sections of the region. The region's prosperity is, therefore, contingent on the behaviour of the Kosi which remains, in fact, a Cinderella for the engineers as well as the Kosi Project Administration. If it behaves well (when rainfall is not too high and the flow is normal), the prospects for crops become bright; but when floods occur, the people in general and peasants in particular suffer. In order to alleviate people's miseries, authorities were prompted to harness the "River of Sorrow" by constructing parallel embankments (western and eastern) tied up to a masonry barrage at Hanumannagar and by diverting large part of the discharge through sluices as well as numerous irrigation canals. This belated venture, however, has proved its worth in years of normal river flow.

REFERENCES

1. Francis, Buchanan, 1928 : *An Account of the District of Purnea*, 1809-10, (Ed. V.H. Jackson), Patna, pp. 13-14.
2. L.S.S. O'Malley, 1911 : *Purnea District Gazetteer*, Calcutta, p.2.
3. Ibid., p. 21
4. K.N. Das, 1967 : The Kosi drainage and its geographical significance, *Trans. of the ICG*, Vol. 4, pp. 27-28.
5. D.N. Wadia, 1953 : *Geology of India*, London, p. 253.
6. A.M. Heron, 1922 : Geological results of Mt. Everest expedition, *Geog. J.*, Vol. LIX, pp. 418-22.
7. S.G. Burrard and H.H., Hayden, 1933 : *The Geography and Geology of Himalayan Mountains and Tibet*, New Delhi, p. 64.
8. J.B. Auden, 1936 : Traverses in the Himalayan, *Rec. GSI*, Vol. LXIX, Calcutta, pp. 143-46.
9. J.B. Auden, 1946 : Blaini-Talchir, *Current Science*, Vol. XV, No. 1, pp. 346-48.
10. D.N. Wadia, 1953 : *op. cit.*, pp. 363-65.
11. Kosi Dam Project, Preliminary Report, *CWINC*, 1953, Delhi, p. 22.
12. H.W. Tilman, 1952 : *Nepal Himalayas*, Cambridge, pp. 222-38.
13. S.G. Burrand, et al., 1933 : *op. cit.*, pp. 1-6.
14. Ibid., p. 204.
15. L.R. Wager, 1937 : The Arun river drainage pattern and the rise of the Himalayan, *Geog. J.*, Vol. LXXXIX, No. 3, pp. 239-48.
16. B.W. Sparks, 1964 : *Geomorphology*, London, pp. 121-22.
17. L.R. Wager, 1937 : *op. cit.*, pp. 241-46.
18. D.N. Wadia, 1953 : *op. cit*. p. 22.
19. Ibid., p. 199.
20. N.E. Odell, 1922 : Observations on the rocks and glaciers of Mt. Everest, *Geog. J.* Vol. LIX, p. 302.
21. Rainfall characteristic of the Kosi Catchment, Govt. of Bihar, 1962, RVPD Kosi Wing, Investigation and Research Division, Birpur, p. 2.
22. Ibid, p. 8.
23. S.G. Burrand, et al. 1933 : *op. cit.*, p. 203.
24. A.R. Bhaskaran, 1962 : Landslides : Their origin and control in the Kosi Catchment, *JSWCI*, Vol. 10, Nos. 1-2, pp. 80-82.
25. K.N. Das, 1968 : Soil erosion and the problem of silting in the Kosi Catchment, *JSWCI*, Vol. 16, Nos. 3-4, pp. 62-64.
26. J.A. Dunn, 1942 : *Mem. G.S.I.*, Vol. LXXVIII, Calcutta, p. 6.
27. M.S. Krishnan, 1953 : The structural and tectonic history of India, *Mem. G.S.I.*, Vol. 81, Calcutta, p. 82.
28. James Rennell, 1783 : *Bengal Atlas*, Plates V & VI, Calcutta.
29. L.S.S. O'Malley, 1911 : *op. cit.*, pp. 3-4 (Adapted from 'Tent Life in Tiyarland' by J. Inglis, 1892).
30. R.R. Diwakar, (Ed) 1958 : *Bihar Through the Ages*, (Chap. II: Geography of Bihar contributed by Prof. P. Dayal), Patna, pp. 30-31.
31. F. Buchanan, 1928 : *op. cit.*, p. 9.

32. Arthur, Geddes, 1960 : The alluvial morphology of the Indo-Gangetic Plain : Its mapping and geographical significance, *IBG. Trans. and Papers*, Vol. 28, pp. 263-65.
33. B. Wolman, Leopold, Luna, M. Gordon & John Miller : *Fluvial Processes in Geomorphology*, San Francisco, p. 291.
34. Ibid., (1964) 268. Also see : L.B. Leopold and T. Maddock, (1955) : Flood control problems with reference to Kosi River, *JSWCI*, Vol. 3, No. 4, pp. 172-73.
35. K.N. Das, 1968 : *op. cit*, pp. 64-65.
36. T.N. Sheshadri, 1956 : *Kosi River and its Behaviour*, Paper No. 1 of 1966, Kosi Project Deptt.; Patna, p. 21.
37. C.C. Inglis, 1943 : *Rapid Westerly Movement of the Kosi River* Annual Report (Technical for 1941-42) p. 2 (cyclostyled copy received from Director, *CWPRS*, Khadakvasla, Poona).
38. Ibid., p. 2
39. K.N. Das, 1968 : Westward shift in the courses of the Kosi, *NGJI*, Vol. XIV, Part 1, p. 24.
40. C.C. Inglis, 1943 : *op. cit.*, pp. 2-4.
41. K.N. Das, 1967 : *op. cit*, p. 29.
42. Arthur Holmes, 1966 : *Principles of Physical Geology*, London, pp. 743-44.
43. K.N. Das, 1968 : *NGJI*, *op. cit.*, p. 31.

FOUR

P. B. Saxena and *Satya Prakash*

A STUDY OF THE MORPHOMETRIC DETERMINANTS OF THE STAGE OF CYCLE OF EROSION IN THE NAYAR BASIN (GARHWAL HIMALAYAS)

Hypothesis

THE STAGES of landscape brought up by the river cycle of erosion are precisely characterised and determined by the progressive returns of maximum relief ratio (above 75%) maximum amount of dip/slope (45-100%), high dissection index (above 30%) high hypsometric integrals (above 60%), high percentage of unconsumed upland (above 30%) and with low circularity index.

With the advancement of the cyclic landscape the trend of relief ratio, amount of dip/slope, dissection index, hypsometric integrals and percentage of unconsumed upland tends to decrease to minimum while circularity index increases.

The penultimate stage of the landscape finally achieves maximum circularity with minimum returns of other variables progressively.

Geoidentity

The Nayar Basin which is one of the important drainage basins in the Himalayan Ganga System,[1] consists of 1997 km^2 area extending from 29° 45′ to 30° 15′ lat. and from 78° 34′ to 79° 12′ longitudes. The Nayar river rises from the Dudhatoli mountain 3116 metre high and ends in the Ganga at Viyasghat mentioned as Sapta Samudrik (in the Kedarnath 10th Skand Puran), i.e., the collection of seven rivers water. The whole basin is generally uneven and irregular. The Dudhatoli Khatil-Anticline divides the basin into two intra-basins of the Purvi Nayar and the Paschmi Nayar which meet together at Satpuli, where the valley has been broadened and the basin seems to be under youth stage.

The systems of the Purvi and the Paschmi Nayars consist of almost sub-parallel drainage pattern,[2] because they flow parallel to the back-bone of the Dudhatoli Anticline on its both sides. The altitude of the valley ranges from 900 m to 3600 m. Maximum summits are observed in between 1800 m to 2100 m. In general the slope ranges from 10° to 50°. The basin of the Purvi Nayar is narrowly scarped and more dissected than the Paschmi Nayar. The complete Nayar basin is drained up to fifth order streams. The first order streams are generally noticed seasonal and perennial. The first and second order streams generally develop dentritic drainage system and the third and fourth order streams develop sub-parallel drainage system. In the fifth order region, where both the Purvi and Paschmi Nayars meet together forming the Nayar itself ingrown—paired broad benches have been developed. The Purvi Nayar consists of nine prominent knick points with receding waterfalls from source to the confluence. First, second and third knick points occupy their positions on almost the similar distance but fourth, fifth and sixth are being closer to each other and the rest seventh and eighth recede together. Thus the bench curves of the Purvi Nayar overall depict waning slope forming a concave structure of the valley. In comparison to Paschmi Nayar, the Purvi Nayar has deeper, steeper and narrower valley structures.

Morphometric Evaluation of the Valley

The quantitative analysis of the basin confirmed with field observations depict very meaningful results. Geologically the basin exposes glacio- and glacio-fluvial surfaces of landscape. The Dudhatoli anticline consists of granitic rocks. The intrusive crystalline rocks are observed at the outcrops exposed to the surface. The trend of the strikes of the Garhwal-nappe tends from north to south in the Dudhatoli range and the Krol nappe is observed in the southwest of the basin near Satpuli range. The Krol-nappe lies in the direction of northwest to southeast. Beneath this nappe main Himalayan range with granite rocks covers the whole Landsdowne range. At present the area comes under the fluvial landscape but the existence of glacial morains indicates its preformation of the glacial period. Whenever the outcrops have been exposed in the valley, mica, slate, flint, sandstone and granitic rocks are observed.

When the data of mean annual rainfall and temperature were fed in Peltiers' scale 2,[3] of different plots of mean annual rainfall and temperature to find out the intensity of different morphogenetic processes, the following field observations were recorded to confirm the quantitative results of the morphometric parameters.

(i) Frost action—Maximum,
(ii) Chemical weathering—Moderate,
(iii) Weathering region—Maximum-Mechanical and Moderate Chemical,
(iv) Mass Movement—Moderate,
(v) Fluvial erosion—Maximum,
(vi) Morphogenetic region—Fluvial Landscape with High Mechanical and Moderate Chemical Weathering,
(vii) Dominant processes—Weathering and running water,
(viii) Drainage—sub-parallel and dendritic.

The morphology of the intrabasins of the Nayar valley have been found tapered, curvelinear and irregular along the Garhwal-nappe from north to south direction in the Dudhatoli range and Krol-nappe. The valley has been broadened towards south in between the ends of the Dudhatoli anticline[4] and the Viyasghat. The intrabasins of the Nayar tributaries obtained circular, semi-circular and parabolic shape after the confluence of Eastern and Western Nayar due to having least effect of the north-south trend of the Dudhatoli anticline. All the morphometric parameters used to evaluate the terrain absolutely follow the trend and more of the geological formation of the region. The morphometric results confirmed with field observations have been analysed below.

Circularity Index

Circularity index[5] deals with the ratio of the drainage basin area (A_b) to the area of a circle having the same perimeter.

It is a quantitative approach to compare precisely, area of the actual form of the drainage basin to the area of the circle encircled by the perimeter of the same drainage basin. High, medium and low values of basin circularity are correlated with early, middle and youth stages of the drainage basins. The calculated average circularity index for selected intrabasins of the Nayar river stands less than 1.5 covering 50 per cent of the drainage basins of the Nayar valley (Fig. 4.1). About 20 per cent area of intrabasins of the Nayar valley constitutes the middle youth covering nine intrabasins. The standard value of the circularity index proposed for the middle stage ranges 1.5 to 3.0 (Fig. 4.1). About 30 per cent area of the valley covering 12 intrabasins consist the youth stage, where the values of circularity index are observed more than 3 in Fig. 4.1. It is generally observed that the youth stage starts from the confluence of the Purvi and Paschmi Nayars where the intrabasins consist of more or less parabolic shape.

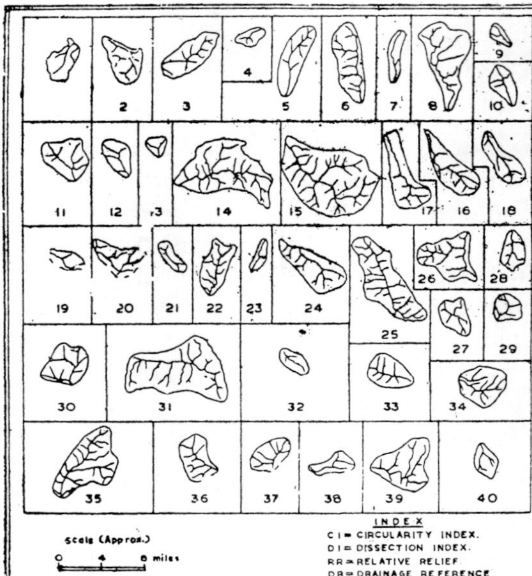

DR	CI	DI	RR
1	3·25	4·0	1·0
2	3·12	6·0	1·5
3	3·90	7·0	2·5
4	3·25	6·4	2·0
5	3·75	7·0	1·5
6	3·05	5·0	2·5
7	3·65	6·5	1·7
8	1·95	10·0	4·5
9	1·62	12·0	3·5
10	2·10	15·0	3·5
11	1·75	12·0	3·5
12	1·82	15·0	4·0
13	1·52	10·0	4·0
14	2·5	12·0	4·0
15	2·75	15·0	3·0
16	1·75	14·0	4·6
17	1·52	13·0	4·5
18	1·63	12·0	4·5
19	1·56	10·0	4·5
20	2·75	20·0	4·5
21	3·27	4·0	6·0
22	3·4	6·0	4·5
23	1·42	18·0	4·5
24	3·25	7·0	2·5
25	3·9	5·5	3·0
26	1·43	17·0	5·5
27	1·27	18·0	5·0
28	1·2	16·0	5·5
29	0·91	19·0	5·5
30	0·99	22·0	5·5
31	1·42	20·0	5·5
32	1·47	19·0	6·0
33	1·46	19·0	5·5
34	1·47	16·0	5·5
35	1·01	19·0	3·0
36	1·42	20·0	5·5
37	1·25	21·0	6·0
38	1·31	19·0	6·0
39	1·29	17·0	5·5
40	1·47	18·0	5·0

FIG. 4.1 : *Morphology of Drainage Basins*

Stages	Drainage basins	C.I.	Percentage of the total area
Early	19	Below 1.5	47.5
Middle	10	1.5 − 3	25.0
Youth	11	Above 3	27.5

Dissection Index

Dissection index deals with the relationship of the real area to the projected area between contours and is calculated as proposed by Slauctajas (936)* but this scheme does not give a vivid picture of dissection of a particular region in detail. Dissection index, therefore, is expressed in terms of ratio between maximum relative relief (R.R.) and maximum absolute relief AR. Dissection is directly related to the stage of cycle of erosion. Very high values of dissection closely correspond to youthful stage whereas low values are related to penultimate stage.[6]

It is observed that the values of dissection index varies from 16 to 22 in early youth, 10 to 15 in the middle youth and below 10 in the youth stage (Fig. 4.2). The trend of the values of dissection index in the valley tends to decrease from early stage to youth stage.

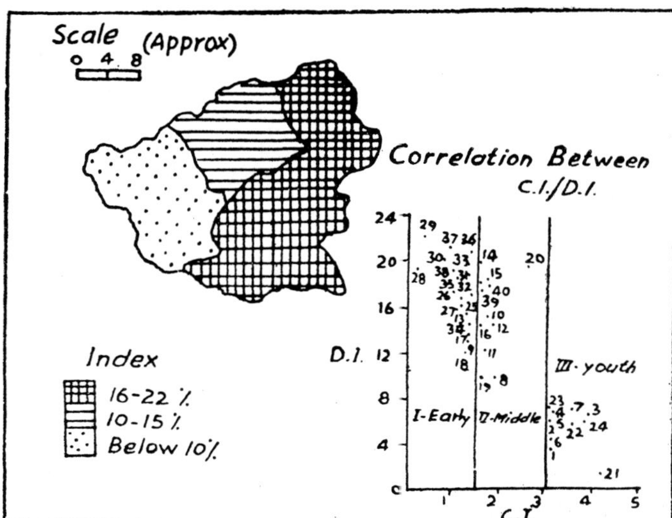

FIG. 4.2 : *Dissection Index, the Nayar Basin*

* $DI = \frac{R_a - P_a \times 100}{P_a}$, where R_a stands for real area, P_a for projected area.

Stages	D.I.	No. of basins	Percentage of the total area
Early	Above 15	19	47.5
Middle	10-15	10	25.0
Youth	Below 10	11	17.5

Hypsometric Integrals

The hypsometric and erosion integrals calculated from the percentage hypsometric curves, give accurate knowledge of the stage of the cycle of discussion. The following hypothetic standards have been recognised for determining the stages.

% of hypsometric integrals	Stages
30	Old
30-60	Mature
60-80	Youth
80-100	Middle
100	Initial

The hypsometric integrals[7] of the Purvi Nayar basin and Paschmi Nayar basin stand at 80 and 65 per cent relatively. They indicate youth and pre-youth stages in respective regions. The Purvi Nayar basin keeps high integral value because about 50 per cent area of the basin is above 3000′.

Stages	% of hypsometric integrals	No. of basins	% of total area
Early	90-100	18	45.0
Middle	80- 90	10	25.0
Youth	60- 80	12	30.0

Unconsumed Upland Ratio

Unconsumed upland ratio shown by highest closed contours is very significant quantitative indicator[8] of conventional geomorphic stage determination, as the hill tops and water divide summits have not been eroded during the initial stage of the cycle of erosion, when valley deeping is most active. With the change of time, water divides and summits are narrowed down because of lateral erosion, which occurs due to the valley widening process. D. Johnson (1933) gave the following criteria to determine the stage of the cycle.

% of unconsumed upland	Stges	
	Initial youth :	Full youth
Above 66	Early youth	
33-66	Middle	
33-10	youth	
10- 5	Early	Maturity
5	Full	Maturity

Stages	% of unconsumed upland	No. of basins	% of total area
Early youth	Above 66	19	47.5
Middle youth	33-66	9	22.5
Youth	10-33	12	30.0

Average Slope Index

Based on the concept of chronological approach the study reveals that the landforms undergo progressive changes starting from initial forms through sequential forms to ultimate forms, with the result the hill tops and interfluves are gradually lowered weathering, rain wash and creep leading to decline in the steepness of slope-angles[9] slopes, therefore it may be correlated with stages of landforms development and the following scale may be suggested :

10-20°	Youth
20-30°	Middle youth
30°	Early youth

Stages	Average slope	No. of basins	% of total area
Early youth	Above 30°	18	45.0
Middle youth	20-30°	10	25.0
Youth	Below 20°	12	30.0

In the Nayar basin about 45 per cent area undergoes above 30° slope and 30 per cent area remains below 20° slope. Thus the Early youth stage significantly coincides the limits as propounded in the hypothesis (Fig. 4.3).

Relative Relief Ratio

Relative reliefs being the difference in heights between the highest and lowest points per unit area are directly related to the degree of dissection. The more the intensity of dissection (particularly vertical erosion) the greater the relative reliefs. Since the high medium and low dissection indices are related to youth, maturity and old stages, the same principles may hold good in case of relative reliefs. The relative relief of the study regions have been calculated by using Smith's method in a grid square of one square mile and have been presented in the following table :

Stages	R.R. categories	No. of basins	% of total area
Early youth	Above 500′	19	47.5
Middle youth	300-500′	10	25.0
Youth	100-300′	11	27.5

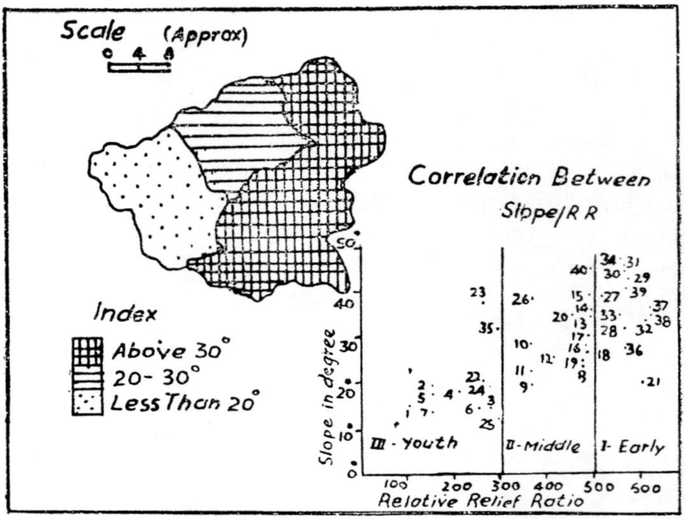

FIG. 4.3 : *Slope Structure of the Nayar Basin*

On the basis of the observations given in the table it is clear that the area under early youth stage of the river basin occurs about 47 per cent above 500′ covering maximum intrabasins[10] of the river Nayar. About 27.5 per cent area of the Nayar basin under youth stage (covering 11 intrabasins) occurs and the rest 25 per cent area of the basin is confined by the middle youth stage. Significantly Relative Relief categories correlates other indices as mentioned above to justify the area delimitation of cyclic stages of the landscape as depicted in Figs. 4.4 and 4.5.

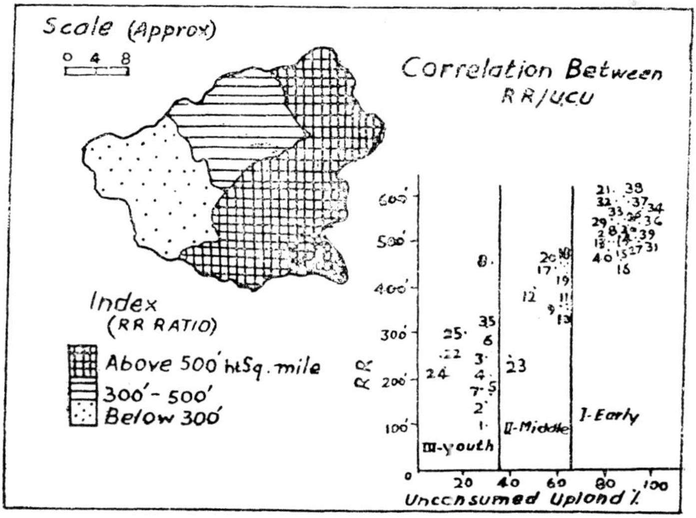

FIG. 4.4 : *Relative Relief Ratio of the Nayar Basin*

Area-delimitation of Cyclic Stages of the Landscape

After studying the observations of the said indices, it is concluded that the river Nayar passes through early, middle and youth stages. On the basis of the geomorphic and its structure valley has been delimited under the following geomorphic tracts :

1. The upper Nayar Dudhatoli Khatil water divide—(Early youth) comprises higher values of dissection index, hypsometric integrals and lower value of circularity index.

   In the basin the maximum inconsumed upland is observed along the Dudhatoli anticline and its environs on account of moderate chemical weathering and the hard structure of Garhwal-nappe. The sub-parallel drainage pattern of the tract has formed the parallel gulley formations, the recession of waterfalls tends to

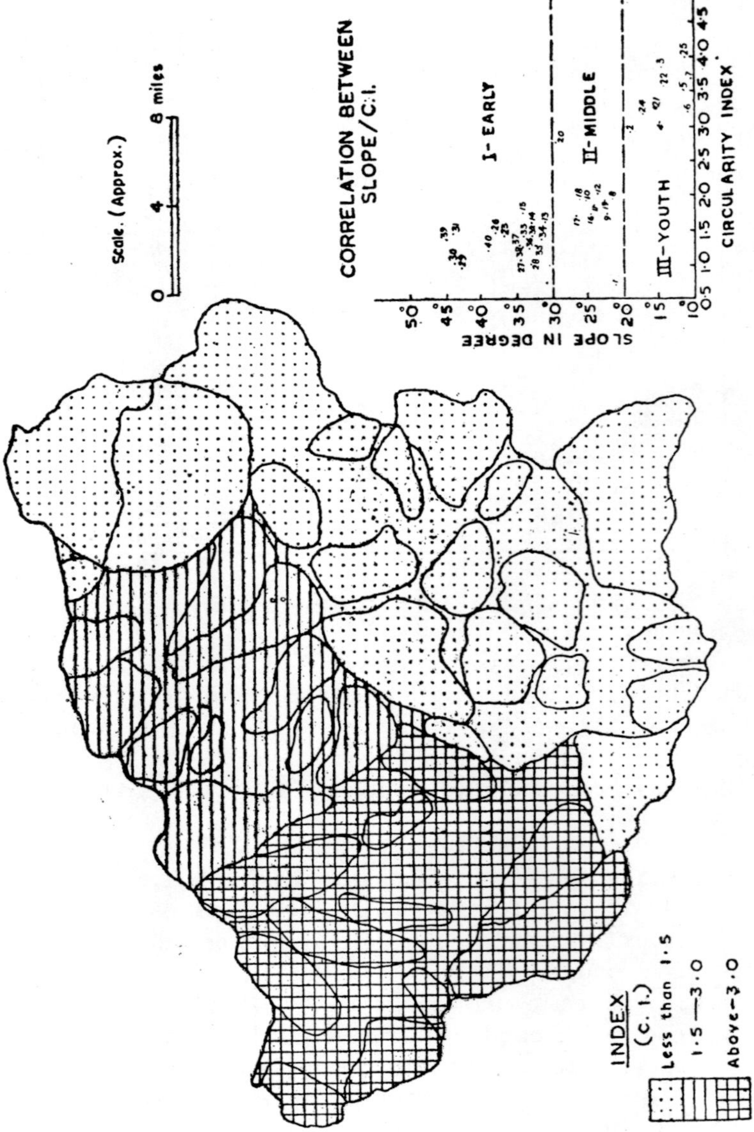

FIG. 4.5 : *The Nayar Basin—Circularity Index*

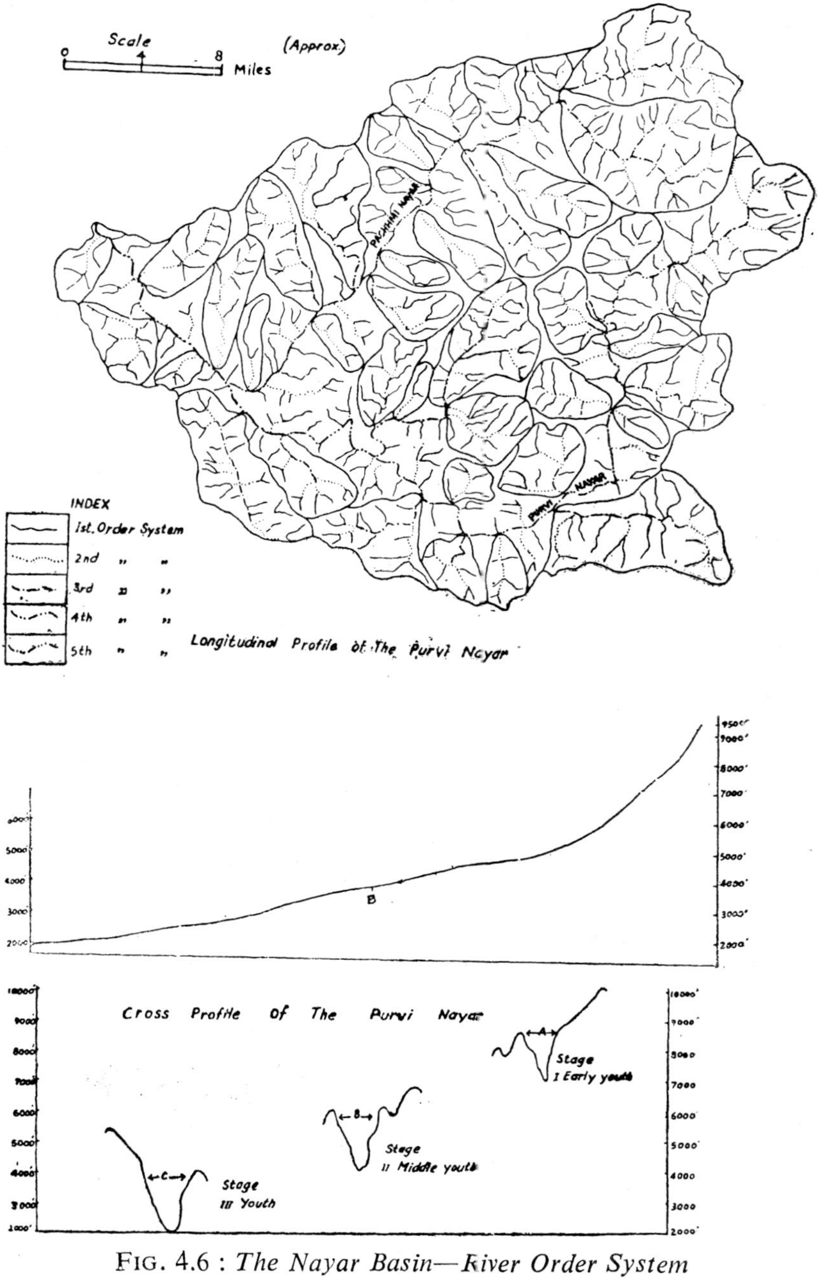

FIG. 4.6 : *The Nayar Basin—River Order System*

incise the landscape to form intrenched meanders, deep channels and gorges. The longitudinal profiles of the river consists a degraded trend with steep slope above 30°.

2. The Purvi-Paschmi Nayar Binsar interfluve—(Middle youth)—consists of landscapic features evolved by lateral and vertical erosions simultaneously. In this tract the waterfalls are gradually receding at considerable space. Thus the knick points are forming waning bench curves, consequently the formation of alluvial terraces starts. The incised meanders of the perennial and non-perennial rivers have been found engrown. The trend of the denudation of interlocking spurs initiates, with the result the stacks and stumps of the spurs are noticed far away the sides of the river benches. The longitudinal profile of the river shows concavo-convex trend with moderate slope ranging 20° to 30°.

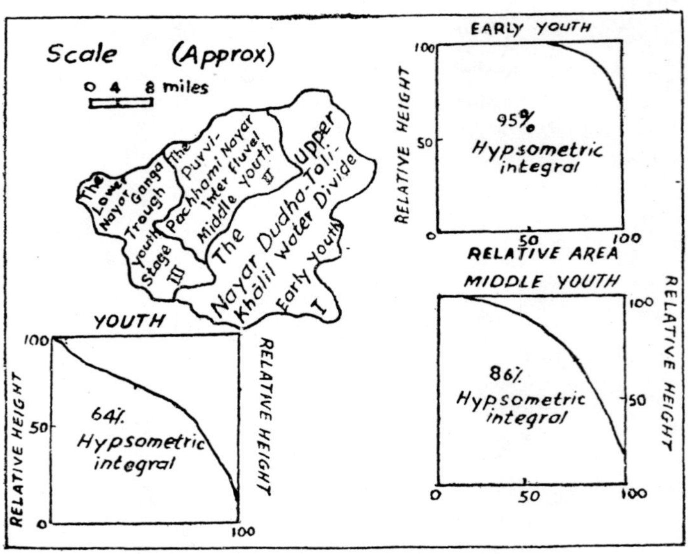

FIG. 4.7 : *Area Delimitation of Cyclic Stages*

3. The lower Nayar Ganga Trough—(Youth stage)—constitutes the parabolic intrabasins with paired river benches. The fifth order system exists with alluvial deposits in the form of linear concentric and triangular shaped terraces. The river profile becomes gentle graded to the local base level of the Ganga. With the result river consists rapids on 20° slope.

Summarisingly, the morphometric parameters held a lot in determining the stages of landform development but it is not an end of the problem rather field observation, field measurement and information of geological structure are also necessary as quantitative and statistical methods offer no substitute for original thought about the problem but they do provide a valuable method of guiding thought and reasoning.

REFERENCES

1. P.B. Saxena and Satya Prakash, 1977 : A study of drainage morphometry for resource planning in the Alaknanda Valley (Garhwal Himalayas) *Indian Journal of Regional Science*, Vol. IX. No. 2.
2. Arthur L. Bloom, 1979 : *Geomorphology : A Systematic Analysis of Late Cenozoic Landforms*, pp. 268-277.
3. L.C. Peltier, 1950 : The Geographical cycle in pereglacial region as it is related to climatic Geomorphology, *A.A.A. Geogr.* 40, pp. 216-230.
4. P.B. Saxena, 1980: A study of land sliding and its land depletion in the Alaknanda Valley (Garhwal Himalayas) 'Pub. *Himalayan Geology*, Vol. IX.
5. $CI = \frac{A_b}{A_c}$ derived by V.C. Miller : A quantitative Geomorphic Study of Drainage Basin Characteristics in the Clinch Mountain Area, Virginia and Tennessee, Technical Report No. 3 (Microfilmed).
6. O.D. Von Engelin, 1942: *Geomorphology:* New York, the Macmillan Company, p. 655.
7. C.A.M. King, 1966 : *Techniques in Geomorphology*, London, Edward Arnold, pp. 319-321.
8. A.N. Strahler, 1950: Equilibrium theory of erosional slopes approached by frequency distribution analysis : *Geol. Soc., American Bull.*, Vol. 57, pp. 571-98.

9. A.N. Strahler, 1956 : Quantitative slope analysis, *Geol. Soc. American Bull, 9*, Vol. 67, pp. 570-96.
10. P.B. Saxena, 1979 : Water resources and potential utilization programme in the Susma valley, Dehra Dun (Garhwal-Himalayas) *Indian Journal Science*, Vol. VI., No. 2.

APPENDIX

Nayar Basin

Name of the drainage Basin	Circularity Index	Dissection Index	Slope Angle (Degrees)	Relative Relief (Ft.)	Hypsometric integral (%)	Unconsumed upland (%)
1	2	3	4	5	6	7
Tairi gad	3.25	4	10.5	100	62	30
Bhal gad	3.12	6	19	150	63	31
Mol gad	3.90	7	15	250	62	31
Guin gad	3.25	6.4	16	200	63	32
Than gad	3.35	7	15	150	62	32
Khar gad	3.05	5	15.5	250	64	32
Ghat gad	3.65	6.5	10.5	170	67	29
Ir gad	1.95	10	22	450	81	35
Chair gad	1.62	12	22	350	82	63
Kutli gad	2.10	15	25	350	83	64
Bidoli gad	1.75	12	23.5	350	81	63
Chhaya gad	1.82	15	24	400	82	37
Syoll gad	1.49	16	31	500	95	82
Syoll gad	1.48	18	32.4	500	95	83
Dhailyuli gad	1.55	18	34	500	95	84
Bali gad	1.75	14	25	450	85	86
Kalgadinadi gad	1.52	13	26.5	450	84	55
Pingad	1.63	12	25	450	83	67
Kalgadi gad	1.56	10	23	450	83	66

(*Contd.*)

Appendix—Contd.

1	2	3	4	5	6	7
Si gad	2.75	20	30	450	83	59
Nadoli ka khala	3.27	4	15	600	97	85
Bengani gad	3.40	6	16	250	74	13
Unika gad	1.42	18	37	250	73	43
Bhisgadhi gad	3.25	7	17	250	79	15
Medi gad	3.90	6.5	12	300	78	25
Kali gad	1.43	17	39	550	96	87
Balsa gad	1.27	18	34	500	94	87
Gherigad	1.20	16	33	550	95	86
Kale gad	0.91	19	44	550	95	85
Kail gad	0.99	22	44.3	550	96	87
Khatal gad	1.42	20	45	550	97	90
Lachhi gad	1.47	19	33	600	96	87
Panchras gad	1.46	19	34	550	96	85
Dhundi gad	1.47	16	46	550	97	85
Sari gad	1.01	19	30.2	300	77	31
Gwarikhal gad	1.42	20	33.2	550	96	87
Nansyan gad	1.25	21	34.5	600	95	90
Chanchariya gad	1.31	19	34.2	600	96	90
Pasol gad	1.29	17	40.5	550	96	92
Bagru gad	1.47	18	43.5	500	95	91
NAYAR	4.80	—	—	—	—	—
NAYAR	2.96	—	—	—	—	—
Nayar as a whole	1.33	—	—	—	—	—

FIVE

Nageshwar Prasad

SOME ASPECTS OF MEANDERING STREAMS OF THE BARAKAR BASIN AND THEIR SINUOSITY INDEXES

MEANDERING STREAMS OF THE BARAKAR BASIN

FOLLOWING the general but natural rule that no straight river occurs in a state of nature (Davis, 1913), the Barakar Basin (23° 41′ 30″ N to 24° 31′ 30″ N Lat. and 85° 10′ E to 86° 54′ E Long.), in its own case also, is characterized by meandering and sinuous courses of streams. The Basin is drained mainly by Karkara, Saghar, Harnaro, Keso, Pachkara, Baretu, Irga Khakho, Usri, Beri, Rajoya and Barki Jhar from the north, and Kewta, Kolhuwatari, Barsoti, Khero, Bakra, Chirki, Khudia and Pusai from the south including the main stream Barakar (a tributary of the Damodar) in the basin area of about 7026.1 km^2 in the Chota Nagpur Highlands of Bihar (India). Barsoti is its principal tributary from the south and the Usri from the north. Some of the streams are meandering and all are sinuous. Thus meandering is a natural state of affairs, obtained by gradual stages, as the channel departs from the straight line (Tanner, 1968). Straight channels are so rare as to be almost non-existent (Leopold and Wolman, 1957).

Fig. 5.1 shows some of the meandering streams of the Barakar Basin. It just provides a bird's-eye view of main streams located at different sections of the Basin. The rivers Barakar, Barsoti, Irga, Usri and Khudia, etc., flow on the different planation levels of varying altitudes and lithological characters. But the tortuosity of the course of the meandering streams is seen everywhere. The universality of sinuosity of streams may be

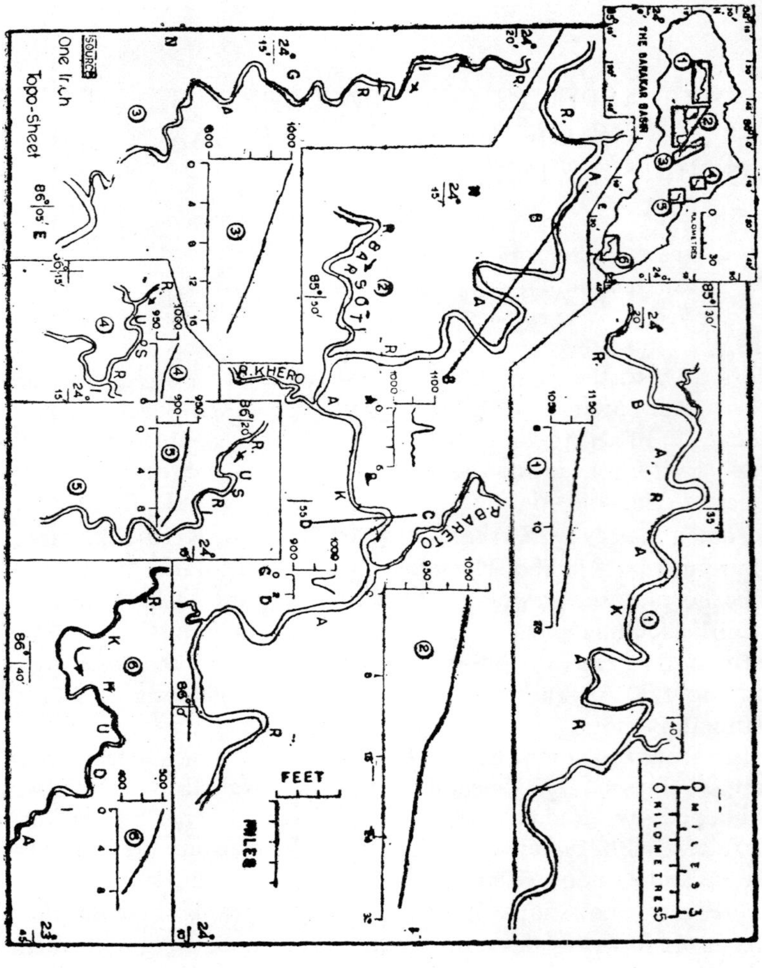

FIG. 5.1 : *Meandering Streams*

observed throughout the Basin. Almost all the streams of the Basin are more or less in sinuous state in general. We may, therefore, conclude that in the Barakar Basin also, meandering has been a natural state of affairs (Bagnold, 1960; Leliavsky: 1959; Leopold and Langbein, 1966; Leopold and Wolman, 1960; Melton, 1938; Scheidegger, 1961; Thornbury, 1954; Wooldridge and Morgan, 1965).

What is here most important to note is that the meandering streams of the Barakar Basin have established that meandering is not limited to large rivers which are mature in the Davisian sense (or supermature, or "old"). In the Barakar Basin, streams of all sizes meander at all altitudes. This holds true even in case of those streams which have many small waterfalls and rapids present throughout meandering reaches. Meandering streams, in general, have gradients between about 24 cm/km and about 11 m/km in case of the Barakar river. The river Barsoti has a gradient of 3.12 m/km and that of 2.14 m/km in the Usri river (Table 5.1).

It should be noted here that the Barakar Basin presents a variety of landscapes characterized by scarps, rugged topography, long, deep and terraced river valleys; waterfalls and rapids, residual hills, gullies etc. The Basin consists of crystalline gneisses and granites.

Another striking feature is that these meandering streams have not developed on wide flood plains; rather they flow in comparatively deep valleys (Fig. 5.1). They are actually entrenched, which flow through a gorge or chasm cut in bedrock. These meandering streams seem to be commonly ascribed to epeirogeny. It is because of this that the river courses in the Barakar Basin are meandering while "the strike of the jointed gneiss probably does not turn as sharply and frequently as the meanders" (Ahmad, 1969). Here it may be suggested that these streams are most probably antecedent in nature. The region has been subjected to peneplanation several times and hence the course of the rivers would have been marked by meandering and the meanders have been able to degrade the courses sufficiently fast to keep pace with an uplift rising across its path. Thus, these meandering streams of the Barakar Basin seem to be the feature of rejuvenation. The epeirogenetic uplift during the Tertiary times has been responsible for initiating a new denudational cycle.

Now we would like to note how much sinuosity these meandering streams did develop.

SINUOSITY INDEXES

"Meandering is applied to sinuous channels that exhibit a certain degree and regularity of sinuosity" (Brice, 1962). Of all the existing sinuosity indexes, the most common is that which is derived by dividing the length of a reach as measured along the channel by the length of a reach as measured along the valley. According to this index a stream having sinuosity of 1.0 is straight, just above unity to 1.3 is sinuous, and more than 1.3 is meandering. A second and more recent index is derived by "dividing the length of the channel in a given curve by the wavelength of the curve". (Leopold and Langbein, 1966). A third index of sinuosity is "derived by dividing the thalweg, or the line of maximum depth, by the length of the valley" (Leopold, Wolman and Brush, 1960). These indices do not consider the role of surface configuration of topography in directing the course of the stream.

Noting some inherent flaw in these existing sinuosity indices, Mueller (1968) has derived a new sinuosity index very recently and in that he has given due importance to the channel and surface configurations in making the sinuous bends of the rivers. Mueller's insistence on including the role of the topographic factors in the computation of sinuosity indexes is remarkable, as it gives a new dimension to the concept of sinuosity. According to him, the tortuosity of the course of a meandering stream is the outcome of both topographic and hydraulic factors and is expressed by a ratio called the index of sinuosity.*

**Index of Sinuosity*

$$\text{HSI (Hydraulic Sinuosity Index)} = \frac{CI - VI}{CI - 1} \times 100$$

$$\text{TSI (Topographic Sinuosity Index)} = \frac{VI - 1}{CI - 1} \times 100$$

$$\text{CI (Channel Index)} = \frac{\text{CL (Channel Length)}}{\text{AL (Air Length or Straight Length)}}$$

In the Barakar Basin, measurements have been taken in the full course of the Barakar river, the Bārsoti and the Usri for the sample study; although all the rivers show some amount of sinuosity, as noted above. Both the hydraulic sinuosity index (HSI) and the topographic sinuosity index (TSI) for the meandering bends of these important rivers have been determined from the one-inch Topographical Sheets. The data have been tabulated in Table 5.1.

The study reveals the following points:

(1) The Hydraulic Sinuosity Index (HSI) is not high in most of the river reaches. The HSI ranges from 0 to 23 for all the measured river-bends of the Basin. It is highest (23.10%) at Bend II of the Barakar river. Normally the percentage of the HSI is not high.
This shows that the rivers have not developed flood-plains in general. It is because of the fact that the rivers pass through the hilly tracts and their valleys are comparatively constricted. The low percentage of the HSI indirectly suggests that the meandering streams do not belong to the initial denudational cycle; but the area has been rejuvenated and the role of the topography has been increased thereby.

(2) In the majority of the cases the channel deviation is controlled topographically. It is evident from the high

$$\text{VI (Valley Index)} = \frac{\text{VL (Valley Length}}{\text{AL (Air Length)}}$$

$$\text{SSI (Standard Sinuosity Index)} = \frac{\text{CI}}{\text{VI}}$$

N.B.: (1) CI is an index of total sinuosity, both hydraulic and topographic.

(2) VI is an index of total topographic sinuosity.

(3) Valley Length (VL), along a stream, is the length of a line which is everywhere midway between the base of the valley-walls. It will equal Channel Length (CL) wherever the valley-walls descend directly to the water's edge, and will be less than CL wherever a floodplain has developed.

(4) Air Length (AL) is the shortest distance between two points of a meandering bed of a river (for big rivers) and the shortest distance between the source and mouth of the streams in case of the small rivers.

TABLE 5.1: **Sinuosity Indexes (Mueller's Model)**

Streams	Valley Length (VL)						Channel Length (CL)		Air Length (AL)	
	Left Bank		Right Bank		Average Length					
	Miles	Kms.	Miles	Kms.	Miles	Kms.	Miles	Kms.	Miles	Kms.
1	2	3	4	5	6	7	8	9	10	11
BARAKAR RIVER										
Bend I (A-B) 595-424 M	9.70	15.52	9.70	15.52	9.70	15.52	9.70	15.52	7.80	12.48
Bend II (B-C) 424-422 M	5.10	8.16	5.00	8.00	5.05	8.08	5.30	8.48	4.20	6.72
Bend III (C-D) 422-370 M	14.70	23.52	14.43	23.09	14.57	23.51	14.63	23.41	8.70	13.92
Bend IV (D-E) 370-344 M	13.57	21.71	13.47	21.55	13.52	21.63	13.77	22.03	9.00	14.40
Bend V (E-F) 344-326 M	24.33	38.93	24.53	39.25	24.43	30.09	24.53	39.25	15.50	24.80
Bend VI (F-G) 326-262 M	35.73	57.17	35.13	56.21	35.45	56.69	36.53	58.45	21.70	34.72
Bend VII (G-H) 262.214	M24.33	38.93	24.53	39.25	24.43	39.09	25.13	40.25	20.40	32.64
Bend VIII (H-I) 214-103	M46.03	73.65	44.63	71.41	45.33	72.53	46.33	74.93	37.60	60.16
Total	173.49	277.59	171.42	274.28	172.46	275.93	176.42	282.32	91.20	145.92
BARSOTI RIVER From Source to Mounth (457-284 M.)					33.63	53.81	33.63	53.81	24.70	39.52
USRI RIVER										
Bend I (from source to Kharagdiha) 395-358 M.					7.41	11.86	7.41	11.86	6.25	10.00
Bend II (from Kharagdiha to Mouth) 358-214 M.	41.8	66.88	41.00	65.60	41.40	66.24	42.00	67.20	34.00	55.60
Total					48.81	78.10	49.41	79.06	40.25	65.60

N.B. :

A=Source of Barakar; B=Confluence point of Barakar and Kewta near Padma C=Spot height 422 M.(1385′) D=Near Barhi; E=Near Tilaiya; F=Confluence point of Barakar and Keso; G=Confluence point of Barakar and Irga; H=Confluence point of Barakar and Usri; I=Mouth of Barakar.

Streams	C.I.	V.I.	H.S.I. %	T.S.I. %	S.S.I	Gradient in Metre/Km.	Average Width in Metre
	12	13	14	15	16	17	18
BARAKAR RIVER							
Bend I (A-B) 595-424 M	1.24	1.24	0.00	100.00	1.00	11.02	10
Bend II (B-C) 424-422 M	1.26	1.20	23.10	76.90	1.05	0.24	130
Bend III (C-D) 422-370 M	1.68	1.67	1.47	98.53	1.01	2.22	290
Bend IV (D-E) 370-344 M	1.53	1.50	5.66	94.34	1.02	1.18	320
Bend V (E-F) 334-326 M	1.58	1.57	1.72	98.28	1.01	0.46	190
Bend VI (F-G) 326-262 M	1.68	1.63	7.35	92.65	1.03	1.09	260
Bend VII (G-H) 262-214 M	1.23	1.20	13.04	86.96	1.03	1.19	310
Bend VIII (H-I) 214-103 M	1.24	1.20	16.67	83.33	1.04	1.48	650
Total	(Av.) 1.94	1.89	5.32	94.68	1.03	1.74	270
BARSOTI RIVER From Source to Mouth (457.284 M)	1.36	1.36	0.09	100.00	1.00	3.20	
USRI RIVER							
Bend I (from source to Kharagdiha) 395-358 M.	1.19	1.19	9.12	100.00	1.00	3.12	150
Bend II (from Kharagdiha to Mouth) 358-214 M	1.21	1.19	9.52	90.48	1.02	2.14	190
Total	(Av.) 1.20	1.19	9.32	95.24	1.01	2.63	170

C.I.=Channel Index=$\frac{CL}{AL}$, V.I.=Valley Index=$\frac{VL}{AL}$, H.S.I.=Hydraulic Sinuosity Index=% of $\frac{CI-VI}{CI-I}$

T.S.I.=Topographic Sinuosity Index=% of $\frac{VI-I.}{CI-I}$; S.S.I.=Standard Sinuosity Index=$\frac{CI}{VI}$

topographic sinuosity indices (Table 5.1) in most of the river bends. The higher TSI indicates greater, irregularity of the initial surface.

(3) The Standard Sinuosity Index (SSI) increases generally towards the lower reaches of the rivers. The exceptions are at Bend II (1.05) and at Bend V (1.01) of the Barakar river. Similarly at these bends, the HSI is also irregular, i.e., this breaks the increasing tendency towards the lower reaches. The explanation is apparent. At Bend II, this is not the natural meandering but it is the case of stream-piracy. At Bend V, the SSI is undesirably low, apparently because of narrow valley (190 m width). Here the Barakar flows between ridges in a deep valley, although the gradient is comparatively very low (0.46 m/km)

(4) No doubt, the Table 5.1 shows the general increasing tendency of the SSI (1.00 to 1.04 in case of the Barakar and 1.00 to 1.02 in case of the Usri) towards the lower reaches. But at the same time there is a general increasing tendency of the gradient also (1.18 m/km at Bend III while 1.74 m/km at Bend VIII). These figures, of course, do not indicate strong hydraulic factors of sinuosity; because the natural rule follows that the relative increase in the SSI coincides with the relative decrease in the gradient of the stream.

(5) As the river progresses in the cycle of erosion, the role of the hydraulic sinuosity increases and the role of the topographic sinuosity decreases, but the index of the total sinuosity or the SSI is less than desirable as was deemed essential in the Mueller's analytical tool in the study of regional landforms and hydrology.

Thus concluding all these we may say that the streams of the Barakar Basin are sinuous of the Low SSI, and they are in the maturity or in the middle stage between youth and old stage of the cycle of erosion. The entrenched character of the meandering streams leads to the conclusion that the region has experienced rejuvenation which has increased the dominating influence of the topographic sinuosity, and thereby slackened the development of the floodplains reducing the amount of the Hydraulic Sinuosity Index (HSI).

SUGGESTED READINGS

Ahmad, E., 1969: Ranchi to Rajrappa—Some geomorphic observations, *Geographical Outlook*, Vol., 6, p. 1X.

Bagnold, R.A., 1960: Some aspects of the shape of River Meanders, *U.S, Geol. Surv. Profess. Paper*, 282-E, pp. 135-144.

Brice, J.C., 1962: Streams, *Laboratory Studies in Geology*, No. 207, Freeman, London, p. 8.

Davis, W.M., 1913: Meandering valleys and underfit rivers, *Ann. Ass. Am. Geog.*, Vol. 3, p. 4.

Leliavsky, S., 1959: *An Introduction to fluvial hydraulics*, Constable and Co., London, pp. 92-191.

Leopold, L.B.; Bagnold, R.A.; Wolman, M.G. and Brush, L.M. Jr., 1960: Flow resistance in sinuous or irregular channels, *U.S. Geol. Surv. Profess. Paper*, 282 D, p. 133.

Leopold, L.B. and Langbein, W.B., 1966: River Meanders, *Sci. Amer.*, 214(6), pp. 60-70.

Leopold, L.B. and Wolman, M.B., 1960: River Meanders, *Bull. Geol. Soc. Am.* 71, pp. 769-794.

Melton, F.A., 1938: Underfit meanders of floodplain streams, *Proc. Geol. Soc. Am.*, p. 324.

Mueller, J.E., 1968: An introduction to the Hydraulic and the topographic sinuosity indexes, *Annals Ass. Am. Geog.*, Vol, 58, pp. 371-385.

Scheidegger, A.E., 1961: *Theoretical Geomorphology*, Springer Verlag, Berlin, pp. 176-190.

Tanner, W.E., 1968: *The Encyclopedia of Geomorphology* (Ed., by R.W. Fairbridge), Reinhold Book Corporation, New York, p. 959.

Thornbury, W.B., 1954: *Principles of Geomorphology*, Johan Wiley and Sons, New York, pp. 618.

Wooldridge, S.W. and Morgan, R.S., 1965: *An outline of Geomorphology*, Longmans, London, pp. 142-43

SIX

Nageshwar Prasad and *S. N. P. Gupta*

STREAMLINE SURFACES OF THE BARAKAR BASIN
A Perspective of Landscape Evolution

INTRODUCTION

THE STREAMLINE surfaces of the Barakar Basin (Fig. 6.1) indicate the amount of denudation still to be achieved (Monkhouse and Wilkinson, 1966). Each line outlines a class correlation with existing planation surface between 103 metres (338′) and 610 metres (2000′) and above. Most of the area of the Basin, as a whole, is in upland, valleys are narrow but deep and relief is not very great over most of the area. The Basin (23°41′ 10″ to 24° 31′ 30″ N Lat. : 85°20′ to 85°54′ E Long./ 7026.1 km^2) consists of crystalline gneiss and granites and is supposed to represent an originally solidified crust. There is no evidence of well-defined Caledonian cycle earth deformation in the Basin, but the Hercynian revolution probably resulted in small linear fractures as the Gondwana coal bearing basins. The epeirogenetic uplift during the Tertiary times is supposed to be responsible for initiating a new denudational cycle which has carved out the present landscape of the Barakar Basin.

METHODS AND OBSERVATIONS

The streamline surfaces of the Barakar Basin have been prepared by plotting intersections of actual contours within the main streams (Dury, 1952). Streamline contours of 553 m (1750′), 457 m (1500′), 381 m (1250′), 305 m (1000′) 228 m (750′) and 152 m (500′) have been drawn by joining points where main streams cross actual contour lines.

The dissection that has been registered along the scarps and the Parasnath-Tundi areas, comes out in the map giving an idea of even further dissection. The area of active dissection seems to agglomerate along the margins of these regions. The surface seems to be stepping down to north particularly in the scarp-zone and the Parasnath-Tundi area, and to the southeast in general. Erosion by the rivers Rajoya, Usri, Irga, Bareto, Kewta, Barsoti, Chirki, Khudia, etc. along with the main stream Barakar, is directly marked in the map (Fig. 6.1). Narrow valleys, gorges, rapids and waterfalls have developed in these sections frequently.

DISCUSSION AND CONCLUSION

The comparison between two maps physiography (Fig. 6.2) and the streamline surfaces (Fig. 6.1) clearly shows the areas, deeply dissected by rivers and extensive interfluves surviving in between the valleys. The study brings out the following facts to be noted (Prasad, 1979):

(1) The rivers of the Barakar Basin are active in denudation of the area. Rapid denudation is associated with gorges and narrow valleys of the Basin, the intensity of runoff, amount of load discharge, channel gradients and ruggedness of the surface, etc.

(2) The streamline surfaces mark somewhat new physiographic zones, different from the present relief features. These have extended to the probable lower areas and minimised the upper surfaces. The Table 6.1 gives clarification to this point:

TABLE 6.1: **Present and Probable Physiographic Units**

Present Physiographic units	Area in km^2	Probable physiographic units of streamline surfaces	Area in km^2
The Scarpzones	338	The Scarp Zones	189.6
The lower Hazaribagh Plateau	3730	The Lower Hazaribagh Plateau	3332.3
The Outer Erosion Fringe	2643	The Outer Erosion Fringe	3019.6
Damodar Trough	315	Damodar Trough	484.5
	7026		7026.0

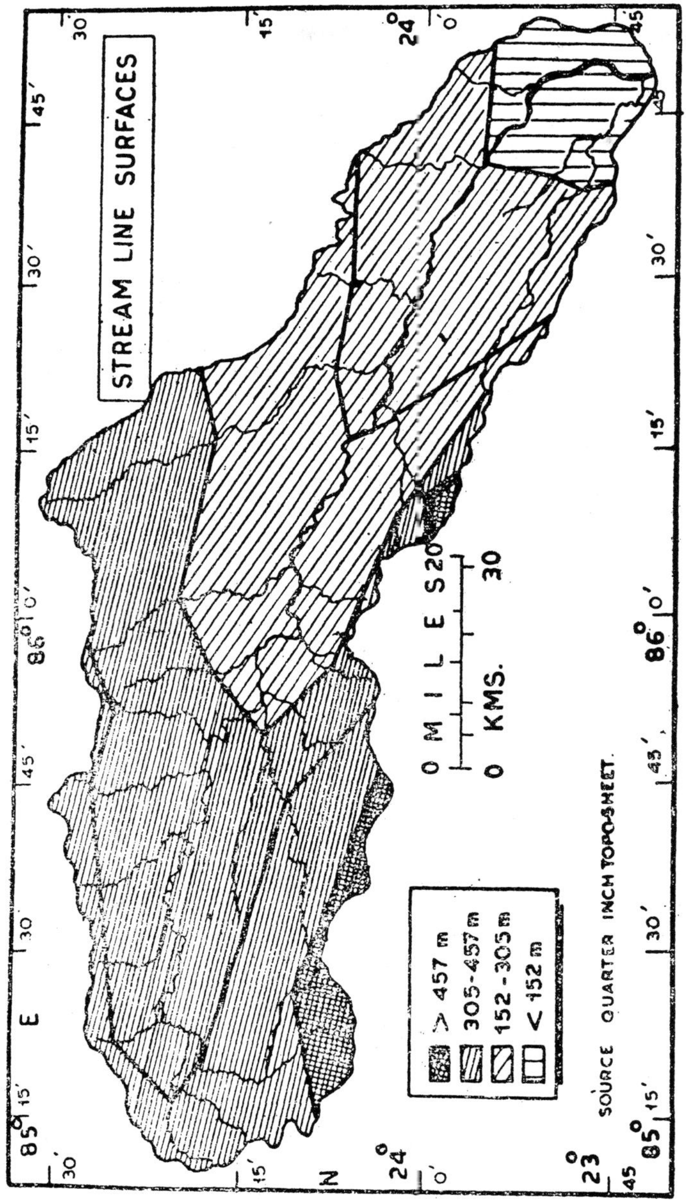

FIG. 6.1 : *The Barakar Basin—Streamline Surface*

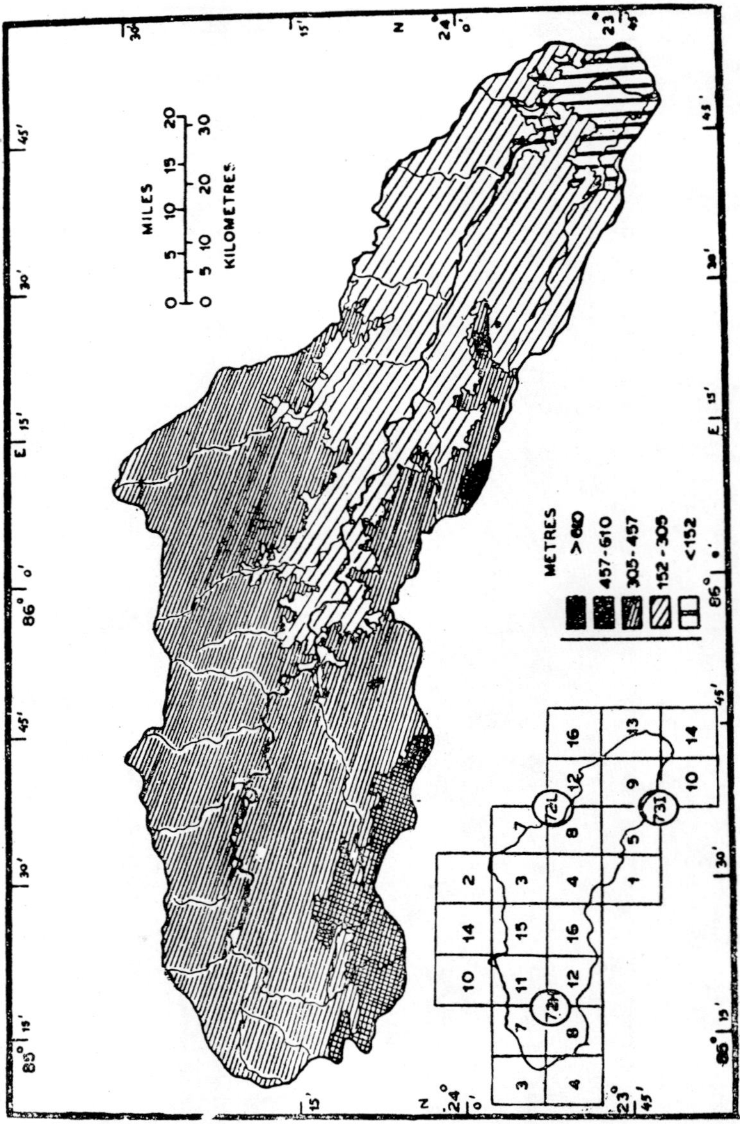

FIG. 6.2: *The Barakar Basin—Physiognomy*

(3) The streamline surfaces give an idea of the direction of denudation. The scarp zone in the west and the Parasnath-Tundi area in the mid-southeast seem to be stepping down northwards, while the tendency of the whole Basin is east and southeastward.

(4) It seems that the lower denuded areas are receding westward rapidly specially after the uplift of the area during the Himalayan orogeny which has given a new strength to drainage making it capable to erode down surface rapidly.

(5) The Barakar, Kewta and Barsoti rivers in the scarp zone are very active resulting in apparent scarp-retreat. It seems that the scarp zone would have extended for northward formerly. The hills rising above 2187′ (667 m) between Gorhar *nala* and Khero may be a proof. It has probably been left isolated by retreat of the scarp zone in course of active erosion by the tributaries of the Barsoti. It is also expected that the Mahabar Hills and the surrounding areas will be disconnected physiographically from the scarp zone by headward and lateral erosion by Kewta and Barsoti.

(6) Similarly in the middle southeast the Khudia and Chirki rivers are forking up the area between Parasnath and Tundi Hills. This has resulted in narrowing the Tundi Hills subsequently. If this would have consisted of softer rocks, the stream line surface of 750′ (228 m) and 1000′ (305 m) would have receded up to the Parasnath Hills. But the hills owe their preservation to their hardness. In spite of this structure the forking up of the rivers Khudia and Chirki will isolate the hills in course of time, and then it will remain as a mere monadnock on the lower eroded surface.

(7) In the mid-north the Usri and Irga are very active. The streamline of 1000′ (305 m) joins these valley-tips surviving a major interfluve in between. Similarly a number of interfluves are surviving in the Basin. These will be denuded by the active rivers in course of time, isolating some hills and hillocks, monadnocks and tors, as existing in the present time also by the former retreat of streamline surfaces.

(8) The map (Fig. 6.1) also clarifies that most of the interfluves of large sizes are located in the middle and mid-east portion of the Basin below the 305 m stream-line. This shows that in these areas the account of denudation is larger than in other areas. This is mainly due to the increasing water volume and thereby increased eroding capacity in the middle reaches of the Basin.

(9) It is no doubt that the damming of the river Barakar at Tilaiya and Maithon has slackened the intensity of retreat to some extent, but it is an evident fact and a natural phenomenon in the Basin which prevailed in the past, is active in the present time and will characterise denudation in future.

(10) Thus, concluding all these we may say that the stream-line surfaces of the Basin show that the area is under rather rapid denudation by a number of active rivers. However, a greater amount of denudation is still to be achieved. Most of the area of the Basin is in upland, valley are narrow but deep, and relief is not very great over most of the Basin. Therefore, the Basin is, most probably, between youth and maturity, better say, it is progressing toward maturity.

SUGGESTED READINGS

Dury, G.H., 1952: Methods of cartographical analysis in geomorphological research, *Indian Geog. Soc. Silver Jubilee Souvenir Volume*, Madras, pp. 136-140.

Monkhouse, F.J. & Wilkinson, H.R., 1966: *Maps and Diagrams* Methuen & Co. Ltd, London, p. 66.

Prasad, N., 1979: Geomorphology of the Barakar Basin, Unpublished Doctorate Thesis, Ranchi University, Ranchi, pp. 183-86.

SEVEN

O. P. Singh

MORPHOMETRIC EVALUATION OF LANDFORMS IN PALAMAU UPLAND

PROLOGUE

IT IS a great complexity and indecision regarding the definition and the approach in the study of landforms. Geomorphologists like Russell, Kesseli, Hammond, Smith and others have advocated for the geographic approach in landform study. Russell[1] has pointed out that geologists are mainly concerned with 'structure, process and time' while geographers are related to 'what, where and how much'. Kesseli[2] has remarked that a study of assemblages of geomorphic landscape is the best means of establishing true geographer's geomorphology. He laid three concepts, i.e., (i) concept of descriptive landform analysis, (ii) concept of landform type and (iii) concept of landform type region. Hammond[3] found it necessary to coin a new word 'landsurface form' in order to avoid unnecessary dualism between the explanatory (genetic) and descriptive (surface form) branches. Smith[4] also favoured non-genetic study of landforms for geographers.

Some geomorphologists like Dury[5] and Strahler[6] gave more emphasis to the mathematical symbolization of landform characteristics, especially in the field of drainage network. Zakrzewska has also pointed out that "descriptive (preferably quantitative) landform analysis, as an objective statement of facts, should be the first step in any type of landform study, especially in geography, where facts of the earth surface are always in focus".[7]

In the light of the above development in the landform study, the present paper is an attempt to go on intermediate path between the descriptive and quantitative approaches of study. The study has been dealt into four steps as suggested by Weaver,[8] i.e., (i) Formulative basic criteria, (ii) Changing the scale of observation, (iii) Dismantling the landforms, and (iv) Reassembling the landforms.

Methodology

For the analysis, One Inch Topo Sheets of Survey of India have been used. The whole area (8511.08 sq. km) of Palamau Upland has been divided into unit square of 10 sq. km each and within these units, all the morphometric attributes like absolute relief, relative relief, dissection index, drainage frequency, drainage density and slope have been enumerated, classified and reproduced in the form of maps and diagrams by suitable cartographic devices and their certain categories have been designed. For reassembling the landforms, all the morphometric attributes have been superimposed on a single frame and the most coinciding boundaries have been selected to delimit the landform categories. Where this superimposition is not very striking, the boundaries have been marked on the most suitable lines. In this way, the entire terrain has been classified and their micro-details have tabulated. Finally, the boundaries are checked up by the help of some additional information collected from the field, such as recent changes which took place due to erosion intensity along the course of major rivers etc.

The Region

Palamau Upland (23°39′ 40″ N—24° 33′ 50″ N and 83° 20′ E—85° 56′ 10″ E), a' 'broken and divided country' of Palamau District, Bihar is characterised by rugged terrain, poor soil, scrub forest and poor agriculture experiencing tropical monsoon climate. After feeding the data of mean annual rainfall and temperature in Peltier's scale of morphogenetic region and processes, the study region exhibits moderate morphogenetic region characterised by moderate chemical weathering, minimum wind action, but maximum alluvial erosion by the North Koel and its tributaries.[9] Physiographically, the area (8511.08

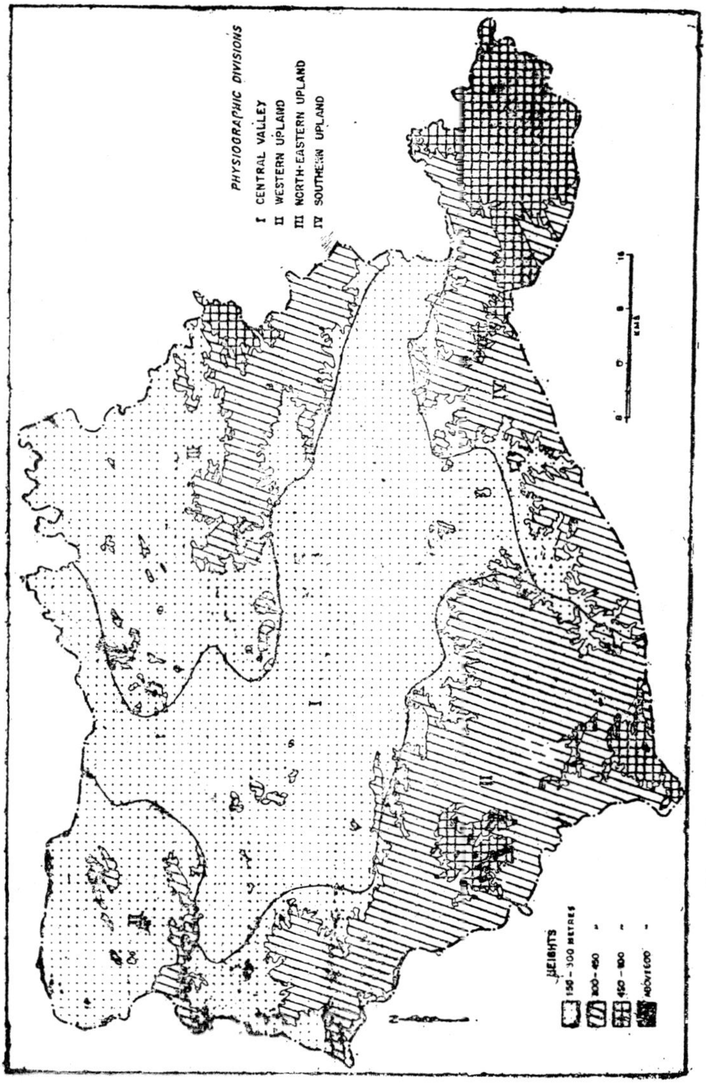

FIG. 7.1 : Major Physiographic Units and Altitudinal Zones

sq. km) with an average elevation of about 340 metres, is a part of Peninsular India, exhibiting, as it does, two major relief features, the broad valley (31.23% area) in the centre and the uplands (68.77% area) surrounding it. Representing the three geological formations, i.e., Archarans, Vindhyans and Gondwanas, the present surface configuration of the study region is a function of different phases of uplifts, tectonics and the process and magnitude of erosion in consonance with the resistivity of rocks.

Morphometric Analysis

Out of the various morphometric attributes, some important ones are selected for the present study, such as absolute relief, relative relief, dissection index, drainage density and slope. In general, the class values of different morphometric attributes are the same as suggested by Singh[10] with some subdivisions depending upon the nature of relief.

Absolute Relief

In the first phase of the morphometric analysis, the area has been divided into nine altitudinal zones (Fig. 7.1). The areaheight break-up (Table 7.1) indicates that a large proportion (29%) of the total area falls in the altitudinal zone of 300-375 metres followed by 375-450 metres (23.22%). Beyond these height-groups, the areal shares of the altitudinal zones decrease so much so that it is only 0.24 per cent for above 750 metres. The scarp and hillocks fall in the height group of 600-750 metres and above.

The absolute relief analysis comprises of profiles and areaheight relations, for which all types of profiles, hypsographic curves, altimetric frequency and spot-height frequency histograms have been separately drawn.

The serials profiles (Fig. 7.2) indicate cross-sectional picture of the whole region. The composite profiles carry only the highest sections of each serial profiles. The projected profiles drawn for the study area (Fig. 7.3) well indicate the high magnitude of the dissection of the region and vertical differences between the hill-tops and valley floor. Many of the hill-tops abruptly project above the nearby surface rendering

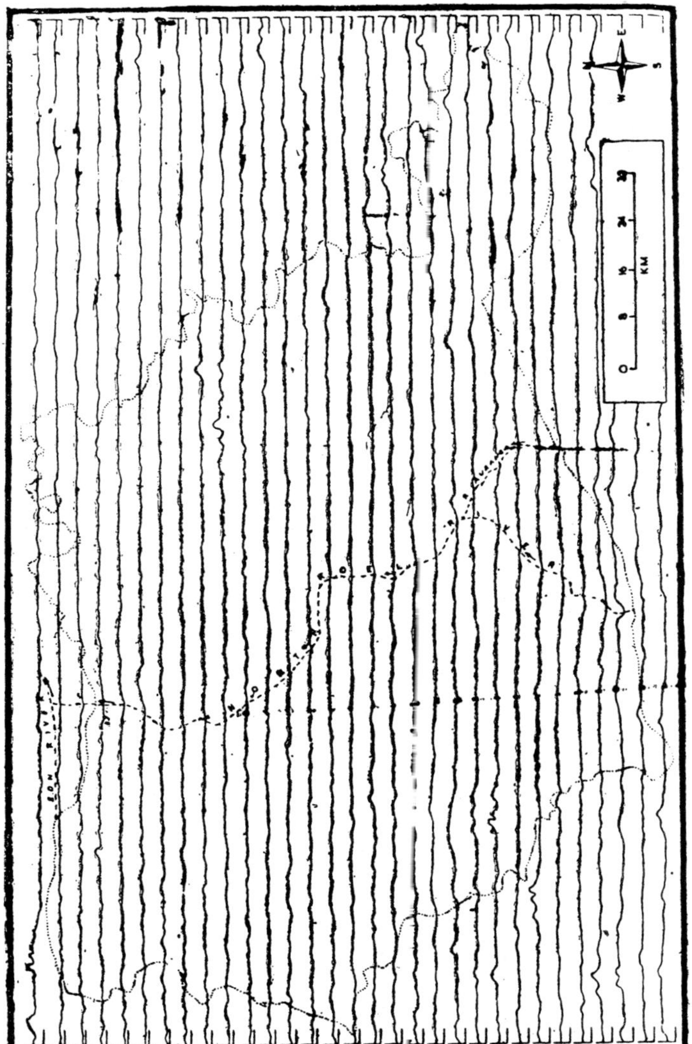

FIG. 7.2 : *Serials Profiles*

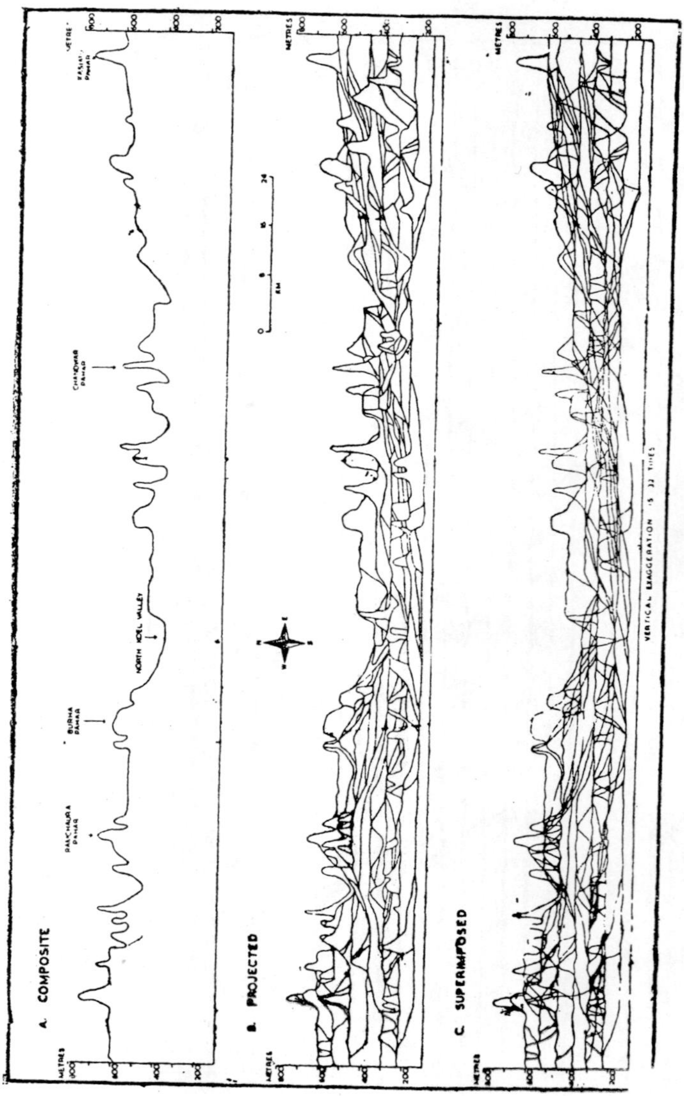

FIG. 7.3 : *Profiles*

higher relative relief. The superimposed profiles represent the clustering of the majority of hill-tops at four different levels which represent four erosion surfaces in the study area at the height groups of 150-250 metres, 300-400 metres, 500-600 metres and 700-800 metres. A minor surface is also visible in the height group of 425-475 metres (Fig. 7.3).

TABLE 7.1 : **Area-Height Break up**

Height Group (metres)	Total Area %	Area West of the North Koel %	Area East of the North Koel %
150–225	8.66	56.94	43.06
225–300	17.63	48.12	51.88
300–375	29.00	48.42	51.58
375–450	23.22	56.99	43.01
450–525	10.71	46.07	53.93
525–600	7.16	26.05	73.95
600—675	2.77	43.48	56.52
675—750	0.62	87.80	12.20
Above—750	0.24	100.00	—
Total	100.00		

The hypsographic curves present a generalisation of the surface character along with the areas incorporated in different height groups. The notable fact from the Fig. 7.4(A) is a sharp decline above 635 metres and below 260 metres which is probably due to small coverage by hillocks and valley bottom commonly occurring into the respective height groups. The dominance of the area between 300-375 metres is due to the wide coverage of partly the Central Valley and partly the gentle sloping low lying area of the Uplands.

Two separate frequency histograms using the spot-heights and highest points in the grid square of 10 sq. km each have been drawn, Fig. 7.4 (B), 7.4 (C). Both the histograms indicate four erosion surfaces in the study area which very well correspond to the results drawn through the superimposed profiles (Fig 7.3).

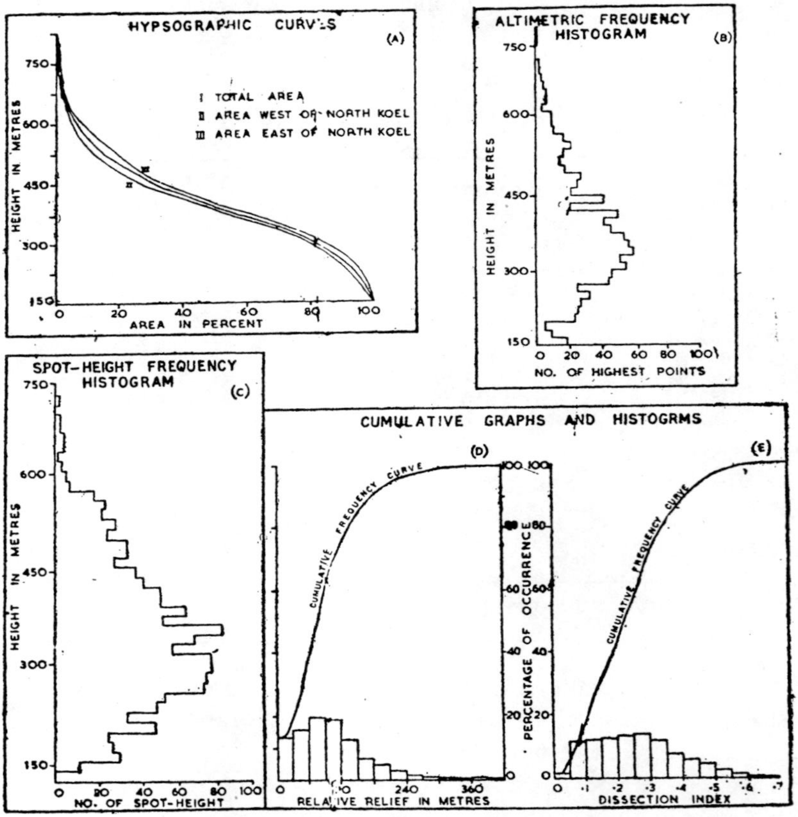

FIG. 7.4 (A) : *Hypsographic Curves*
FIG. 7.4 (B) : *Altimetric Frequency Histogram*
FIG. 7.4 (C) : *Spot-Height Frequency Histogram*
FIG. 7.4 (D) : *Cumulative Graphs and Histograms*
FIG. 7.4 (E) : *Cumulative Graphs and Histograms*

Relative Relief

The frequency distribution of relative relief computed in the grid of 10 sq. km each, exhibits a marked unevenness when grouped in the class interval of 30 metres, as the coefficient of variation stands as high as 66 per cent. The whole area may be divided into six categories which are as follows :

(A) Extremely Low Relative Relief (0—15 metres),
(B) Moderately Low Relative Relief (15—30 metres),
(C) Low Relative Relief (30—600 metres),
(D) Moderate Relative Relief (60—120 metres),
(E) Moderately High Relative Relief (120—240 metres),
(F) High Relative Relief (240 metres and above).

The above categories are represented with the areal shares of 1.07, 8.88, 15.74, 46.29, 24.77 and 3.25 respectively. The extremely low, moderately low and low categories are mainly confined to the Central valley whereas others are observed on the Uplands (Fig. 7.5).

TABLE 7.2: **Frequency Distribution of Relative Relief**

Relative Relief (metres)	Frequency %	Cumulative frequency %	Categories
0-15	2.13	2.13	Extremely Low
15-30	11.18	13.31	Moderately Low
30-60	15.65	28.96	Low
60-90	19.71	48.67	Moderate
90-120	19.38	68.05	
120-150	12.67	80.72	Moderately High
150-180	6.92	87.64	
180-210	4.90	92.54	
210-240	3.09	95.63	
240-270	1.60	97.23	High
270-300	1.17	98.40	
300-330	0.96	99.36	
330-360	0.21	99.57	
360-390	0.32	99.89	
390-420	0.11	100.00	

Mean=103 metres, Mode=69 metres, Median=92 metres, C.V.=66 per cent.

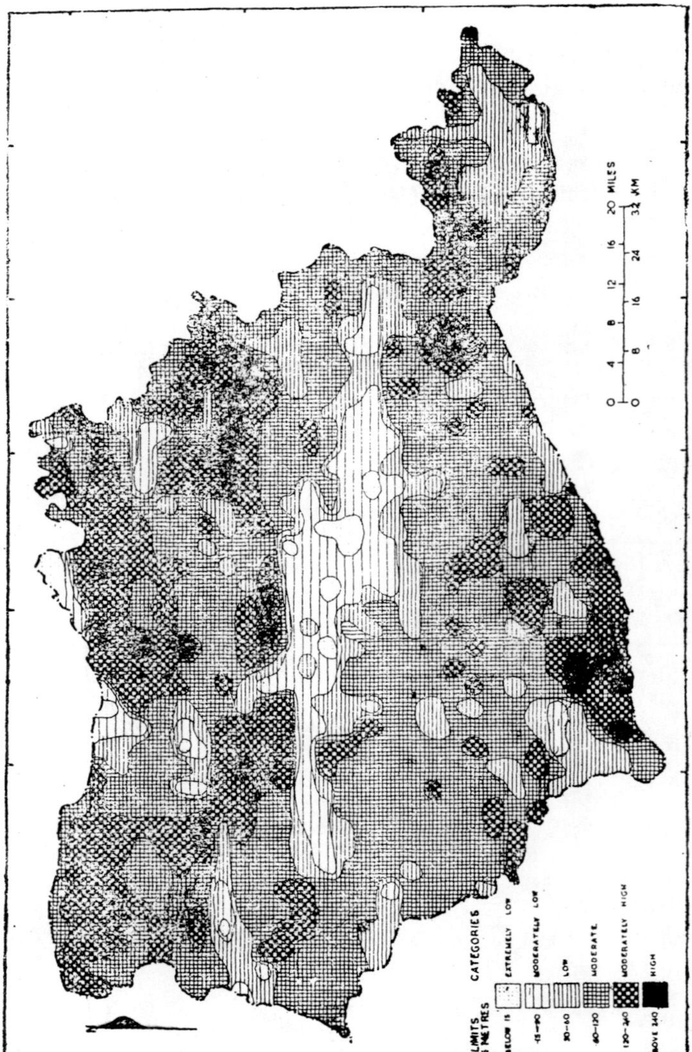

FIG. 7.5 : *Relative Relief Categories*

The two variables, i.e., Absolute and Relative Relief, correlated statistically show that the coefficient of correlation (+0.56) is not perfect looking into the age of the study area. It might be attributed to the impact of various upheavals particularly the Tertiary Movements, responsible for Himalayan Orogeny and upliftment of this area causing its rejuvenation. The positive value of the correlation coefficient suggests that the value of Relative Relief increases in accordance with the Absolute Relief as is also clear by the Regression Lines (Fig. 7.6).

Dissection Index

The concept of Relative Relief is not entirely satisfactory as a criterion of the nature of relief "Equal relative altitude are not always of equal importance, since their absolute altitude may differ. The picture gained from relative altitude only is static, for it fails to take into account the vertical distance from the erosion base".[11] On the basis of this fundamental fact, the dissection index, i.e., a ratio between the relative relief and absolute relief which always varies between zero (complete absence of dissection) and one (extreme case, vertical cliff at sea-shore) has been taken into account (Table 7.3).

TABLE 7.3 : **Areal Coverage of Dissection Index**

Dissection Index		
Dissection Index Values	Area (sq. km.)	Area %
Below 0.1	771.16	9.06
0.1—0.2	1998.31	23.48
0.2—0.3	2763.41	32.47
0.3—0.5	2620.37	30.79
Above 0.5	357.83	4.20
Total	8511.08	100.00

The area has been classified into four major classes of dissection index which are as follows :

(1) Low Dissection Index (0.01—0.1)
(2) Moderate Dissection Index (0.1—0.3)
(3) High Dissection Index (0.3—0.5)
(4) Very High Dissection Index (Above 0.5)

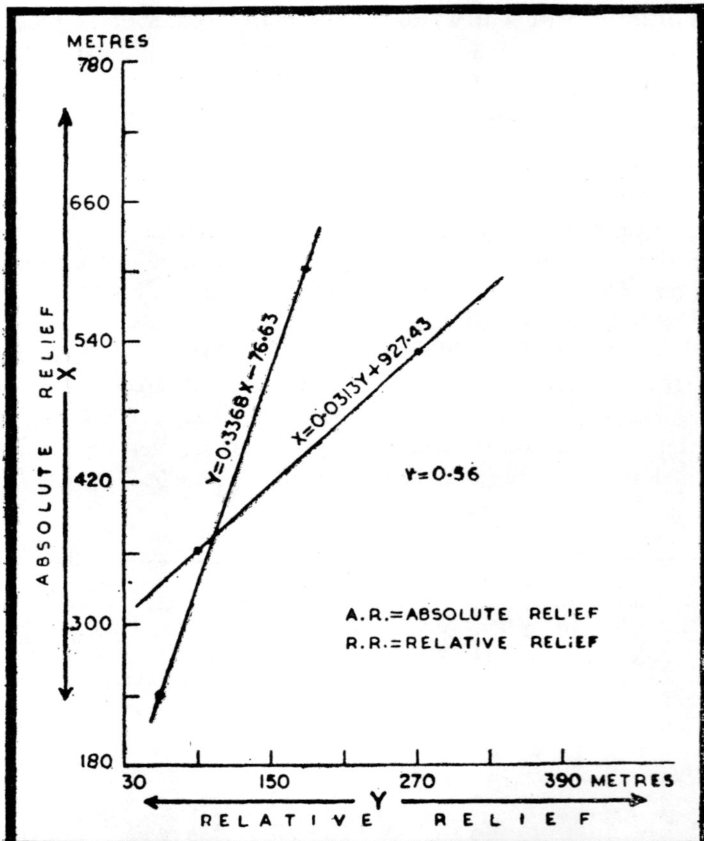

FIG. 7.6 : *Regression Lines A.R. vs. R.R.*

The dissection index categories—low, moderate, high and very high, share the total area in the ratios of 9.06 per cent, 55.95 per cent, 30.79 per cent and 4.20 per cent respectively. Areas of low category are mainly confined to the Central Valley (59.96 per cent of its area) whereas very high category is well marked in the northeastern part (Fig. 7.7).

Drainage Frequency

Another phase in dismantling the landforms goes with the study of drainage characters. The drainage frequency of the area has been determined by counting the number of stream segments in the unit of grids of sq. km each. The value ranges from 0 to 25 streams/10 sq. km. The drainage frequency plot (Fig. 7.8) suggests positively a skewed distribution.

The classification of total frequency range may lead to the delineation of four drainage frequency areas (Fig. 7.8).

(1) Areas devoid of drainage channels.
(2) Areas of poor drainage frequency, (1—10 streams),
(3) Areas of medium drainage frequency (10—20 streams),
(4) Areas of high drainage frequency (Above 20 streams).

The area, in general, have poor drainage frequency (49.93 per cent of the total area), followed by medium drainage frequency (47.93 per cent of the total area). Areas devoid of drainage channels are observed in the form of three detached small patches in the North Koel Valley where the relative relief is less than 15 metres with more or less level surface (below 2° slope). The areas of high drainage frequency, covering 1.74 per cent of the total area (Table 7.4) occur in eleven small patches due to local lithological variations.

TABLE 7.4: **Areal Coverage of Frequency and Density of Channel**

Drainage Frequency			Drainage Density		
Number of Streams/10 sq. km.	Area sq. km.	%	Density km./sq. km.	Area sq. km.	%
Nil	33.99	00.40	0-1	1350.64	15.87
Below 10	4249.80	49.93	1-2	6873.06	80.75
10-20	4079.82	47.93	Above	287.38	3.38
Above 20	147.46	1.74			
	3511.08	100.00		8511.08	100.00

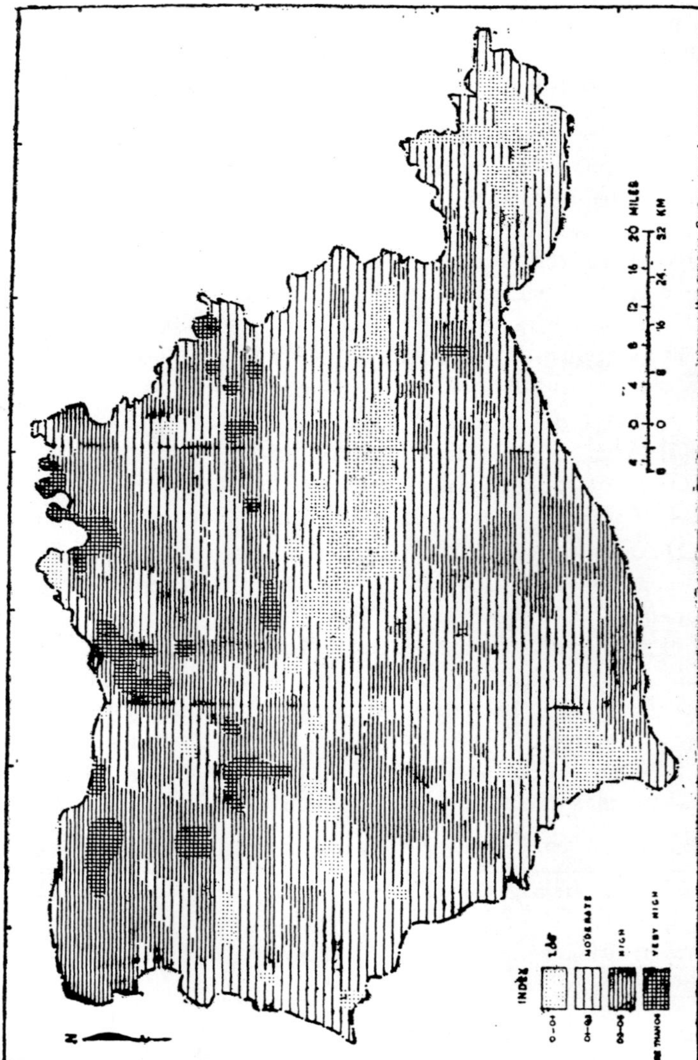

FIG. 7.7 : *Dissection Index Categories*

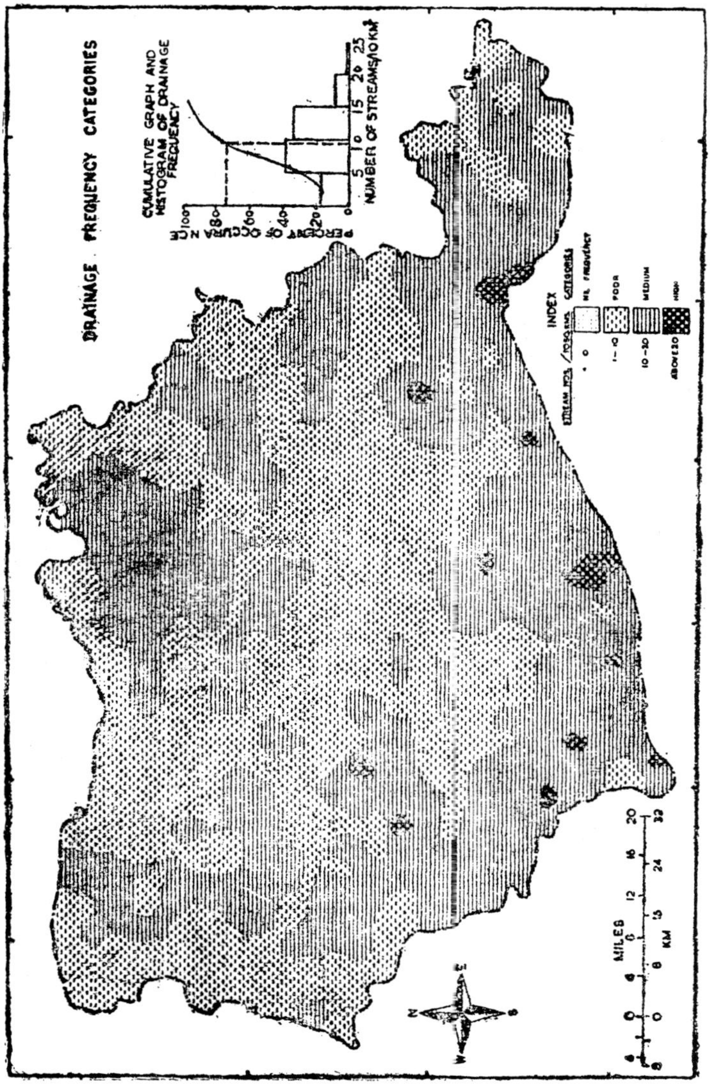

FIG. 7.8 : *Drainage Frequency Categories*

Drainage Density

Another part of the drainage study comprises the drainage density, simply a channel length per unit area. The three categories of drainage density have been observed in the study area (Fig. 7.9).

(1) Extremely Low Density (below 1 km/sq. km)
(2) Low Density (1-2 km/sq. km)
(3) Medium Density (Above 2 km/sq. km)

About 80 per cent of the total area is covered with low drainage density whereas others are insignificant. The low density is due to structural control as the predominance of granites and gneisses of Archaean Formations and poor supply of rain water have restricted the branching of the rivers and their tributaries.[12]

While correlating the drainage density with absolute relief and relative relief, the author comes to the point that there is no good interdependence as is obvious from the lower value of their correlation coefficients[13] (0.238 and 0.173 respectively), although its dependence on absolute relief is better as is also observed by Singh[14] in his research area. The observed anomaly between the two correlation coefficients may be attributed to the exclusion of the ephemeral seasonal streams, small enough to be counted and measured.

Average Slope

The last aspect of landform study is the slope, the loss or gain in altitude per unit horizontal distance in a direction. The area has been analysed by Wentworth's method which indicates the relative position of slope in the area. This technique is very useful quantitative generalization for the study area.

On the basis of the slope values ranging from 0° to 14°, the area may be divided into following regions.

(1) Level (below 2°—5°)
(2) Gentle (2°—5°)
(3) Moderate (5°—10°)
(4) Moderately Steep (10° and above)

The method used here for dismantling the area into slope zone could provide results only 14° slope, although the area

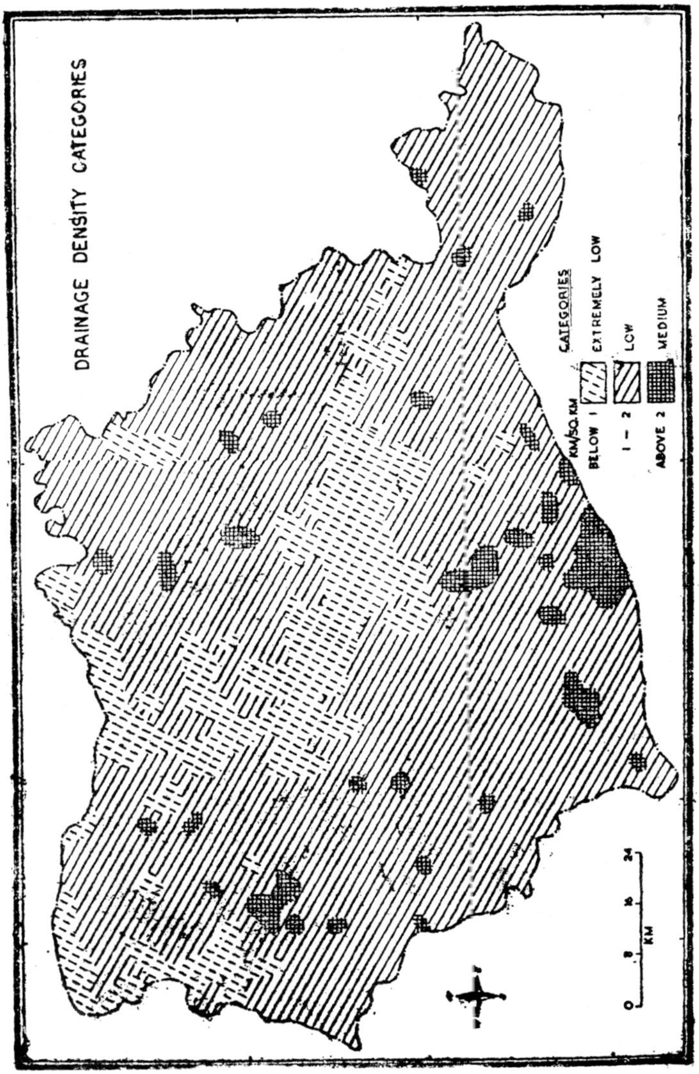

FIG. 7.9 : *Drainage Density Categories*

is not devoid of steep scarps and hillocks. The level, gentle, moderate and moderately steep slope categories cover 22.54, 50.02, 25.96 and 1.48 per cent of the total area respectively (Table 7.5).

TABLE 7.5 : **Areal Distribution of Slope**

Slope Value	Area	
	sq. km	per cent
0°—2°	1918.56	22.54
2°—5°	4257.61	50.02
5°—10°	2209.58	25.96
10° and above	125.33	1.48
Total	8511.08	100.00

As far as the areal distribution and the spreads are concerned, 2°—5° slope appears much more pronounced in areal expanse (Table 7.5), almost all of which lie mainly on the uplands in which Balumath Tabeland and Dhurki-Bhandaria Highlands are more significant.[15] This category is followed by the slope zone of 5°—10°, covering three main blocks, i.e., Bhandaria-Dhurki Highlands, Bhaunathpur Upland and Auranga-Amanat Interfluve. Below 2° of slope covers the lower reaches of North Koel and the Amanat rivers (Fig. 7.10).

Reassembling and Landforms

This is the last step of Weaver. The conclusion drawn in the above analysis may now be tabulated to mark some distinct boundaries of landform units. The diverse topographic characteristics, i.e., the occurrence of extensive valley in the centre, higher degree of undulations in the east, higher altitudinal surface in the west and table-type surface with higher altitude in the south, lead to the delineation of four first order landform units (Fig 7.11).

(I) Central Valley (RM, DM, TL, SG,)
(II) Western Uplands (RM, DM, TL, SG,)
(III) North-Eastern Uplands (RM, DM, TL, SG,)
(IV) Southern Uplands (RM, DM, TL, SG,)

The Central Valley, covering about 31.23 per cent (2657.92 sq. km) and lying between 150-450 metres height is occupied by the North Koel and its tributaries. The Western Uplands, lying between 150-750 metres or above and accounting for about 32.15 per cent (2735.69 sq. km) of the total area are characterised by more dissected and highest altitudinal surface. The North-Eastern Uplands, (20.60 per cent of the total area), situated near Auranga Hills and Hazaribagh Plateau is characterised by river basins, detached hillocks and highly dissected surface and scarps. The southern part of the study area, lying between 200-675 metres, is characterised by table-type features which are not found in other three units because it is closed to 'Pat' region.

The above four first order regions have further been subdivided into as many as eight second-order and eighteen third-order regions where the criterian have been mostly the apparent dissimilarities[16] (Table 7.6).

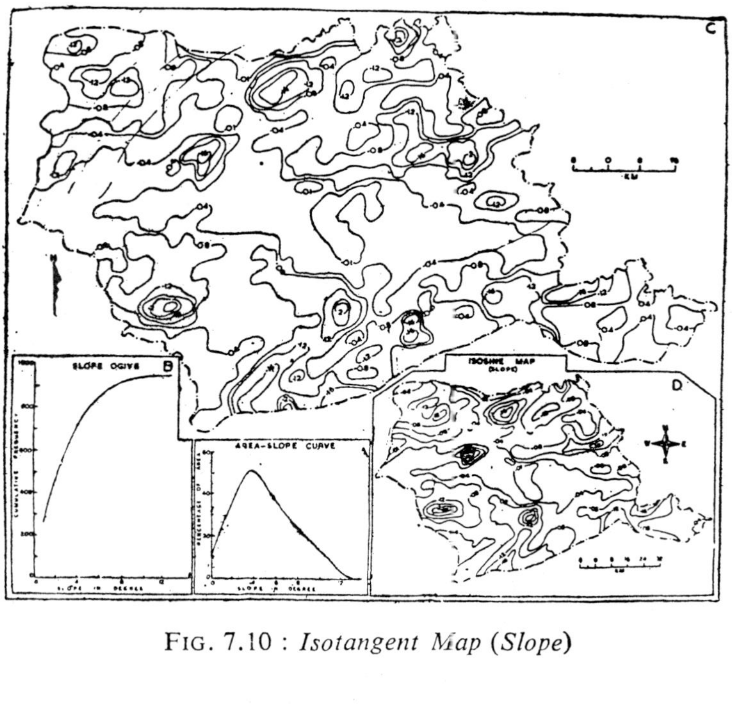

FIG. 7.10 : *Isotangent Map (Slope)*

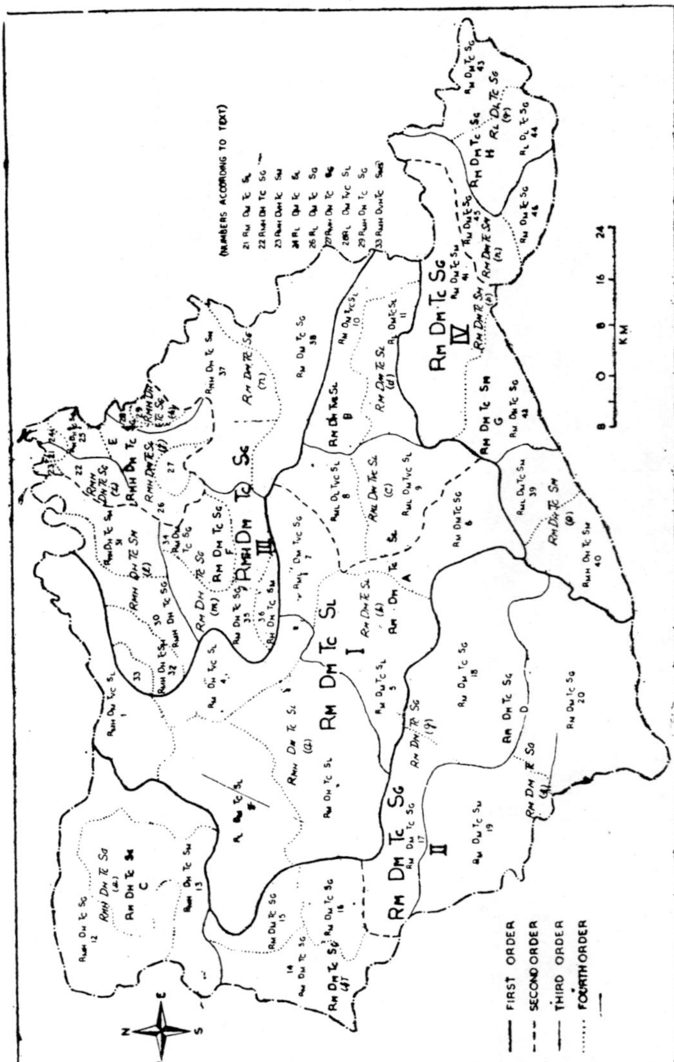

FIG. 7.11 : *Morpho-Units (Composite)*

TABLE 7.6: **Landform Regions of Palamau Upland**

I. *Central Valley*
- A. North Koel Valley (R_m, D_m, T_l, S_l)
  1. Lower North Koel Valley (R_{mh}, D_m, T_l, S_l)
  2. Middle North Koel Valley (R_m, D_m, T_l, S_l)
- B. Amanat Valley (R_m, D_m, T_{el}, S_l)
  3. Lower Amanat Valley (R_{ml}, D_m, T_{el}, S_l)
  4. Middle Amanat Valley (R_m, D_m, T_l, S_l)

II. *Western Uplands*
- C. Bhaunathpur-Dhurki Uplands (R_m, D_m, T_l, S_g)
  5. Bhaunathpur Uplands (R_{mh}, D_m, T_l, S_g)
  6. Dhurki Uplands (R_m, D_m, T_l, S_g)
- D. Ranka-Bhandharia Highlands (R_m, D_m, T_l, S_g)
  7. Ranka Highlands (R_m, D_m, T_l, S_g)
  8. Bhandhria Highlands (R_m, D_m, T_l, S_g)

III. *North-Eastern Uplands*
- E. Chhatarpur Basin (R_{mh}, D_h, T_l, S_g)
  9. Upper Batra Catchment (R_{mh}, D_h, T_l, S_g)
  10. Upper Batana Catchment (R_{mh}, D_h, T_l, S_g)
  11. Upper Surhur Catchment (R_{mh}, D_h, T_l, S_g)
- F. Dissected Zone (R_m, D_m, T_l, S_g)
  12. Denuded Scarp Zone (R_{mh}, D_h, T_l, S_m)
  13. Nawa Dissected Zone (R_m, D_h, T_l, S_g)
  14. Manatu Dissected Zone (R_m, D_m, T_l, S_g)

IV. *Southern Uplands*
- G. North Koel Amanat Interfluve (R_m, D_m, T_l, S_m)
  15. Barwadih Upland (R_m, D_h, T_l, S_m)
  16. Satbarwa Upland (R_m, D_m, T_l, S_m)
- H. Balumath Tablelands (R_m, D_m, T_l, S_g,)
  17. Bariatu Tableland (R_e, D_l, T_l, S_g)
  18. Harhanj Tableland (R_m, D_l, T_l, S_g)

Legends

Relative Relief : (R) (i) Extremely low (R_{el})=less than 15 metres (ii) moderately low (R_{ml})=15-30 metres; (iii) low (R_l)=30-60 metres; (iv) moderate (R_m)=60-120 metres; (v) moderately high (R_{mh})=120 metres – 240 metres; and high (R_h)=more than 240 metres.

Dissection Index (D) : (i) Low (D_l)=below 0.1; (ii) moderate (D_m) =0.1-0.3; (iii) high (D_h)=03-0.5; (iv) very high (D_{vh})=above 0.5.

Drainage Density (T) ; (i) extremely low (T_e)=below 1 km/sq. km; (ii) low (T_l)=1-2 km/sq. km; and (iii) medium (T_m)= above 2 km/sq. km.

Slope (S) : (i) level (S_l)=less than 2°; (ii) gentle (S_g)=2°-5°, (iii) moderate (S_m)=5°-10°; and (iv) moderately steep (S_{ms})=above 10°.

The above eighteen third-order landform regions may be subdivided into further orders but they become so micro in area and their morphometric characteristics so common, that they have not been described here in order to avoid monotony and repetitions.

EPILOGUE

The classification of landform units of Palamau Upland on the basis of morphometric attributes is so vivid and lucid that they give a complete panoramic physiographic glimpse of the region and almost every details of relief, dissection, drainage etc. are well represented on the maps. The micro landform units, so determined, can be used as ideal regions for soil conservation, afforestation, canal formation, mining and agricultural purposes.

REFERENCES

1. R.J. Russell, 1949 : Geographical Geomorphology, *Annals, A.A.G.*, Vol. 39, p. 10.
2. J.E. Kesseli, 1946 : A neglected field : Geomorphology, *Annals A.A.G.*, Vol. 36, p. 93.
3. E.H. Hammond, 1964 : Classes of land surface form in the forty-eight states of U.S.A. Map Supplement No. 4., *Annals, A.A.G.*, Vol. 54, p. 108.
4. G.H. Smith, 1935 : Relative Relief of Chio", *Geographical Review*, Vol. 25, p. 272.
5. G.H. Dury, 1963 : *The Face of the Earth*, Pelican Books, p. 39.
6. A.N. Strahler, 1952 : Hypsometric (area—altitude) analysis of erosional topography, *Bull. Geol. SOC, Ame.*, Vol. 63, pp. 1117-1142.
7. B. Zakrzewska, 1967 : Trends and methods in land form geography : A review article, *Annals, A.A.G.*, Vol. 57, p. 121.
8. G.D. Weaver, 1965 : What is landform ?, *The Professional Geographers*, Vol. 17, pp. 11-13.
9. S. Singh, and O.P. Singh, 1976 : Morphometric controls on rural settlements of Palamau upland : A quantitative approach, *The National Geographer*, Vol. XI, p. 124.
10. R.L. Singh, 1974 : Morphometric analysis of terrain, *N.G.S.I., Bull.* No. 22. p. 2.
11. Nir Dov, 1957 : The ratio of relative relief and absolute altitudes of Mt. Carmel, *Geographical Review*, Vol. 47, p. 565.
12. O.P. Singh, 1980 : Geomorphology of drainage basins in palamau upland in *Recent Trends and Concepts in Geography* Edited by R.B. Mandal, Concept Publishing Company, New Delhi, pp. 229-47.

13. O.P. Singh, 1977 : Palamau upland : A study in landforms and settlements, unpublished Ph.D. Thesis, Department of Geography, B.H.U., p. 73.
14. K.N. Singh, 1968 : Southern uplands of eastern U.P.: landforms and settlement distributions, unpublished Ph.D. Thesis, Department of Geography, p. 107.
15. O.P. Singh, 1976 : Slope studies of Palamau upland : A quantitative approach, Indian Geographical Studies, *Research Bulletin*, No. 6, p. 18.
16. O.P. Singh, 1976 : A quantitative approach to the classification of morpho-units : A case study of Palamau Upland, *National Geographer*, Vol. XI, p. 83.

EIGHT

Kalyan Mal Hironi

EVOLUTION OF THE ANTECEDENT DRAINAGE PATTERN
A Study of Banas

IN THE present paper the author has investigated the evolution of a transverse valley of the Banas near Raj Mahal in Rajasthan. Raj Mahal is 7 km west of the Jaipur-Kota highway towards northwest of Tonk township. At this place the Banas after traversing a vast alluvial plain has cut across three parallel ridges of Alwar quartzites. How the river has cut across these transverse hills of bold resistant quartzites? This has been a point of discussion from the very beginning. A.M. Heron (1938) while studying Geology of whole Rajputana has remarked about the development of Banas at Raj Mahal. Heron pointed out that the course of the Banas river is a case of antecedent development of drainage. Its conclusions have further been substantiated by later workers like Chatterji and Bagchi (1957). The present paper also aims to investigate the problem thoroughly and supports the theory of Heron.

GEOLOGY

The isolated hill at Raj Mahal is composed of rocks of Aravalli and Delhi supergroup. The main rock formations mat within the area are gneiss, schist and quartzites (Fig 8.1)

The River Banas

The Banas is the major river of eastern Rajasthan, which originates from the Kumbhalgarh hills in the Central Aravalli range. From Kumbhalgarh it flows towards east and after flowing for about 16 km it crosses an elongated range and

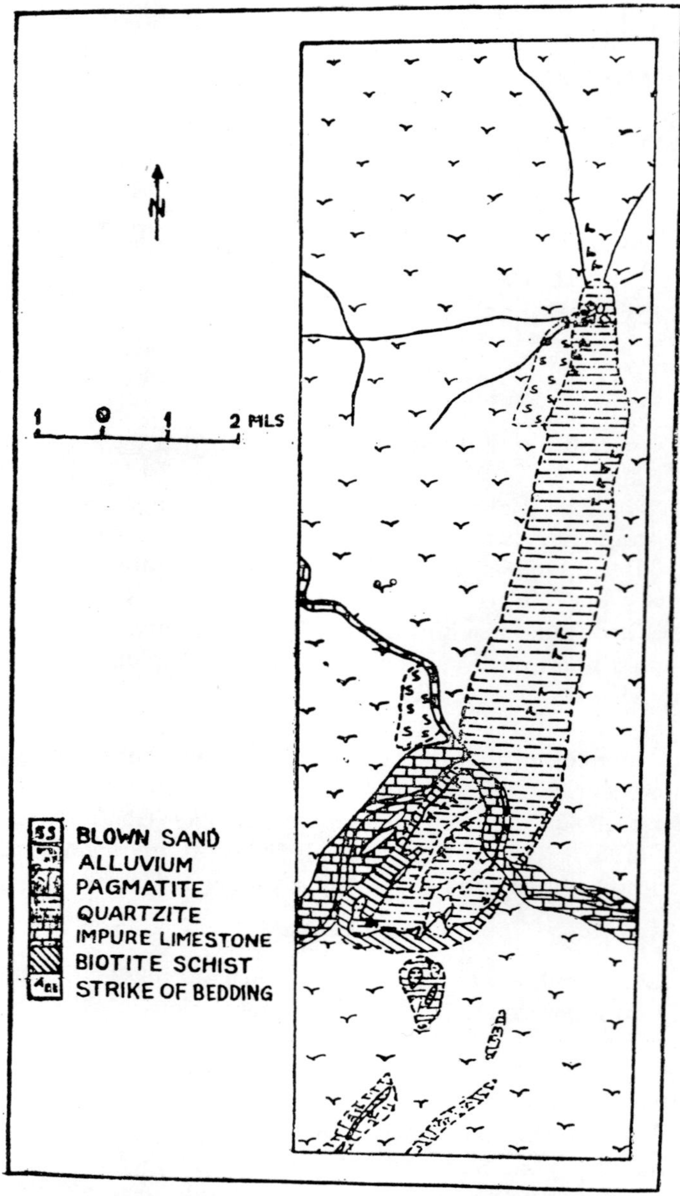

FIG. 8.1 : *The Raj Mahal Area—Geology*

then enters into the last plateau of Mewar. Near about 10 km northwest of Nathadwara, the river takes a northeastern direction and this direction continues till Tonk, from Tonk it takes an easterly direction and then it takes sudden turn towards south and crosses Sawai Madhopur and Karoli line of hills. Near Rameshwarm it joins the Chambal, where the Banas joins Chambal the area is called 'Panchnand'.

The course of Banas throughout its long profile is meandering. It appears that the river once might have been flowing on the peneplain and the peneplain has been uplifted and the river has established its course on the same meandering course; as a result of upliftment, the course of Banas is incised.

The section of the Banas near the Raj Mahal area is very interesting and significant from geomorphological point of view. Here the Banas at Raj Mahal wandering through the extensive and wild alluvial plain, breaches a great isolated flat ridge of the Delhi quartzites; in a gorge, one can infer that in the case of Banas its course across the ridge was in existence before the Aravalli peneplain was wrapped in the cretaceous period and cut down vertically as the plain arched (Heron, A.M., 1938). The course of Banas here is a good example of antecedent river action. The Banas had its valley over the extensive peneplain and was flowing freely over this peneplain. Later on the peneplain was uplifted the river continued its development as the peneplain arched. Chatterji and Bagchi (1957) are also of the same opinion that the Banas is a good example of antecedent river action. Here the river has cut across two ridges of quartzites, one near Bisalpur and another near Raj Mahal village. At the both sides the quartzites ridges are steep facing in each other. The height of these ridges is about 300 metres from the river bed. By studying this section in the field one tempted to say that the river has cut down its course vertically as the plain of Aravalli was uplifted. The course of Banas here is meandering; steep slopes and cut banks are common and significant fluvial features of the section.

SUGGESTED READINGS

Heron, A.M., 1923: *Geology of Western Jaipur*, Records of GSI—Part IV, Vol. LIV.

James, F.W. and Parizek, E.J. 1956 : Influence of underlying structure on stream courses and valley profiles in the Georgia Piedmont, *Annals of American Association Geographers*, 46, pp. 129-139.

Manjrekar, B.S., 1961-62 : A report on the preliminary investigation of the garnet deposits near Raj Mahal (Tonk District) and Sarwar (Ajmer Distt.) Rajasthan.

NINE

R. P. Verma and *P. K. Roy*

PALAEO-DRAINAGE STUDIES AND ITS UTILITY IN FLOOD PROBLEMS
A Case History in Lower Luni Basin

LUNI RIVER, the only integrated drainage flowing through the semi-arid tract of Western Rajasthan, west of Aravalli Range, is important fluvial geomorphic feature. The drainage network of Luni system, as evidenced from the geomorphic studies, study of aerial photographs and satellite imageries is in an advance stage of degradation. The fact became more pronounced and of far reaching importance after the devastating floods of July, 1979 experienced in this basin.

Luni river originates in the highlands of Aravalli range around Ajmer and flows from NE to SW upto Tilwara (Barmer Dist.) from where it takes an acute elbow turn and flows in as almost southerly direction finally draining into the vast Rann of Kutch. During high floods, possibly the water drains into the Arabian Sea. It is striking to note that during the present, all along its course no tributaries join the main Luni channel from the north, whereas the entire western slopes (in broad sense) of NE-SW trending Aravalli range are drained in the Luni river. Thus Luni river also broadly defines the boundary of an arid tract to the west and semi-arid tract to the east.

Detailed studies of Luni basin in the area between Dundara in the east and south of Tilwara in the west suggest an integrated network of drainage joining the main channel during a more humid past which at present are totally or partially obliterated due to onset of aridity in this tract during the Quaternary times. The relics of these old courses are manifested in different ways and with the help of different studies a detailed map of disorganised courses of Luni system has been prepared.

In the foregoing paragraphs, the different aspects on which study has been conducted are discussed.

Photogeomorphic Studies

Detailed study of aerial photographs on 1 : 30000 scale were carried out to study and demarcate the disorganised courses of Luni river. On the basis of different photo-elements, it was found that there existed a well integrated drainage pattern which is not traceable on the ground during the present times which drained the area north of present Luni Channel. Also few channels coming from the south were also easily picked up with the help of a.p.s. The important disorganised channels picked up are:

(a) Abandoned course of Jojri river which drains the highlands of Jodhpur was picked up from east of Patau to the north of Balotra. The course is almost non-existent on the ground except for in the form of small bank sections in a very isolated manner in some stretches.

(b) Few major disorganised channels north of Balotra one coming from Newai and other one from Bedarlai were picked up. These both channels meet near Pachpadra Town and the entire Pachpadra salt works is situated in the channel once drained the area north of Pachpadra.

A cursory look at the map suggests that the highlands of Shergarh were drained by this network of channels and Rann of Thob is also an abandoned course of this network.

(c) Third abandoned channel picked up with the help of the study of a.p.s. is seen draining near Gol and meeting the present Luni channel near Mekarans.

(d) Apart from these abandoned courses, which once drained the northern areas, a major disorganised course of Mitri river is picked north of Rakhi which is now a waste Rann of Sanwarla and drains in the present channel of Luni river south of Silor.

The zone north of Balotra (between Balotra and Pachpadra) is a zone of major confluences. In this zone meet the older courses of Jojri, the channels draining area north of Pachpadra and also some older disorganised courses of Luni which look off from the northern bank of the river. This whole tract during floods is inundated and here lies the important salt work of Pachpadra. So on the basis of these studies, it is now possible to say that origin of Pachpadra salt works is due to choking of earlier drainage and concentration of salts in stagnant drainage.

(e) Study of a.p.s suggests that Luni river had a series of channels off taking from its right and left bank, draining in the main channel at point in the down stream after travelling for some distance. Such was the elaborate net of Luni channel itself as seen west of Balotra, north of Jasol, around Kher Temple, west of Bagundi, south of Tilwara and north of Kitpala, Sinli. In the last case, a vast looking flat carved out by the shifting course of Luni from east to west (broad directions) is seen northwest of Variya and north of Sinli. It is still possible to pick one of the disorganised courses in this featureless flat.

During the present times many rivulets originating from the relief around Jasol, Nakodaji, Variya die before meeting the present course of Luni.

Thus, if all the abandoned and disorganised courses of Luni system picked up from aerial photographs are joined, then it will be amply clear that the entire highlands of Shergarh and Jodhpur were being drained in the Luni river by a well integrated drainage network and the confluence of which during present days is defined by the Pachpadra salt work and the area around. These courses due to some major climatic shift towards aridity during Quaternary times resulted in the total disintegration.

Geomorphological and Quaternary Geological Evidences

To substantiate the findings made on the basis of aerial photostudies, geomorphological and Quaternary geological studies undertaken in this area amply proved the point.

Extensive development of older fluviatile sequence in all these zones with considerable thickness of quaternary fluviatile deposits in the form of gravel to pebble conglomerate with alternate silt-stone/clay-stone is an evidence to suggest that this tract once experienced very dynamic fluvial activity in which the discharge was high carrying large amount of fairly graded sediments from long distance. Such thick fluvial sequences encountered in the well sections where no channels now exist confirm the argument in favour of disorganisation of drainage. These fluvial deposits and the flat older aggraded alluvial plains coupled with aerial photostudies are conclusive proof for a well integrated drainage course meeting the Luni channel from the northern areas in the past. Also the floodplains of these one time active channels in all probabilities also merged with each other making the older alluvial plain an extensive unit found almost uninterrupted in large tracts.

In some of the areas few bank sections are preserved in the older courses which show fluvial sediments and some form of valley. The best of these is seen near Bedarlai, where the left bank of the disorganised channel is still defined by 1-2 m scarp of a sequence of river borne cobble and boulder conglomerate in a fine metric. The linear course with a well defined depression is also traceable from little north of Bedarlai to Pachpadra salt works.

The older courses and shifting of main Luni channel between south of Tilwara and north of Sinali is evidenced by a moisture charged vast slightly low lying flat. Other courses which earlier took off from main Luni channel, however, have not left any appreciable mark.

Disorganised Courses and Floods

The entire Luni basin from Pipar and Bilara downstream faced one of the worst floods in the living memory during July 1979 in which lakhs of cattle heads perished and many human lives were lost. Many villages were totally wiped out. In this context it becomes pertinent to study the nature of floods and the role played by the network of disorganised courses.

A detailed account of the damages done by the floods

shows interesting facts that major damage has been done in only those zones in which the river revitalised its older abandoned and disorganised courses. It is evident from the fact that the entire road between Kalyanpur and Balotra got extensively damaged by flood water as the road runs almost in the disorganised bed of Jojri river. Similar situation is encountered between Jethantri in the east and Tilwara in the west (a tract of about 30 km) in which the entire railway tract (Jodhpur-Barmer Metre Gauge) was washed by the flood waters. A detailed examination of the disorganised courses of Luni river in this part will show that there were many courses, which took off from the northern bank of Luni, and met the main channel in the downstream segment. These courses, which were totally disorganised before floods, and on which extensive farming was being practised got revived during the floods perhaps due to heavy demand on the main channel and wiped out all the structures, farms and wells which came in its way. This situation was repeated in this stretch between Jethantri and Tilwara causing most extensive damage to the railway tract, totally eating away the fertile land and burrying the wells.

Also got revival was the abandoned and disorganised course of Luni river south of Tilwara, the water which flowed through the vast flat north of Sinli and finally drained in the main channel of Luni near Bajawa (west of Sinli).

Extensive damage was also caused to the road between Pachpadra and Balotra which as discussed elsewhere, is the confluence point of the disorganised courses of Jojri and the courses which once drained the tract north of Pachpadra.

The railway line between Tilwara RS and Tilwara fair RS also got totally washed by flood waters as it fell right across an older disorganised course which got revived during the floods.

It will, thus be clear from the foregoing discussion that what has been termed as the flooding by Luni river, is not necessarily flooding but a simple case of revival of its own older courses which were once well integrated. This fact assumes more importance when planning about flood relief measures in this basin, because this flooding is not a simple case of a river toppling its banks. What is needed in this

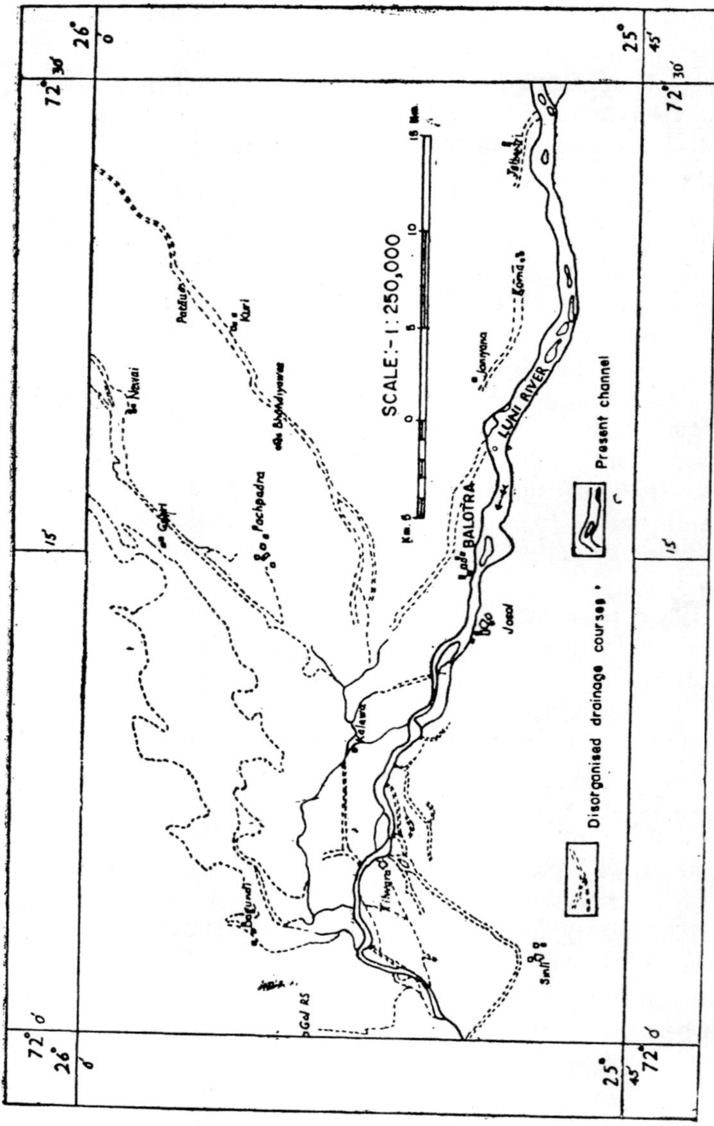

FIG. 9.1. *Map Showing Disorganised Drainage Courses in Lower Luni Basin Barmer District, Rajasthan*

case is a careful study and demarcation of all the disorganised courses before planning any flood control measures. Also there is a need to keep this fact in mind when planning any engineering structures in such areas.

CONCLUSION

Luni river system, which is in an advanced stage of disorganisation, makes an interesting case history to understand the role of aerial photostudies and geomorphology in studying the problems connected with floods in such tracts.

SUGGESTED READINGS

Ghose, B., 1971 : Geomorphological aspects of the formation of salt Basins in Western Rajasthan, *Proc. Symp. Problems of Indian Arid Zone*, Jodhpur, 1964.

Ghose, B. and Singh, S, 1965 : Observation on geomorphology and land use of a Kitnod Village, *Geographical Review of India*, Vol. XXVII, No. 3.

Singh, S.; Pandey, S. and Ghose, B., 1966 : Geomorphology of Siwana development block in Barmer District, *Annals of Arid Zone*, Vol.5, No. 2.

Singh, S.; Pandey, S. and Ghose, B., 1971 : Geomorphology of the middle Luni basin of Western Rajasthan, India, *Annals of Arid Zone*, Vol. 10, No. 1.

TEN

B. V. Singh and *A. Singh*

ISOSTATIC READJUSTMENTS AND THEIR REJUVENATION
Fluvial Geomorphology in the Chambal Basin

INTRODUCTION

THE PRESENT paper deals mainly with the analysis of isostatic readjustments and rejuvenation—fluvial geomorphology in the Chambal Basin. The area of study falls in the semi-arid zone of Central India, consisting of south-eastern and north-western flanks of Rajasthan and Madhya Pradesh respectively. The Basin covers about 78,750 km^2, in the Vindhyan tectonic area of the Deccan Plateau, bounded by the graticular of 74°—79°—20′ E and 22° 30′—26° 56′ N.

The Chambal, the biggest tributary of the river Yamuna, rises in the Northern slopes (Janpao spur) of the Vindhyas at an altitude of 840 m near Mhow, District Indore (M.P.) (Fig. 10.1). After traversing a course of 960 kilometres, it confluences with the Yamuna near Muradganj, District Etawah (U.P.).

An uplift or subsidence of the land causes imbalance in the initial units of geologic materials and topographic features which are subjected to sculpturing or deposition to maintain isostatic readjustments. The cretaceous and other intermittent upheavals of anticlinal part of the Vindhyan ranges favoured the isostatic readjustments in the Chambal Basin.

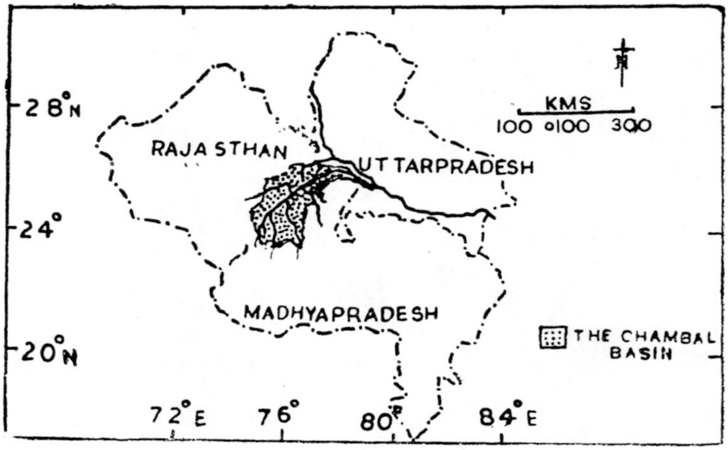

FIG. 10.1 : *Location of the Chambal Basin*

Isostatic Readjustments or Evolution of Landforms (Tectonic Features) in the Chambal Basin

Tension Theory of Origin

The occurrence of the Chambal rift valley, ponded fault basins (trough-faulting), fault-line springs and Vindhyan strikes etc. are the outcome of tensional theory of origin. These uplifted (Vindhyan ranges or scarps) and down-dropped landforms (weak parts between Vindhyan and Aravalli scarps of resistant rocks), are the formations resulting tension due to tectonic movements.

Evidences of Tension

The faulting in the Chambal Basin has produced a series of depressions and uplifts, running from south-west to north-east direction in the region. In the Vindhyan country of the Basin, there are seven (including one Chambal rift valley) tectonic depressions (Fig. 10.2). The Chambal rift valley (a major depression) has a south-west to north-east straight extent of more than 200 km. The average width of each depression is about 2 km. Besides, a series of ground water flows, linearly situated along the Lower Chambal Valley, reveal that some sub-surface disturbance has taken place in the zone.

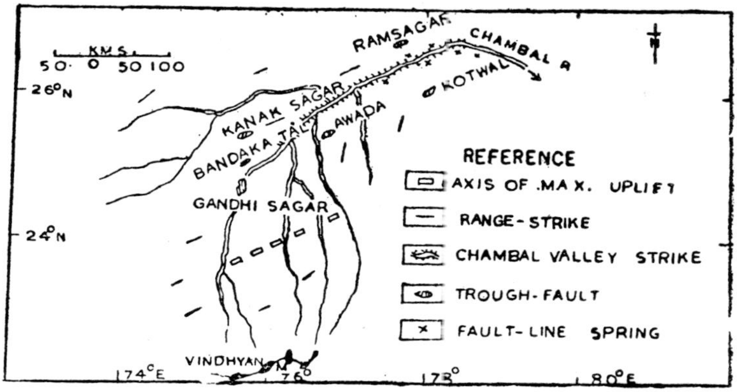

FIG. 10.2 : *Major Trough-faults and Strikes in the Chambal Basin*

The study of the drainage pattern, in the support of the dynamic rejuvenation, establishes the general history of the drainage development. The study of the drainage divides suggests that the area initially was traversed by south-west to north-east flowing parallel and rectangular Chambal tributaries (Fig. 10.3) separated by uplifted and fractured parallel interfluves of the Kota, the Kanjarda, the Kauntel, the Jhalawar, the Umatwara and the Morena plateau. Besides, the maximum tectonic movement (in the post-cretaceous) in the central part of the upper Chambal Basin evolved "some vertical uplift and rejuvenation in its streams. That some youthful landforms . . . cannot be denied". . . they show an unmistakable sign of having given way or yielded to tensional and vertical forces, resulting in a number of normal faults".[1] (Fig. 10.4).

The upper part of the Basin is a 'block' in which the structural disturbances have taken place. The curving south-east boundary of the Aravalli is a fault which brings the Vindhyans against the Aravalli system. This great fault, having a throw of 1500 m is roughly with the course of the river Chambal from Kota (Rajasthan) to Usaithghat (M.P.). It is notable that this is the only course of the river Chambal

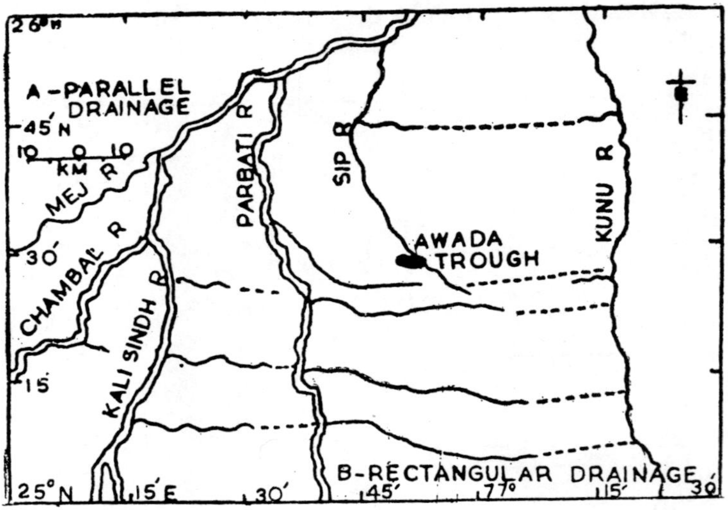

FIG. 10.3 : *Landscape Pattern due to Faults*

which runs roughly in a straight line. Occurrence of scarps in the alluvial deposits also favoured the view of tensional faulting. Alluvial scarps locally called *Puthas** are specially found in the Chambal valley from the mouth of the river Kunu up to Usaithghat (Distt. Morena). These faulted scarps have a range from a few feet to 100 feet in height.

Major Factors which Determine the Intensity, Type and Degree of Isostatic Readjustments Evolution of Landforms

1. The periodicity, i.e., continuity of tectonic forces or intermittence of tectonic forces govern the isostatic readjustments in a different way. The upward and downward earth movements were of intermittent nature, which affected the Vindhyan uplands at its fringes by slight movements of secular upheaval, faults and depressions. In the process of fault there have formed lower Chambal strike valley which

*The elevated grounds (scarps) composed of alluvium (sands) are locally called *Puthas*.

indicates 3 to 4 intermittent slight movements during the Cretaceous, Mid-Miocene, Pleistocene and recent ages.

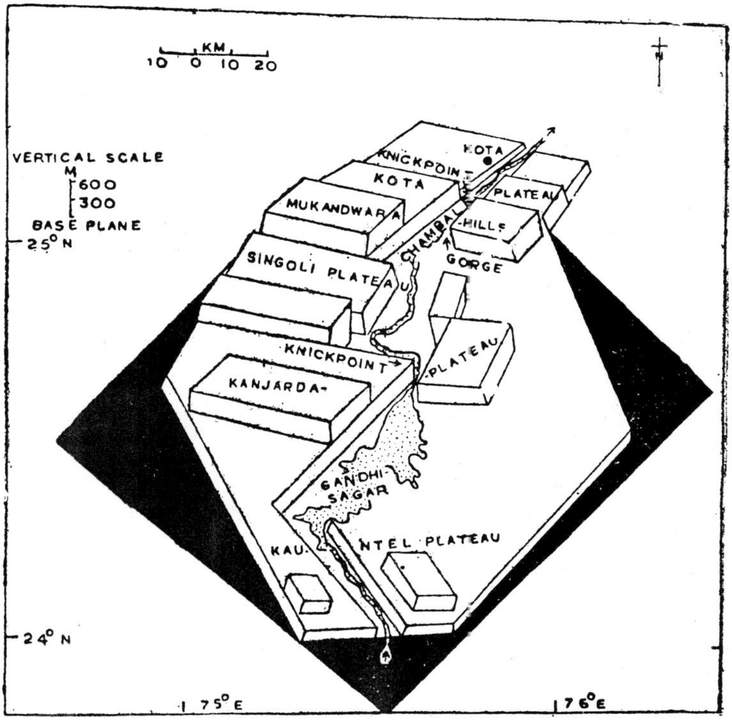

FIG. 10.4 : *Post-Faulting Landscape in the Upper Chambal Basin*

2. The height of uplift and depth of fissures or rift valley governs the elevation of the knick points, the degree of slopes, the height of river terraces, the rate of valley bottom erosion and the height of the summit plains (Fig. 10.5). Hence the elevation of knick points in the Chambal valley is higher (375 m near the boundary of Malwa Plateau) and variable (225 m on the fringe of the gorge near the town of Kota) than the knick points which are found in the valleys of the Chambal tributaries (e.g. the knick points in the Kunu valley at 250 m near the boun-

dary of District Shivpuri and 225 m near the 'Terhati'.* (District Morena). Similarly, the elevation of river terraces, the degree of slopes and the rate of valley bottom erosion is higher in the area of superposition of Chambal valley than the other valleys of the Chambal system. The height of the summit plains is closely associated with the height of uplift. Hence there is a great variation in the vertical distance of the summit plains as well as in the uplift of the land surface, e.g., the highest uplift area of the Malwa plateau, revealed the highest summit plain at the elevation of 480 m—600 m, while the lowest uplifted area on the northern flanks of the Vindhyans in the lower Chambal Basin, produced the lowest summit plains at the elevation of 180—305 m.

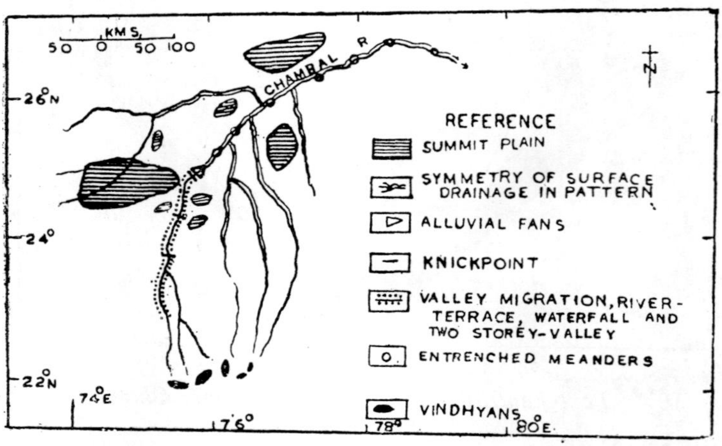

FIG. 10.5 : *Typical Examples of Rejuvenated Landforms*

In general, the region, as it is previously stated, was slightly uplifted during the movements of secular upheaval, hence the Vindhyans having the low elevation resulted the location of knick points, river terraces and summit plains

*A depressed dissected tract in the Vindhyans of district Morena is locally called Terhati.

etc. at moderate height in the region.

3. The Physical structural differences of rocks help the preservation of the rejuvenated morphology. For example massive sand stones and granites of the gorge valley areas of the upper Chambal valley, have preserved the two and three cycle—valleys as well as paired and single terraces. While the channels in the schists, slates and alluvial areas in the lower valley, are poor preservers of such morphology (except the ravines).
4. The tableland features of the Vindhyan area are partly responsible of the occurrence of summit plains, parallel drainage of rivers and deep cutting of river valleys in the Chambal valley system. The severe gullying, initiated by rejuvenation in due to the uplifted edge of the tableland.
5. In the Vindhyan areas, the major cause of soil erosion in the whole valley system, is due to the lack of maturity in the base level of the Chambal Valley (caused due to evolution of landforms) hence, extensive areas of morphostructures, morphosculptures and ravines are the Chief characteristics of the rejuvenated landforms. The soil erosion and down cutting, not only in the sense of having been affected by different processes at different times, but also in the sense of being eroded by several types of denudational processes at the same time.

The calculation of down cutting of a process is possible with the help of a quantitative formula. It has been attempted that treats the process. The soil erosion concerns itself mainly with the effects of waste transport in the valley profile, namely, valley wash. It is expressed as a removal of waste from each point on the valley profile. It is as follows:

$$R=KV_j\ (1.0+d_j\ 0.6)^*$$

where R=amount removed from point j in a given time; K=a coefficient whose value is proportional to the local thickness of the waste, V_j=a quasi-velocity at point j, proportional to

*The formula was formulated by Zingg (1940) which is quoted by Frank, A. in an article—Slope process and slope form : A theoretical study, appeared in *International Geog*, 1972, Vol. 1, p. 3

the sine of the local slope angle; d_j= relative distance from the summit to point *j*.

$$(0.0 \leq d \leq 1.0)$$

It is concluded that the rate of removal is higher on the higher and slopy ground of the upper reaches of the Chambal valley while it is lower with the increasing distance from the summit plains in the lower Chambal valley.

It is notable, that, there is general lack of flood plains except 'Kachhars' even in the lower Chambal valley and the banks of the rivers are always higher on both the sides of the rivers, because the long profile of the open river valleys indicate that they have not attained the base levels (Fig. 10.6).

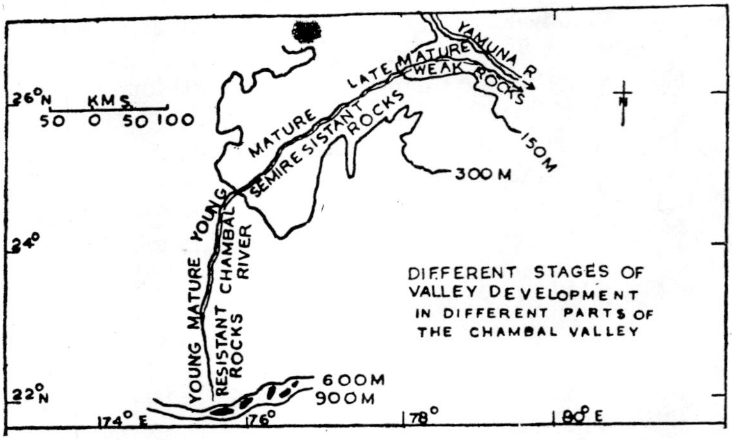

FIG. 10.6 : *The Chambal Cyclic Interruptions*

6. Transverse fissures or river valley faults of the Chambal are responsible for the occurrence of a gorge having a depth of 150 m and a length of 32 km above Kota in the hard resistant sand stone formations (Fig. 10.7). It is also a result of superposition of youthful stage of valley development. Such gorges on a small scale are also formed in the courses of the river Kunu. The Parbati, the Kali Sindh and the Kunwari when they cross the hard resistant rocks like sand stone formations.

7. Shifting in the course of the river Yamuna and subsidence in the Ganga valley, are also important factors in resulting rejuvenation morphology. In the lower reaches, the Chambal flowing on its lower land surface, has been intermittently affected by the fall in the level of the Ganga valley. Consequently it has cut through the banks and the remains of each bank survive as terrace ravines. "Probably the Chambal system is older than the Yamuna which formerly belonged to the Indus system in the tertiary and Holocene periods. Due to subsidence in the Ganga delta and later uplift in the Aravalli-Delhi axis. The Yamuna changed its course, because a tributary of the Ganga and usurped the catchment areas of the Chambal and other streams of the central foreland".

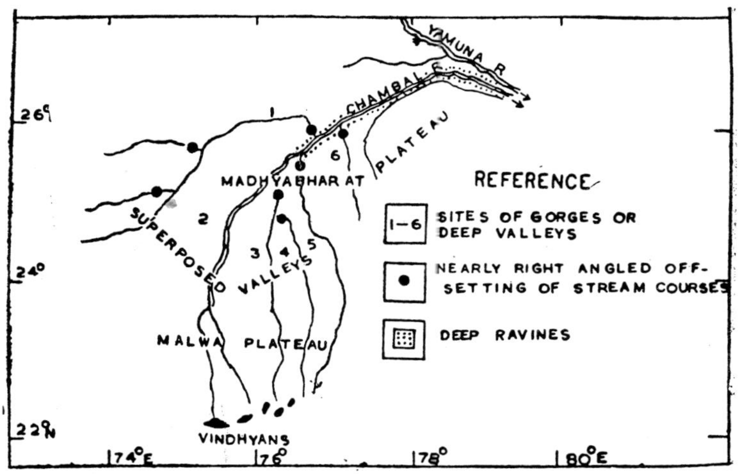

FIG. 10.7 : *Present Topography of the Chambal Basin*

Rejuvenated Fluvial Geomorphology

Rejuvenation of a valley means a morphological unconformity or cyclic interruption. In other words, there occurs a lack of conformity between the geomorphological landscape of the upper parts of valleys and of the lower parts of valleys. As it is observed in the Chambal valley by an eminent British Geographer who says: "From Kota the Chambal runs down stream to the north-easterly direction and occupies a strike valley, where gullying along the banks, it assumes a course of slight recent rejuvenation".[2] Thus the Isostatic readjustments have resulted various rejuvenated fluvial geomorphic landforms which are as follows:

1. Interruption of the cycle or the lack of conformity between the topographic features of the upper Vindhyan valleys and those of the lower basin valleys. In the Pre-Cretaceous, Mid-Miocene and Pleistocene periods, intermittent diastrophic movements evolved generally three cyclic interruptions followed by three youthful ages of valley developments in the lower parts of the upper valleys in the Chambal valley system (Fig. 10.5).
2. Superposition of youthful age of valley development on previously established channels is a notable sign of dynamic rejuvenation. It is confirmed by an eminent British geographer: "The lower Chambal may be regarded as occupying a strike valley in the Vindhyan scarp lands, but above Kota it is probably superimposed, cutting across the strike, and its upper reaches are more nearly consequent on the Deccan lavas of Malwa". (Spate, 1967).
   A common example of superposition of youthful features is in the form of the two storey-valley of Mid-Miocene period in the sandstone formation area of the upper valley. The other example of the termination of previous cycle and superposition of valley-in-valley occurred in the Pleistocene period on the channels where the Chambal crossed the sandstone formation. In the same way the river Banas, revealed the same phenomena of rejuvenated superposition in its upland area as it is stated.
   "This phenomena of succession is amply demonstrated by the Banas which is par excellence a superimposed stream. (Bhattacharya, 1971).
3. Outflows of large tectonic lakes (trough faults), e.g., those of Gandhi Sagar valley lake in Malwa plateau, Kanak Sagar valley lake in Tonk plateau, Awada valley lake in Morena plateau, Pagara and Kotwal valley lakes in Gwalior plateau, etc., are the notable features to reveal the rejuvenated landforms in the Basin.

4. Typical examples of collosal rejuvenated landforms exist in various parts of the basin in the forms of:
   (a) Knick points or the retreating steps causing the interruptions in the profiles of the streams. These benches existed in the valleys of the river Chambal, the Banas, the Parbati and of the river Kunu specially in the section of uplifted blocks near the trough fault or the rift valley.
   "Along the Chambal, there is one bench or knick-point at about 375 m near the boundary of Malwa plateau and the triple scarps. Thereafter, at the terminus of the Singoli scarp near Rawat Bhata the second bench occurs. Further downstream, a last Knick point at a height of 225 m is found where the Chambal emerges from the gorge section near the town of Kota. . . the Chambal was interrupted three times, giving rise to successive rejuvenation of various erosion surfaces". (Sharma, H.S., 1972)
   (b) Peneplanation in different ages is also an evidence of interrupted cycle of erosion and rejuvenation, e.g., Pre-cretaceous (at the elevation from 480 m to 600 m) are marked in the peneplaned Akoda hill, the Satur plateau (about 16 km west of Bundi town) and the Palki Mesa. Mid-miocene erosion surfaces (at the elevation from 360 m to 450 m) or peneplanation is marked in the Bundi-Sawai Madhopur and in the Mukandwara hills. And pleistocene erosion surfaces, dissected Vindhyan scarp lands rich in sandstones in Morena plateau as well as Karauli plateau (at the elevation from 180 m to 305 m) are noted in the transitional plain (between uplands and alluvial plains) and ravines.
   (c) Valley-in-valley or two storey-valleys are preserved in the cross profiles of the river Chambal, the Kali Sindh, the Parbati, the Kunu and of the river Banas in the resistant rock areas specially in the upper valleys.
   (d) Accelerated rates or down-cutting in the Chambal valley from Malwa plateau to the town of Kota and accelerated rates of lateral erosion in the form of extensive ravines formation in the weak rock areas of

the lower Chambal basin. Severe gullying initiated by rejuvenation, is due to the edge of 'tableland', hence most of the erosion and ravine formation has occurred during the last 400 years.

(e) Evolution of waterfalls and rapids in the Chambal valley near Bhansorgarh.

(f) Entrenched or ingrown meanders are evolved in the Chambal rift valley from Kota up to the confluence of the Chambal with the river Yamuna. These meanders are formed in the soft alluvium of the flood plain along the banks of the Chambal.

(g) Valley terraces are produced by surges of erosion along the banks of the river Chambal, the Kali Sindh, the Parbati, the Kunu etc. where they cross the rising fault-blocks. Hence they reflect the periods of rejuvenation.

(h) Symmetry of surface tributaries in deep-cutting of valleys in the resistant upland area and lateral erosion and ravines in the weak rock areas. The paralleled drainage, developing upon the upland area of uniform resistance, imply a notable lack of parallel topographic unconformity or structural control. In the upland area, all the tributaries are subparallel to the river Chambal and they confluence with the river Chambal nearly at right angles The river Parbati, the Kunu and the Banas (Singh, A., 1978). Confluencing with the Chambal rift valley nearly at right angles, are forced to rejuvenation and readjustment with the rift valley. Therefore 30 m, 25 m and 20 m of valley trenching in the lower Banas, the Kunu and the Parbati valley is attributable to this cause respectively. On the other hand, in the lower Chambal basin, where the paralleled topographic features and structural control dominate, none of the tributaries of the Chambal system confluence with the river Chambal and they drain in the same direction at lower topographic levels and have developed with the respect to new base levels of erosion.

Thus the plateau has its youthful or rather rejuvenated landforms and youthful gorge-like river courses. The vertical scale from initial Vindhyan flats to the level of the Chambal rift valley is mainly responsible for the cutting of deep valleys, ravine formations and head-ward eroding streams in the region.

(i) Ground water outflows or spring-like water flows occur in the floor of the north-eastern rift valley due to fault margin and the river reversal.

(j) The rate of fall (vertical) due to the formation of the Chambal strike valley, is also responsible for the rejuvenated landforms. From its source (840 m) to its confluence with the river Yamuna, the Chambal has a descent of about 720 m of this 450 m are in the first few km, 120 m up to Kota and 150 m up to the confluence with the Yamuna.

(k) The rejuvenation and reversal of the Chambal valley caused normally small alluvial fans at the base of a scarp near Kota. These are resulted from sinking of a land at the base of a scarp and with the alluvium derived from the upstream block where continued erosion along the base of the scarp may prevent the development of extensive alluvial fans.

The quantitative analysis of the drainage system establishes some mathematical relationships between elevation and length ratio of streams. In the upper Chambal valley system, the ratio of length of the stream in respect to the elevation of the region is lower, i.e., there is about 220 km average length of the stream and about 440 m average elevation of the area. Thus there is a ratio of 1:2 between the length of the streams and the elevation of the area. On the other hand, in the lower Chambal valley system, the ratio of length of the streams in respect to the elevation of the region is higher, i.e., there is 400 km average length of the streams and about 200 m average elevation of the area, hence the ratio is 2:1 between the two.

The System of Isostatic Readjustments/Geomorphic Regions

Many difficulties involved in defining and measuring individually the features of isostatic readjustments and rejuvenation morphology of the Basin. And the equally serious problem came in fitting them into a meaningful tectonic disturbance/

geomorphic feature ratios. Very little work has been done so far for categorizing the Basin or a country in terms of this ratio. Therefore, the authors made an attempt to form the isostatic readjustments/gemorphic feature ratio region on the basis of the quantitative formula.*

The following five geomorphic regions are worked out in (Fig. 10.8).

Region of the Highest Geomorphic Features

The region, falling in the South-West part of the Basin, consists a number of plateau like the Kauntel, the Jhalawar, the Kanjarda, The Singoli and the Kota. The total area of the region is about 33750 km^2. The region is the highest in the sense of elevation of topography as well as of geomorphological landscape. The average height of the land is 480 m from mean sea level. It is mostly a Vindhyan hilly area. The rocks are resistant, hence the fluvial rejuvenational features are bolder than those in the other parts of the basin.

It is a faulted mainly Malwa plateau in which 'blocks' troughs and the river valley faults evolved side by side, e.g., the Gandhi Sagar trough between the ranges of the Kanjarda

*$X=(2U+3S_1+3S_2+2R_1+2P_1)+(4Y+3M+2L)+(4K+4T+4G+4V+4E+2F_1)+(5S_3)+(5R_2)+(2P_2)+(3H+2S_4)+(2W_2)$

The major groups of variables used in formula for evaluating a geomorphic region index are as under:

1. Five variables of terrain (U=uplands + S_1=Scarps + S_2=Summit plain + R_1=Ravines + P_1=plains)
2. Three variables of valley stage (Y=youthful + M=Mature + L=Late mature)
3. Six variables of typical rejuvenated landforms (K=Knick-point, + T=river terraces + G=gorges + V=valley-in-valley + E=Entrenched meanders + F=waterfalls
4. One measure of range strike (S_3=Vindhyan strike)
5. One measure of rift valley (R_2=Chambal rift valley)
6. One variable of Ponded troughs (P_2=Ponded troughs)
7. Two measures of rock resistants (H=Hard rocks + S_4=Soft rocks)
8. One measure of spring like water flows (W_2=Water flows)

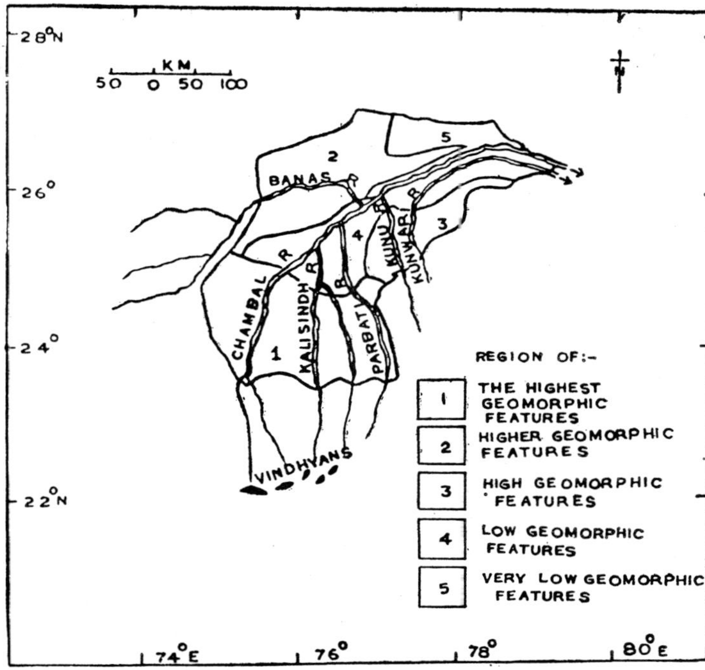

FIG. 10.8 : *Isostatic Readjustments/Geomorphic regions of the Basin.*

S.No.	Isostatic readjustments/geomorphic regions	Geomorphic feature index	Average height of the Region (in Metres)
1.	Isostatic readjustments of the highest geomorphic features	42	480
2.	Isostatic readjustments of higher geomorphic features	33	380
3.	Isostatic readjustments of high geomorphic features	31	370
4.	Isostatic readjustments of low geomorphic features	18	244
5.	Isostatic readjustments of very low geomorphic features	16	100

plateau, the Chambal gorge-fault between the ranges of Kota plateau etc. There is a subsidence of the Singoli plateau between the uplifted Mukundwara hills and the Kanjarda plateau.

Superposition of streams and the vertical corrosion in the area is the Chief characteristic of the region. All the streams are flowing parallel to the north and there are a number of parallel interfluves among them.

The most striking features formed by the rocks in this area, is the evolution of scarps. The region is deeply eroded and extensive erosive surfaces are evolved over the pile of Vindhyans, drained by the headwaters of the Chambal, the Kali Sindh, and the Parbati.

The major fluvial geomorphic landforms of the river valleys are in the form of knick points, gorges, ponded troughs, summit plains, two-storey-valleys etc. Besides most of the valleys are flanked by terraced slopes; the paired terraces of the river Kali Sindh from Tropic of Cancer up to Sarangpur is a typical example.

The region has rich black soils, forests and water resources and proper planning of these resources will help in growing a variety of food and industrial crops. The high potential of large streams like the Chambal, the Kali Sindh and the Parbati when dammed and utilized for generating power, may increase the economic growth of the region. Similarly the vast forest resources may rise the supply of timber, forest produce, fuel and fodder. Teak and Sal of Vindhyans may well be developed manifold for industrial growth.

The Region of Higher Geomorphic Features

It is spreading in the north-west part of the basin. It falls mainly in the Lower Banas plain. The total area of the region is about 15000 km^2. And the average height is 380 m from mean sea level. The region is drained by the river Khari, the Moshi, the Morel, the Golwa and the Mej.

The Bundi and the Karauli Hills, which are folded and faulted, break the monotony of the plain. They give rise to an undulating topography and sharper geomorphic features, with boulders, blocks and depressions. These hills lying in the middle of the area, look like the backbone of the area. The rocks are semi resistant.

The major landforms of the area are in the forms of range strikes, summit plains, scarps, down cutting of valleys, river terraces and troughs which are weaker than those of the previous region. Besides, there is undulating topography coupled with ravines, hence the alluvial deposits have thinner layers.

At present, the region has poor scrub forests and backward agricultural economy. The checking of deforestation and development of plantation may increase the forest resources and timber industries. Similarly the checking of soil erosion will improve the soils, food and industrial crops of the area.

The Region of High Geomorphic Features

The region falls in the South-West part of the district Morena (M.P.) the uplands of Morena and Gwalior spreading in the area, are responsible for the development of geomorphic landscape. The total area of the region is 7500 km^2 and the average height is 370 m. It is drained mainly by the river Kunu, the Kunwari and the Asan.

The major geomorphic landscape is in the form of dissected scarp lands, river terraces, summit plains, troughs and ravines.

The region is rich in sandstones. The total forest area is over 18 per cent of its area. Its proper utilization may increase the timber and various allied industries. At present the means of transportation and agriculture is backward in this zone. The checking of soil erosion and ravines will certainly improve the means of transportation and agricultural landscape.

The Region of low Geomorphic Features

It is a down-dropped triangular tract among the upland area, which is surrounded by the Bundi hills in the north-west, Morena plateau in the east and Kota plateau in the south-west. The total area of the regions is 11350 km^2 and the average height of the region is 244 m.

The region is known as the Harawati plain. It is drained by the river Chambal, the Kali Sindh, the Parbati and the Banas. The rocks are weak, hence the lateral erosion and formation of ravines is the chief characteristic of the region.

It is a part of the Chambal rift valley area where submergence of land took place. Therefore, the deposition and alluvial fans are noted.

It is an agricultural area, but the problem of soil erosion is very harmful. The checking of soil erosion with the help of artificial measures like bunding of land, etc., will increase the production of food and agricultural crops, because it is a fertile plain and well watered by the Chambal, the Kali Sindh, and the Parbati.

The Region of very low Geomorphic Features

It falls in the north-east part of the Basin, i.e., in the Morena and Dholpur plain. The total area of the region is 11150 km^2. And the average height of the region is 100 m. It is drained mainly by the river Chambal and the Kunwari.

It reveals the lowest geomorphic features index (16). It is predominantly a low land with alluvial topography. The evolution of the Chambal rift valley mainly caused the fluvial geomorphic landscape of the area (Singh, B.V. 1968). The major rejuvenational geomorphic features are in the form of lateral erosion, deep ravines and spring like ground water flows.

Due to the previous alluvial deposition and topographic uniformity, none of the tributaries in the region confluences with the river Chambal and they drain in the same direction at lower topographic levels and have developed with respect to new base levels of erosion. In other words the aggradation in this area is ceased and the rivers are engaged in degrading their previous deposits.

The ravines have contributed to its inaccessibility which cause its socio-economic backwardness. The soil erosion, deforestation and dacoity have spoiled its natural assets. The checking of soil erosion and deforestation will improve its forest and agricultural resources. Densely planting in the whole ravine belt will certainly check the soil erosion and various allied problems.

CONCLUSION

It is concluded that the Isostatic readjustments or the structural disturbances (evolution of landforms) in the Vindhyan strata in the form of downward (Chambal rift valley) and upward movements were of tensional tectonic nature during the cretaceous, Mid-Miocene and Pleistocene period (intermittent upheavals) causing the interruption and rejuvenation and produced the classic geomorphic features, in the form of superimposed valleys, knick-points, valley terraces, waterfalls, gorges, summit plains and ponded lakes in the upper Chambal valley region and lateral erosion, extensive ravines, and entrenched meanders and fault-like springs in the lower Chambal valley area.

It is further concluded that the intermittent subsidence in the Yamuna valley is also equally responsible for the rejuvenated morphology in the Chambal valley system.

In the upland area, all the tributaries are subparalleled to the river Chambal and they confluence with the Chambal valley at right angles. On the other hand, in the lower chambal basin, where the parallel topographic features and structural control dominate, none of the tributaries confluence with the river Chambal and they drain in the same direction at lower topographic levels and have developed with respect to new base levels of erosion.

It is established that even at present the major cause of soil erosion and extensive ravines in the whole valley system, is due to the rejuvenation in the Chambal valley system evolving lack of maturity in the base level of the Chambal. Thus the uplands have their youthful or rather rejuvenated landforms and youthful gorge like river courses. While the plains, having matured or late matured stage of river development, reveal the rejuvenated morphology in the form of ravines, entrenched and ingrown meanders, and a few spring-like ground water flows in the Chambal valley. Thus the evolution of fluvial geomorphic landforms reveal that the river Chambal is a rejuvenated stream.

The geomorphic regions, which are evolved, are essential for the integrated area planning.

SUGGESTED READINGS

Bhattacharya, A.N. 1971 : Udaipur-Gwalior region an article appeared in India, *Regional Geography*, National Geographical Society of India, Varanasi, p. 524-25.

Chhibber, H.L., 1954 : The development of landforms in the Himalayas, Pres. Address, Geog. Geol. Sec. Indian Sci. Cong. Hyderabad.

Dey, A.K., 1968 : *Geology of India*, (Ed.) Nat. Book Tru., India (New Delhi).

Kaushik, S.D., 1965 : Geomorphology of rejuvenation in the Garhwal Himalayas, *Indian Geog. J.* 40 (1-2), pp. 13-19.

Mehdiratta, R.C., 1967 : *Geology of India : Pakistan and Burma*, Atma Ram & Sons, Delhi, p. 21 and 31.

Mukhopadhyay, S.C., 1977 : Geomorphology of a part of Lower Kharkai Basin, *Geog. Review of India*, 39 (248).

Singh, A., 1978 : The lower Chambal basin : A study in medical geography unpublished, Ph. D. Thesis, p. 19-20.

Singh, B.V., 1969 : *Land uti. Distt. Morena*, unpub. Ph. D. thesis, U. Agra.

Shafi, M., 1968 : The Problems of waste lands in India, *The Geographer*, Aligarh, Vol. XV.

Sharma, H.S., 1972 : Erosion surfaces in the Lower Chambal Valley, India, *Inter Nat. Geog.* Canada, Vol. I. pp. 62-68.

Sharma, H.S., 1972-73 ; Geomorphic evidences of rejuvenation in the Lower Chambal valley, *Studies in Geog.* U. Rajasthan, Vol. IV.

Spate, O.H.K., 1957 : *India and Pakistan*, Methuen & Co. Ltd., London.

ELEVEN

Surendra Singh

TYPES AND FORMATION OF SAND DUNES IN THE RAJASTHAN DESERT

INTRODUCTION

In the Rajasthan desert, there are two major and one minor zones of the sand dunes. The westernmost zone of the dunes runs through Jaisalmer, Ramgarh, Tanot, Pugal and Mahajan and constitutes high to very high dunes with reactivated crests and flanks. To the east of this zone, another zone of medium to high and very high dunes runs through Chohtan, Barmer, Pachpadra, Shergarh, west of Nagaur, Churu and Rajgarh. The above two dune zones are separated by a vast rocky tract of Jaisalmer-Pokran-Bap-Bikaner sector but they again meet in the north of Bikaner district and extend into Ganganagar district. The third small dunes zone into discontinuous patches passes through Sanchor, Jalor, Saila, Didwana, Sikar and Jhunjhunu and joins the two major zones near Rajgarh. These three dune zones constitute the dunes of different types and origin and they are unevenly distributed. In certain parts of Rajasthan desert, some investigations on the dynamics of sand dunes have been carried out by Pandey *et al.*, (1964), Vats *et al.*, (1976), Singh (1977) and Allchin and Goudie (1971). In the present article, an attempt has been made to discuss the origin and distribution of different types of sand dunes in the Rajasthan desert.

ENVIRONMENT

The climate of the Rajasthan desert is characterised by low

and erratic rainfall, extreme temperatures, high evaporation and wind velocity. The aridity index of the arid western Rajasthan is 78 per cent and the aridity increases from south to north-west. The annual rainfall varies from 100 mm in extreme west to 300 mm in the eastern parts of the desert. About 90 per cent rainfall occurs during the monsoon period from June to September. Maximum summer temperature exceeds 40° C and reach up to 49°C in the months of May and June. The mean wind speed over Western Rajasthan from April to July is 20 kmph. The diurnal temperature fluctuations in summer are marked due to which the dust storms are quite common.

METHODS AND MATERIAL

Investigations on the sand dunes were carried out with the help of the aerial photographs of 1 : 25,000 and 31,000 scales, Landsat imageries of 1 : 500,000 and 1 : 250,000 scales, Landsat false colour uncontrolled mosaic of 1 : 1000,000 scale, topographical sheets of 1 : 50,000 and 1 : 126,720 scales and random field checks. The aerial photographs, toposheets and the Landsat imageries were interpreted and the sand dunes of different types were identified, classified and mapped. The final dune map of Rajasthan desert (Fig. 11.1) was prepared after detailed field check. The parabolic and longitudinal dunes were particularly mapped from the Landsat imageries. The small barchan and transverse dunes could not be identified and mapped from aerial photographs and Landsat imageries, these dunes were mapped from the field. Morphological characteristics of the sand dunes were studied from the field as well as in the laboratory. In order to study the origin of the different types of the sand dunes, the relationship between wind direction and dune orientation was established. The sand samples collected from the different slopes of the dunes were analysed to determine source and origin of the dune sands.

RESULTS AND DISCUSSION

The results of the investigations conducted on the wind regime, processes of sand movement and accumulation of sands and the origin of the different types of sand dunes are described below:

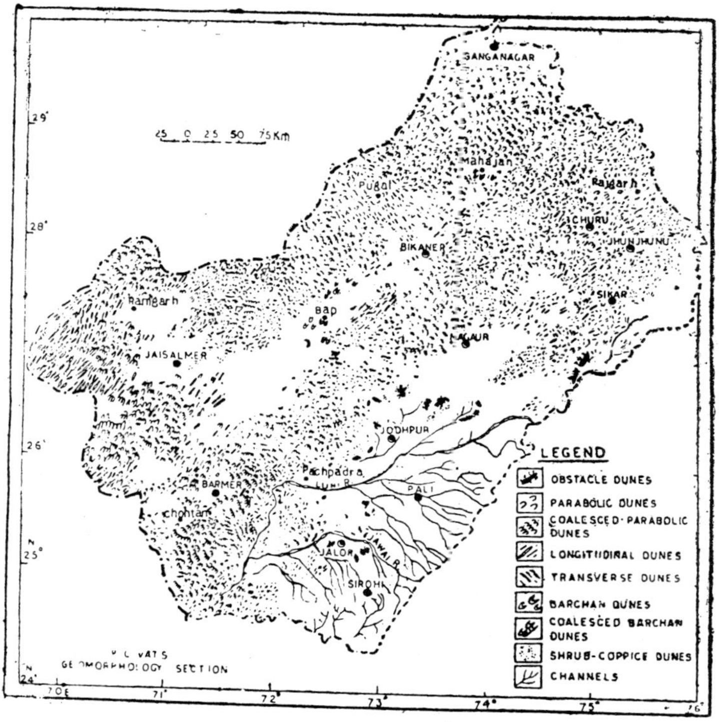

FIG. 11.1 : *Sand Dune map of Rajasthan Desert*

Wind Regime and Dune Orientation

The wind plays a significant role in the deflation and transportation of aeolian materials like sands and dust particles. Analysis of the wind rises from different places and particularly of Dosa, Jodhpur, Barmer and Bikaner have revealed that the dominant wind direction during the summer season is from southwest to northeast but the wind direction in winter season is changed from southwest to northeast and winter winds also influence the aeolian topography. It has been observed that sometimes the local winds also play a significant role in the formation of aeolian features. At Jodhpur 12,

21 and 18 per cent total days of May, June and July respectively have the wind speed range of 20.92—61.15 kmph, at Barmer, 15, 15 and 6 per cent total number of days of the above months have wind speed range of 20.92—61.15 kmph. During these months, the wind speed range of Dosa does not exceed 1.60—19.31 kmph except occasional gales and dust storm.

It has been established that for the movement of the fine grained sands (0.25-0.01 mm) winds with velocity near to 3.5-4 m/sec at the surface are required. The initial wind velocities required for the movement of the sand particles of the different sizes (after Sokolov, 1884) are given below:

Type of sand grains	*Diameter mm*	*Wind velocity m/sec.*
Fine grained	0.1—0.25	4.5—6.7
Medium grained	0.25—0.5	6.7—8.4
Coarse grained	0.5—1.0	9.8—11.4
Very coarse grained	1.0—2.0	11.4—13.0

It may be seen that the winds with different velocities are required for the movement and deposition of the sand grains of different sizes and to create the aeolian features of different shape and magnitude.

Processes for the Transportation and Deposition of Sands

The sediment particles having the diameter range between 0.06 to 2.0 mm have been classified as sand. The sand grains of different sizes are transported and deposited by saltation, surface creep and suspension.

Initially the sand grains are moved by air with a heaping and bouncing process known as saltation. The sand grains of 0.06 to 0.5 mm diameter are transported by saltation. The saltation process starts due to the disruption of the equilibrium in the ecosystem caused by a change in one or more of the component variables. Such disruptions are caused by the numerous changes like reduction of precipitation, increase of temperature, increase of wind velocity, destruction of soil aggregate, reduction of surface roughness and reduction of vegetation cover. Fluid threshold velocity is required to set the sand grains in motion. Impacts from salting particles initiate movement of other particles if their impact threshold velocity is achieved. It has been established that the impact and fluid threshold velocities are almost similar for large amount of sand grains but the impact threshold velocity becomes lower than fluid threshold velocity with increase in size of the sand grains. The impact velocity required for dune sand is 4 kmph at a height of 1 cm which is equivalent to 16 kmph at 1 m height. The initial velocity to set sand in motion is determined by the model diameter of the surface grains and gradually coarse grains are set in motion as the drag velocity increases. The impact threshold velocity for grain size above 2 mm is rarely attained. Due to saltation, the sand grains are initially rolled along the surface in a turbulent air stream, and after rolling a few metres, the bouncing movement starts. Sand grains are lifted up into the air streams a few centimetre above the surface and are carried forward by wind pressure in flat trajectories and are dropped on the ground. By the impact of the descending grains further particles are lifted and thus the saltation process is sustained. The saltation is the dominant process and about 90 per cent sand grains are transported by this process. Due to surface creep process, the sand grains above 0.5 mm diameter are transported along the surface and are deposited in the form of various aeolian features. The impact of the sand grains moved by the saltation also promote the movement of the surface particles by surface creep. Suspension is not very effective process in the transportation of sand grains because vertical velocities required to lift the sand grains is rarely attained. The dust and clay particles below 0.06 mm diameter are transported by suspension process to long distances and are ultimately dropped on the ground.

TYPES AND FORMATION OF SAND DUNES

Based on the interpretation of the Landsat imageries and aerial photographs and the analysis of the physical properties of the dune sands, three major dune types viz. (1) simple, (2) compound, (3) complex have been identified. Major and minor obstacle, parabolic, longitudinal, transverse and barchan, on the basis of their form and number of slip faces have been classified as simple dune types. The barchan and transverse dunes form parallel wavy lines on the Landsat imageries. In the compound dune types two or more same types of dunes are combined by transverse sand ridges. The coalesced barchan and parabolic dunes fall under compound dune types. In the complex dune types two or more basic types of dunes have formed in some areas by the winds from several directions. Since the compound and complex dune types had resulted due to the combination and/or coalescing of simple dune types, the factors responsible for the formation of simple dunes, their distribution and morphological characteristics are discussed below:

Major Obstacle Sand Dunes

Origin

These dunes were formed on the windward and/or leeward sides of the isolated and continuous chains of the hills due to the obstruction in the path of sand laden winds during the prolonged arid phase and they are known as windward and leeward obstacle dunes (Plate 1). The windward obstacle dunes of flat semi-cone shape were formed due to the fixed eddy, decreased wind velocity and a calm zone on the upwind side of the hills. The prevailing southwest winds swept the sand over the hills and deposited it on the leeward side right from the slopes of the obstruction up to few kilometres in the direction of the down wind and the leeward obstacle dunes were formed. At certain places, leeward obstacle dunes had also been formed by the northeast winds which are gently upsloping towards the hill slopes. The isolated hills formed one obstacle dune but the broad hills in chains created several obstacle dunes. These dunes at present are badly dissected, well cemented and vegetated and have been separated from the hills by the gullies of 25 to 50 m depth and 50 m to 100 m width. This indicates that at the time of the formation in pre-Holocene period, these dunes were active and were stabilised in a prolonged humid phase of early Holocene period, around 10,000 B.P.

Distribution and Morphology

These dunes occur scatteringly, mostly in the Central and

northeastern parts of the Rajasthan desert. The windward and leeward obstacle dunes against the isolated hills of rhyolite and granite, and chain of hills of sand stone formations mostly occur in Jodhpur, Nagaur, Jaisalmer and Jalor districts.

The obstacle dunes are of 10 to 40 m and 8 to 15 m in height and their sands are compact and highly calcareous. These dunes are sufficiently degraded and the lime nodules of 8 to 10 cm length and 2 to 5 cm width are exposed at crests and flanks which indicate that these dunes are well cemented and static. The slopes of the windward sides and flanks are 2° and 10° respectively. The orientation of the windward obstacle dunes is N 40°E and of the leeward obstacle dunes is N48°W. The grain size of these dunes varies from 0.03 to 0.25 mm and the sand grains of 0.06 to 0.12 mm are dominant. The mean diameter of these sand grains is 0.14 mm.

Minor Obstacle Sand Dunes (Shrub-Coppice)

Origin

Minor obstacle dunes were formed on the windward and leeward sides of the minor obstacles like shrubs and grasses. The sand carried by winds was caught by shrubs like *Calligonum polygonoides*, *Crotalaria burhia*, *Leptadenia pyrotechnica*, *Aerva pseudotomentosa* and *Capparis decidua* and grasses such as *Panicum turgidum*, *Lasiurus sindicus*, *Cenchrus ciliaris* and *Cenchrus biflorus* and the shrub-coppice dunes in the form of sand mounds, fence line hummocks, low longitudinal and transverse ridges were formed (Plate 2). This process of dune formation formed the dunes on the sandy surfaces. In other deserts of the world, the small shrub-coppice dunes are known as *nebkha* and large one as *rebdon*.

Distribution and Morphology

These dunes are widely distributed almost in all parts of the Rajasthan desert. They occur in the sandy undulating interdune and aggraded older alluvial plains and sandy buried pediments. The shrub-coppice dunes are of different shapes, sizes and orientation. The dunes formed earlier are old, stabilised, degraded and their sands are compact and calcareous. But recently formed dunes are active, their sands are loose and non-calcareous. The sand mounds are of 90 cm to 1.5 m in height and their length and width are almost equal. They do not have any definite orientation and lie quite close to each other. The fence line hummocks of 2 to 5 m in height occur at 120 m to 250 m and 250 m to 500 m apart from each other. The length and width of these fence line hummocks are 300 m to 2000 m and 250 to 500 m respectively. The longitudinal and transverse ridges are 2 m to 4 m in height, 250 m to 1250 m in length and 120 m to 250 m in width and they lie 250 m to 500 m apart from each other. The average orientation of the longitudinal ridges is N 35° E and that of transverse ridges is N 45° W. These ridges do not have slip-faces and the slopes of the windward and flanks are 1° to 2°. About 49 per cent sand grains of these dunes are of 0.06 to 0.07 mm diameter and this percentage is larger than that found in the other dunes.

Parabolic Sand Dunes

Origin

The parabolic dunes were created by the erosional and depositional processes during the prolonged arid phase. The sands were removed by the unidirectional winds through erosional process and created the long scoop-shaped hollows or parabolas with flanks tapering upto several kilometres towards the gentle windward slopes. The sands were subsequently redeposited on the lee of the parabolas by depositional process, created the steep leeward slopes and the parabolic dunes came into being (Plate 3). Their steep depositional faces (slip faces) which form on the dune nose, have foresets consistently dipping downwind. A number of parabolic dunes were formed by the prevailing winds in the same path which were coalesced by the transverse ridges and the coalesced parabolic dunes were formed (Plate 4). These dunes stand out very clearly on Landsat images particularly in band-7 image in white to dark grey and dark tones (Plate 5). Like the obstacle dunes, these dunes were active at the time of their formation but were stabilised during the humid phase. According to Cooke and Warren (1973), the sands are eroded by the winds and the hollows are created. Initially the accumulation of the sands downwind of the hollow forms a blow out dunes, if this dune migrates downwind as the hollow is enlarged, tra ling dunes are left on each flank of the hollow and a parabolic dune is formed. Migration of several dunes down the same path gives the nested parabolic systems. The findings of these authors on the origin of the parabolic dunes are in conformity with the findings of the present author.

Distribution and Morphology

Parabolic dunes occupy more area than the other dunes and scatteringly occur almost in all the districts of the Rajasthan desert. Parabolic dunes 10 to 25 m in height are well stabilised and vegetated. The slopes of the leeward side of these dunes are steep, varying from 14° to 16° and of windward sides are gentle, ranging between 2° to 4°. The sands are calcareous and compact. The average orientation is N 41° E. The coalesced parabolic dunes are also well stabilised, degraded and support good density of vegetation which indicate the advanced development of soil. The coalesced parabolic dunes are of 18 m to 30 m and 40 m to 80 m heights. The number of dunes in dune chains varies from 15 to 35 and the length of these chains ranges between 0.5 to 10 km. The diameter of the sand grains on all the slopes varies from 0.03 to 0.25 mm and 60 to 70 per cent sand grains are of 0.05 to 0.12 mm size. The average length and width of these dunes are 1288 m and 226 m respectively. The orientation varies from N 45° E to N 47° E. The slopes of the flanks, leeward and windward sides are 8° to 12°, 22° to 24° and 2° to 4° respectively. The coalesced parabolic dunes are covered with the fresh sands of 3 to 4 m thickness. The barchan dunes are superimposed on the flanks of these dunes, indicating the renewed aeolian activities.

Longitudinal Sand Dunes

Origin

These are commonly known as sif dunes in North Africa and Saudi Arabia and as sand ridges in Australia. They distinctly appear on Landsat imageries in straight parallel ridges in dark grey to dark tones. The origin of these dunes has been a controversial subject. According to Glennie (1970) the environment with high wind velocity is favourable for the development of the linear or sif dunes. He opines that the higher the wind velocity, the larger the sif dune and greater the dune spacing.

The structures of the longitudinal dunes make a distinctive pattern very different from those of barchan type dunes. In these dunes the slip faces have developed on both sides of the crest and forest dip directions form two clusters of points in nearly opposite directions, rather than a single grouping (McKee, 1979). Bagnold (1941) advocated that the longitudinal dune is a modification of the fundamental barchan form produced by strong cross winds transverse to the prevailing wind direction. The hypothesis propounded by Bagnold is not convincing to the author. In the opinion of the author, the transverse sand ridges coalescing the trailing arms of the parabolic dunes in the lee of the parabolas were removed by the strong prevailing winds resulting into parallel sand ridges which further grew in length downwind by aeolian deposition (Plate 6). The longitudinal dunes have not resulted from the barchan dunes but they are the outcome of the parabolic dunes. In the early Holocene period, the arid phase was terminated and the humid phase started which was responsible for the stabilisation of these dunes. Verstappen (1968) has also reported that the longitudinal dunes in Thar desert were originated from the parabolic dunes. The central curved parts due to funnelling effect were readily removed by the strong winds and the parallel ridges stretching in the direction of the prevailing winds were resulted. King (1960) has also reported that linear dunes were formed due to the separation of the trailing arms of the parabolic dunes from dune noses.

Distribution and Morphology

The longitudinal dunes mostly occur in southwestern parts of Jodhpur district. These dunes are also stabilised, well vegetated and sands are compact. They are 6 m to 15 m and 15 m to 60 m in height and their orientation varies from N 40° E to N 45° E. The slopes of the flanks, leeward and windward sides of the dunes vary from 9° to 14°, 20° to 24° and 2° to 3° respectively. The dunes are at a distance of 600 m to 1200 m from each other. The average values of the length and width of these dunes are 3000 m and 260 m respectively. The dune sands are mixed with lime nodules of 2 m to 25 m length and they are highly calcareous and compact. The sand grains of 0.07 to 0.15 mm diameter are 70 to 90 per cent. The average diameter of these sand grains on leeward slopes is 0.10 mm and on flanks and windward slope is 0.14 mm. The dunes are in alluvial phase but due to renewed aeolian activity, the crests have been reactivated and the sands are piled up on leeward sides and crests. The barchan dunes of 3 to 4 m height are superimposed on the flanks of these dunes.

Transverse Sand Dunes

Origin

The origin of these dunes is least understood and more controversial as to how the transverse dunes could be formed at the right angle to the prevailing winds. Transverse dunes resemble the barchan dunes in being a product of unidirectional winds, with steep slip faces oriented normal to wind direction (McKee, 1979). From the interpretation of the aerial photographs, landsat imageries and field investigations, the author has concluded that these sand dunes were originated due to prodigious sand cover on the ground, abundance sand supply, absence of vegetation and persistant and widespread wind action under arid climatic conditions. The dune building winds due to the above factors created the wave like ridges separated by the trough like furrows at right angles to the prevailing winds and the transverse dunes came into being. These dunes were stabilised during the humid phase which prevailed after the arid phase of pre-Holocene period. The loose sands of 2 m to 3 m thickness are superimposed on all the slopes of these dunes. The most diagnostic feature of the transverse dune structure is the extent of the horizontal or nearly horizontal, parallel laminae formed by the apparent dip of strata. The other noted feature of the dune is the presence of lenticular zones of coarse sand among laminae of fine sand.

Distribution and Morphology

The transverse dunes are distributed scatteringly and occupy a small area of the arid zone. The transverse dunes occur in the southeastern part of Jodhpur district. The high density of these dunes has been found in Pugal area of Bikaner district.

These dunes are of 5 m to 10 m and 10 m to 40 m in height and lie 500 m to 1000 m apart from each other. The orientation of high dunes is N 50° W and that of low dunes is N 45° W. The slopes of the flanks, leeward and windward sides are 12° to 14°, 20° to 22° and 3° to 4° respectively. The average length and width of these dunes are 750 m and 150 m respectively. The sand grains in the dunes are of 0.03 to 0.18 mm diameter which are finer than the grains of other dunes. About 50 per cent sand grains of 0.03 to 0.07 mm size. The average diameter of the grains is 0.11 mm. The sands are highly calcareous and compact. These dunes have undergone changes due to renewed aeolian activities which have deposited fresh sands of 1 to 3 m thickness on the flanks, crests and leeward side of these stabilised dunes.

Barchan Sand Dunes

Origin

The barchan dunes were formed by undirectional uniform winds on the hard surfaces with scanty vegetation and limited sand supply. The prevailing winds deposited the sand on the bare surfaces and they continued to deposit the sand in the lee of the crest till the angle of repose for loose sand was formed and the sand was moved downwind and created the steep face. Due to the absence of vegetation, two elongated ridges on either side of the steep face advanced down-wind up to a few kilometres and the crescent shaped barchan dune with concave leeward side and covex windward side was formed. By this process, several barchan dunes in the same path were originated and grouped together and ultimately resulted into coalesced barchan dunes (Plate 7). The barchan dunes are still being formed on the bare surfaces mostly near the settlements.

Distribution and Morphology

Individual and coalesced barchan dunes are scatteringly distributed in the central, western and southwestern parts of the arid zone. They are generally found near the settlements on the old dunes and in the river beds. Most of the dunes are concentrated in Jodhpur, Bikaner and Churu districts.

These dunes are active, devoid of vegetation and have non-calcareous, loose and single grained sands. The heights of the dunes vary from 2 m to 5 m, 4 m to 10 m and 6m to 15 m. There are 5 m to 15 dunes in the dune chains and the average spacing is 100 m in southwest and 300 m in northwest. The dune orientation varies from N 38° E to 40° E. The average length and width of these dunes are 350 m and 250 m respectively. The slopes of the flanks, leeward and windward sides are 5° to 8°, 14° to 16° and 2° to 3° respectively. The horizontal and vertical sand ripples at 12 cm apart are formed on the dune crests. The sand grain size is 0.03 to 0.25 mm and the grains of 0.18 mm size are dominant. The sand grains of 0.06 to 0.12 mm size are 25 to 50 per cent. The average size of the sand grains on three dune slopes is 0.12 mm.

SOURCE AND ORIGIN OF THE DUNE SANDS

The source and origin of the dune sands has been a controversial subject. Wadia (1962) has reported that the dune sands of this region have been largely transported by southwest winds from Rann of Kutch and they are of marine origin. Contrary to it, the geomorphological investigations on the size and shape of the grains of the dune sands revealed that these sands have been locally derived from the existing landforms by the aeolian process under the prolonged arid climatic phases and are of aeolian origin.

Source of the Dune Sands

The magnitude, shape, orientation and distribution of the sand dunes depend upon the size of the sand grains as the coarse grains drop at a shorter distance while the fine sand grains continue their flight downward to varying distances, depending upon the velocity of the prevailing winds and the violence of the eddies in the leeward side of the dunes. The statistical analysis to compare the mean diameter of the grains of the dune sand from two blocks, viz., Chohtan block which is near to the Rann of Kutch and the Ahor block which is far away from the Rann of Kutch was done and the calculated values of the mean, standard deviation and the coefficient of variation are given in Table 11.1.

TABLE 11.1 : **Analysis of the Size of Sand Grains in Sand Dunes**

Block	Mean (mm)	Standard deviation (mm)	Coefficient of variation %	t
Chohtan block	0.120	0.0025	4.16	2.57*
Ahor block	0.156	0.0270	17.31	

*Significant at 5 per cent probability level.

From the above table, it may be inferred that there is significant difference between the mean diameter of the grains of the dune sands from two blocks. The lower values of the mean, standard deviation and coefficient of variation of the mean diameter of the sand grains indicate that the dune sands of the Chohtan block are finer than that of the Ahor block. The significance of the difference between the size of the sand grains in two blocks has been tested by using t-test, which shows that the difference between the sand grain size is statistically significant at 5 per cent probability level. The significant difference between the size of the sand grains conclusively proves that dune sands have been locally derived by intense aeolian activities from the existing landforms of these blocks and have not been transported from Rann of Kutch as advocated by Wadia (1961). If it had been true that the source of dune sands is Rann of Kutch, then the sands of the Ahor block, which lie far away from the source region of dune sands, should have been finer. The study on the size and distribution of sandgrains at various depth along the different slopes of the dunes in Phalodi and Osian tehsils has also indicated that dune sands are finer in Phalodi tehsil, which is nearer to Rann of Kutch than that of Osian tehsil. The results of this study also confirm that the dune sands have not been transported from Rann of Kutch but have originated from the local rocks and other landforms under the physico-chemical reaction of the weathering and erosion agencies.

The mineral analysis of the dune sands (Table 11.2) shows that the predominant minerals in the sands are quartz,

TABLE 11.2 : **Minerological Composition (in per cent)**

Location	Augite	Appatite	Aegirine	Actinolite	Biotite	Emstatite	Feldspars	Garnet	Hematite & Magnetite	Hornblende	Kyanite	Musconite	Quartz	Rutile	Sphene	Titanite	Zircan	Epidote
Mithari	—	—	—	—	—	—	1.50	3.25	40.20	—	—	—	42.95	—	1.10	—	—	—
Barmer	—	—	2.25	1.10	1.10	4.45	4.33	1.15	20.00	2.25	1.10	1.10	52.77	2.20	—	2.20	—	—
Mokalsar	—	—	—	—	—	1.36	4.55	—	35.33	—	—	1.36	55.98	—	—	—	—	1.36
Bhopalgarh	—	—	—	—	1.35	2.70	24.07	3.20	24.55	2.30	—	—	37.78	0.90	2.75	—	0.90	—
Sardarshahar	2.28	—	—	—	—	—	12.09	—	40.91	—	—	—	42.46	—	—	—	2.94	2.28
Bikaner (Beechwal)	—	—	—	—	—	—	12.75	—	44.12	—	—	—	40.19	—	—	—	2.94	—

hematite, magnetite and feldspar. The Aravallis metamorphic rocks and the igneous rocks of this region consist of almost the similar minerals. This indicates that the sediments reworked by aeolian activities, were transported by the prevailing winds through saltation and surface creep processes to form the sand dunes of the Indian arid zone.

Origin of Dune Sands

The analysis of the size and shape of the sand grains from different parts of the arid zone has indicated that the dune sands are of aeolian origin and not of marine or fluvial origin. It is observed from Table 11.3 that 60 to 75 per cent sand grains in all types of dunes are of 0.06 to 0.12 mm size which reveal that the dune sands were originated under arid phases by aeolian process. The average size of the sand grains varies from 0.12 to 0.14 mm which is much below the average diameter of 0.3 mm for aeolian sands, reported by Bagnold (1941) for Libyan desert. The microscopic examination and the microphotographs of the sand grains and their comparison with the five standard roundness classes of Faber (1960) have shown that more than 75 per cent sand grains are of subrounded to rounded and also well rounded shapes. The roundness in the sand grains has been caused during wind transportation by abrasion such as bouncing off obstructions and wearing off their irregularities and not due to the capacity of winds to pick up the rounded particles as reported by Pettijohn and Lundahl (1943). The values of the roundness of sand grains from the representative sites of the dune country range from 0.4 to 0.9. There is, however, very small percentage of subangular sand grains of more than 0.12 mm size. The microphotographs of the sand grains (Plate 8) also show that there is a difference in the roundness of various sizes of sand grains which increases with the decrease in the size.

The examination of the soil profiles in sand dunes (Table 11.4) also reveals that the sands are of aeolian origin as they constitute only 0.9 to 1.7 per cent silt and 2.2 to 4.0 per cent clay and higher percentage of coarse to fine sands.

TABLE 11.3 : **Percentage Grain size Distribution of Surface Sands in Different Type of Sand Dunes**

No. of types and site of dune	Sand grain diameter in mm						Average diameter
	0.25	0.18	0.15	0.12	0.07	0.06	
1. Obstacle dune (Adhur Bhakari Shergarh)	1.03	3.37	20.84	45.20	28.26	1.31	0.14
2. Parabolic dune (Dabari village)	2.09	8.08	22.79	41.78	21.45	3.87	0.14
3. Longitudinal dune (Jethania village)	0.91	13.66	18.12	23.56	40.73	3.02	0.14
4. Transverse dune (Near Chhatargarh)	—	31.46	11.22	28.46	21.38	7.48	0.12
5. Barchan dune (Shergarh south of Kairaliwala Dhora)	—	5.05	23.78	31.90	37.55	1.72	0.12
6. Shrub-coppice dune (Between Chordiya and Jethania)	—	14.36	23.45	20.91	39.36	1.92	0.12

TABLE 11.4 : Soil Profile in Sand Dunes

Depth in cm	Mechanical composition			
	Coarse sand %	Fine sand %	Silt %	Clay %
0-5	4.0	91.0	1.0	4.0
5-25	5.2	90.0	1.1	3.7
25-50	7.2	88.6	0.9	3.4
50-100	5.6	90.7	0.7	3.0
100-200	3.7	91.0	1.6	3.7
200-300	3.1	92.8	1.7	2.4
300-400	10.7	86.1	1.0	2.2
400-500	6.0	90.0	1.0	3.0
500-600	8.4	87.5	1.0	3.0
600-700	3.4	91.3	1.7	3.6

SUMMARY AND CONCLUSIONS

In the Rajasthan desert, six simple types of sand dunes viz., major obstacle, minor obstacle, parabolic, longitudinal, transverse and barchan have been identified. The factors responsible for the formation of these dunes are winds of sufficient strength and duration, adequate source of sand supply, biophysical barriers, wind deflation and processes of sand transportation. Major and minor obstacle dunes were formed due to the obstruction in the path of the prevailing sand laden winds. Parabolic, longitudinal and transverse dunes had resulted due to the accumulation of the sands by prevailing southwest winds without any obstruction. The transverse dunes came into being due to the accumulation of the sands in the transverse direction of the prevailing southwest winds. The compound and complex dune types were formed due to the combination and/or coalescing of two or more same and different types of simple dunes respectively.

It has been observed that the significant relationship exists between the wind regime and dune orientation. The suspension, saltation and surface creep are three main processes for the transportation and deposition of sands in the form of sand dunes and sand sheets. It has also been established that source of dune sands is not Rann of Kutch but have been locally derived from the existing landforms and they are of aeolian origin.

SUGGESTED READINGS

Allchin, B. and Goudie, A., 1971: Aridity and early man in Gujarat, Western India, *Man*, Vol. 6 No. 2, pp. 245-265.

Bagnold, R.A., 1941: *The Physics of Blown Sand and Desert Dunes* Methuen & Co., London.

Cooke, R.U. and Warren, A., 1973 : *Geomorphology in Deserts*, B. T. Batsford Ltd., London, pp. 318.

Faber, F.J., 1960: Size and shape of sand grains, *Trans. Int. Geol. Cong.* Copenhegen.

Glennie, K.W., 1970: *Desert Sedimentary Environments: Developments in Sedimentology* (V); 14 Elsevier Publishing Co., New York. p. 222.

King, D., 1960: The sand ridge deserts of south Australia and related aeolian landforms Quaternary arid cycles. *Royal Soc. South Australia Trans.*, V. 83, p. 99-108.

McKee, E.D., 1979: *Sedimentary Structures in Dunes in McKee* ED.(ed). *A Study of Global Sand Seas*, U.S Govt. Printing Office, Washington.

Pandey, S.; Singh, S. and Ghose, B., 1964: Orientation, distribution and origin of sand dunes in Central Luni basin. *Proc. Symp. Probl. Indian Arid Zone*, pp. 169-178.

Pettijohn, F.J. and Lundahl, A.C., 1943: Shape and roundness of Lake Beach sands. *Journ. Sedim. Petrol.* Vol. 13, pp. 69-78.

Singh, S., 1977: Sand dunes and palaeoclimate in Jodhpur district. Western Rajasthan, *Man and Environ*, 1, pp. 1-15.

Vats, P.C.; Singh, S.; Ghose, B. and Kaith, D.S., 1976: Types orientation and distribution of sand dunes in Bikaner dist. *Geog. Observ.*, Vol. 12, pp. 69-75.

Verstappen, H.Th., 1968: On the origin of longitudinal (Sif) dunes. *Zeitsch fur Geomorphl*, vol. 12, No. 2, pp. 200-220.

Wadia, D.N., 1961: *Geology of India*, Macmillan, London, pp. 398-99.

TWELVE

Sudershan Mukerjee

STRUCTURAL GEOMORPHOLOGY OF NAGALAND

NAGALAND or land of the Nagas is a narrow rugged mountainous territory in between the Brahmaputra valley and Burma with districts of Kohima, Mokokchung and Tuensang. Geographically lying between 25° 6′ N and 27° 4′ N latitude and 93° 20′ E and 95° 15′ E longitude, on the extreme east of India, the territory has an area of 16488 km^2 with Tirap division of NEFA on the north, Manipur State on the south, broad valley of Assam plains all along its western foothills and Burma on the east.

Apart from a few hundred km^2 of plains in Dimapur the entire country is covered with mountain ranges varying in height from 600 m to 300 m above sea level. The ranges of Nagaland are low, the highest peak being not more than 3000 m whilst majority of the ranges are not more than 1800 m in height. These long parallel unbroken ridges make most of the valleys inaccessible. Yet they broaden out southward to enclose Manipur basin. Among the prominent mountains of the state is the Barail range, that pierces from southwest and runs up to Kohima. To its further east are Kohima Hills and Naga-Patkai Ranges known as Purva-Anchal (Eastern Mountains).

Excluding the Dhansiri, Diyung, Dikhu, Barak and Tizu, there is no worth mentioning river here though small typical annual hill streams are innumerable (Fig.12.1). Though there are five river sub-basins namely Dhansiri, Diyung, Dikhu, Barak and Tizu of which the first three are the sub basin of the Brahmaputra and the other two of Surma and Chenduin respectively, the Barail in the southwest and the Naga ranges in the northeast are the main watersheds.

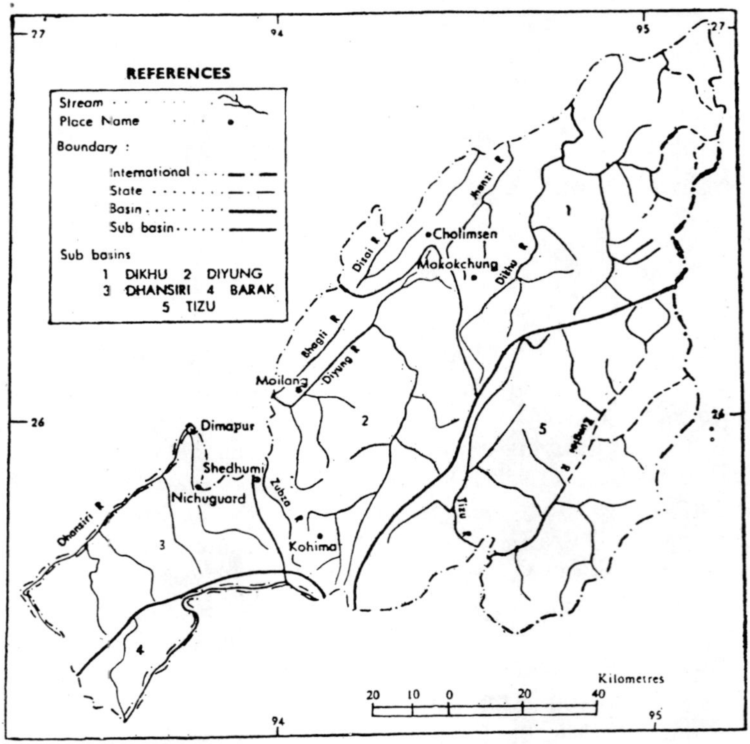

FIG. 12.1 : *Eastern Mountains*

Our knowledge on the structure of this region is very sketchy. However, the region contains Tertiary rocks. The structural history of the region is that it was occupied by geocyncline during the Tertiary era and was gradually filled up and later gently folded. At places in the region the folding has been so intense that the rock beds stand up vertically in absolutely bare hill side. Anyway the geocynclinal sediments were for the most part laid down in a somewhat shallow basin which was sinking throughout the greater part of the Tertiary time and were mainly composed of alternations of shale, mudstone, clay with silt stone and sand stones and occasional conglomerates. The shelf facies also occur here but beneath the alluvium of the plains of Dimapur.

Here the folds are long, individual anticlinal axis running for many kms with more sharply folded anticlines associating with thurst faults. All along the western boundary of Nagaland the thursting is so strongly developed that but little remains of the anticlines and the structure is characterised by a number of nearly parallel strike faults running northeast-southwest, Although many of these follow nearly straight courses for many kms, at points they bend abruptly to make two faults join together. The thrusts are seldom more than a few kms apart, and between them deposits of different geological time outcrop as long narrow strips. The other thrusts or reversed faults into the inaccessible hills on the northeast are less known. A curious feature of the thrusts of the region is that even relatively unconsolidated beds below a thrust are usually disturbed but not through any great thickness. The approximate line of thrust have often developed fault scarps and have given rise to series of deep valleys. Anyway the region has following broad structural divisions—(1) Belt of Schuppen, (2) Patkai-Kohima Synclinorium, (3) Foreland Spur (Fig.12.2).

Belt of Schuppen

A belt of about 20 km broad with its northwestern edge following closely the boundary of Assam valley alluvium and the northeast edge along the Disang thrust is known as Belt of Schuppen (Fig. 12.3). In the belt, eight separate nappes have been demarcated with overthrusts forming a complex pattern, higher ones in places overriding the lower ones. The frontal thrust though known as Naga thrust is seen to have in a longitudinal distance of 100 km, some six different thrusts. The easternmost thrust of the belt is known as the Disang thrust. As regards the stratigraphical throw of the individual thrust it varies greatly, but 3000 to 8000 m is usually for the larger thrusts. The horizontal displacements are not known, they are almost certainly many times the stratigraphical throw. It is suggested that the movement of individual masses might be from 20 to 40 km giving a total crustal shortening for the eight thrusts of anything between 150 km and 300 km.

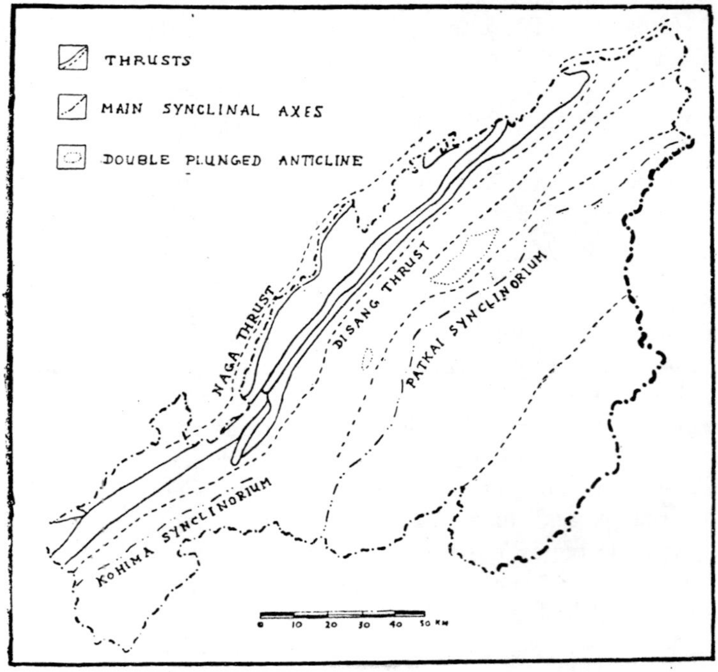

FIG. 12.2 : *Broad Structural Divisions*

The exposed parts of the thrust sheets nearest the plains are composed principally of blue, grey, and mottled clays with a few sandstone beds and is invariably associated with the ground and a few narrow ridges of a slope form of 20/100 to 25/100 gradient. The soft character of these argillaceous rocks is responsible for this type of landform. The exposed parts of the higher thrust sheets are usually made up of arenaceous rocks which are largely very hard, and well cemented, and generally associated, with a high ground of a slope form of 40/100 to 50/100 gradient. However, the thrust sheet along the Disang fault between 25°30′ N to 20°15′ N and another strip in the Desai valley due to bedded sandstones with subordinate carbonaceous shale and clay and coaly shales give rise to a slope-form of 25/100 gradient. The uppermost thrust sheet of the main Disang thrust includes a large thickness of shales metamorphosed into hard, glossy dark blue slates and have evolved a steep slope of a gradient of 40/100 to 50/100 due to their property of resistance to denudation.

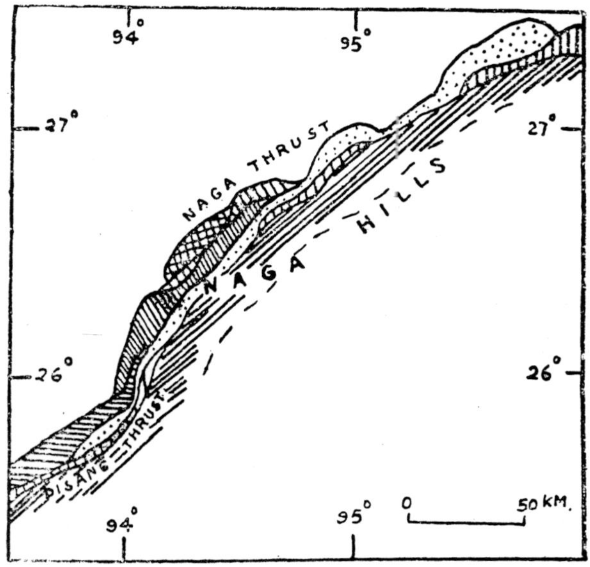

FIG. 12.3 : *Belt of Schuppen (A belt of thrust sheets)*

The drainage system of the belt are more or less a structurally controlled. Taking advantage of the land of shattered rocks of the juxtaposition of hard and soft rocks a valley along the fault line are generally formed. The courses of major rivers like Diyung, Dikhu, Jhanzi in this belt are fault guided. The straight course with gorge like valley of the Diyung and Jhanzi are but thrust fault, while their sharp elbow turn is due to their entrances into Disang thrust plane. The fault generated scarps of the belt are generally fault line scarps which means the line of dislocation, having been brought into relief by work of streams where the differential erosion of rocks of strongly contrasted resistance have played the major role. The fault line scarp along the course of the Diyung river from near Sanis (26°13′ N—24°14′ E) to near-about Moilong (26°4′N—94°4′E) has hard Barail sandstone on the right bank while the bedded sandstones with subordination of carbonaceous shale, clay and coaly shales on the left. Further the fault line scarp developed by river Tsurong or Desai is from (26°41′ N—94°15′ E) to Satsekpa (26°29′ N—94°21′ E) and has sandy shales laminae of clay shales (Surma Series) on the left bank with bedded sandstone of Jenam stage on the right bank. The river Jhanzi forms yet another fault line scarp where hard sandstone of the Barail is on the left while argillaceous Tipam deposits on the right. There is yet another drainage morphology caused by structural variations of the rocks. Due to weak structure of sandy shales, laminae of clay shales, and course ferruginous sands, the river Tsuing has formed an open valley from its confluence with Namsa stream to beyond Boka Pathar (26°28′ N—94°23′ E) a distance of some 15 km. This part of the valley is not only open but is swampy having water for considerable part of the year. Another such swampy patch with an open valley has been formed by the Bhagti river some 5 km downward from its source.

Patkai Kohima Synclinorium

Beyond the belt of Schuppen on the east is the Patkai-Kohima synclinorium. It is a fairly broad area which is broken up occasionally by strike faults. On either side of this synclinorium the dark grey shales of concretionary and carbonaceous origin are overlain by massive sandstones. Throughout a large part of the area there are anticlines, commonly asymmetrical, separated by broader nearly symmetrical synclines. The steeper flanks of the anticlines usually show evidence of overthrusting and in general, adjacent anticlines have their steep flanks in an opposite directions.

The northeastern folded area of the synclinorium known as Patakai synclinorium lies in a country which is difficult of access. Here the highest ground is made by synclines of sandstones while the relatively open ones are separated by faulted and overthrusted shales. On the northwestern flank of this synclinorium there are number of thrust planes, and along some of these the horizontal displacement may have been large. These thrusts could perhaps equally well be included in the belt of Schuppen but for taking Disang thrust as the demarcation line for belt of Schuppen.

To the south-southwest the Patkai synclinorium merges into the Kohima synclinorium where there are broad synclines with narrow, sharply folded and faulted anticlines working as water divide with roughly N-S axes and a slightly southerly pitch. Further south the synclinorium merges imperceptibly into the eastern end of Surma valley.

Excluding the northwestern and southwestern flanks of the synclinorium the whole area has an exposure of either hard, massive and well bedded sandstones or shales metamorphosed into dark glossy slate. Both the rock types due to their resistive character to denudation have given rise to a slope form of the gradient of 30/100 to 40/100. The precipitous Henima scarp and the magnificent cliffs and peaks of Japvo rising almost to 3014 m above sea level are determined by hard sandstones on the anticlines. Among other prominent structural geomorphic features of the region, the dome formation is the foremost. The application of the Himalayan overthrust and the Naga Hills overthrust the two sets of opposed stresses have caused cross folding and development of domes in the central part of the synclinorium in NE-SW axis in the form of big dome (26° 5′ N—26° 18′ N and 94°—22′ E—94°31′E) to the southeast of Mokokchung, circle dome (26°36′ N—26°44′ N and 94°54′ E—94°57′ E) around Wakching and a ridge dome (26°46′N—26°56′N and 95°12′E—95°17′E) on the northeast tip of Nagaland. The domes are double plunging anticlines the length of which in ground plan is not more than three times width. Another prominent structural feature of the region is the formation of rift valley with shutter ridges. Dislocation due to wrench faults caused a rift zone from Katsenyu 25°59′ N—94° 16′ E) to nearabout Philimi (26°4′ N—94° 25′ E) through which the Diyung has its course. Here the wrench faults cut across the grain of the topography and ridges on one side of the fault had been translated to such a distance as to become relocated opposite to the adjacent fault block, to accomplish the character of shutter ridges.

Among other structural geomorphic features developed in

the region is the 'Strike Vale' by the river Zungki. The river Zungki developed most rapidly its valley on the crestal area of the anticline taking advantages of the structural weakness and has formed a broad strike vale. The river Tizu taking advantage of this strike vale flows to the river Chendwin and acquires the distinction of the only river that crosses the Naga range. Strike valleys developed at the headwaters of the Dhansiri, Diyung and some other streams is yet another structural geomorphic feature developed here. Further the river Barak along Nagaland, Manipur border has its course through a strike valley.

Foreland Spur

The basement of ancient crystalline rocks known as Foreland Spur extends for many kms beneath the alluvium of the Brahmputra. Until recent years nothing was known about the structure of the Foreland Spur beneath the alluvium. Recent surveys made in course of prospecting for oil have revealed that basement extends northeast as buried ridges. Above the basement the Tertiary and post Tertiary rocks exhibit low inclination, seldom more than 1/100 to 2/100 gradient, but they are cut into blocks by criss-cross of faults which are revealed by the pattern of valleys eroded along shattered belts of the faults.

The Foreland Spur covers a very little portion of Nagaland and is beneath the new alluvium around Dimapur. The area has received very little attention. Anyway the reactivated fault lines of the ancient time are appear to be exposed by the pattern of the valleys eroded by the rivers Dhansiri and Diphu.

CONCLUSION

Early students of geomorphology expressed landforms as an exclusive expressions of structure. But later on the importance of structural factors got neglected, though the relationship between regional structure and major relief forms has never been challenged. The structural influence was considered as gross, obvious and superficial. Further the structural framework had tended to be seen as something established in the past but now static and unchanging.

The 20th century geomorphological thinking made concentration on cyclic theory where time and process along with structure got an equal footing in formation of landscape. Consideration of time and process is indeed important in the analysis and understanding of geomorphic landscape but not inherently as important as are the structural considerations. Thus a very close and detail examination of structural factors is essential to have a correct interpretation of landforms.

SUGGESTED READINGS

Chatterjee, S.P., 1965 : Physiography, *The Gazetteer of India*, Vol. I, Publication Division, New Delhi, pp. 22-23.

Das, H.P., 1970: *Geography of Assam*, National Book Trust, India, New Delhi, pp. 14-25.

Evans, P., 1933: Tertiary Succession in Assam. *Transactions of the Mining and Geological Institute of India*, Vol. 27, pp. 161-260.

Evans, P., 1964: The Tectonic Framework of Assam, *Journal of Geological Society of India*, Vol. 5, pp. 80-96.

Evans, P., 1965: *Structure of Northeastern Frontier Area of Assam*, Dr. D.N. Wadia Commemorative Volume, Mining, Geological and Meteorological Institute of India, pp. 640-645.

Ganser, A., 1964: *Geology of the Himalayas*, John Wiley and Sons, p. 289.

Kar, Mandira, 1970: Nagaland. An unpublished desertation work, Department of Geography, Calcutta University.

Krishnan, M.S., 1960: *Geology of India and Burma*, Higginbothams (Pvt.) Ltd., Madras, pp. 488-489.

Lepper, G.W., 1933: Geology of the oil bearing regions of the Chindwin-Irrawaddy Valley of Burma and Assam-Arakan, *Proceeding*, *World Petroleum Congress*. pp. 15-25.

Mathur L.P. and Evans, P., 1964: Oil in India, Special Brochure for 22nd *International Geological Congress*, New Delhi, pp. 5-32.

Mukerjee, Sudershan, 1965: Nagaland : A Geographical Sketch. Bombay, *Geographical Magazine*, Vol XIII, No.1, pp. 117-126

Mukerjee, Sudershan, 1975: Landforms of Nagaland, *The Geographical Review of India*, Vol. 37, No. 1, pp. 87-91

Mukerjee, Sudershan, 1979: Slope Analysis of Brahmaputra Basin in Nagaland, *The Geographical Review of India*, Vol. 41, No. 3, pp. 284-290.

Murthy, M.V.N., 1966: Geomorphological Evolution of the Assam Region, *Proceedings of the Symposium and Field Study Meeting*, on the Physical Geography of Eastern Himalayas and Meghalaya Plateau, Department of Geography, Gauhati University, pp. 16-23.

National Atlas of India—Physiographic Regions, Plate No. 41, *National Atlas Organisation*, Calcutta.

Sale, H.M. and Evans, P., 1940: The Geology of Assam—Arakan Oil Region, *Geological Magazine*, Vol. 77, No. 5, pp. 337-363.

Singh, Onkar; and Mukerjee, Sudershan 1971: Purvanchal, India—A Regional Geography, Ed. R.L. Singh, *National Geographical Society of India*, Varanasi, pp. 493-513.

Wadia, D.N., 1965 : Geology. *The Gazetteer of India*, Vol. 1, Publication Division, New Delhi, pp. 144-145.

THIRTEEN

Pramila Kumar

DIASTROPHIC FORCES AND THEIR GEOMORPHIC EXPRESSION IN RANCHI PLATEAU*

INTRODUCTION

RANCHI[1] plateau is a part of a region commonly known as Chota Nagpur plateau. It attains an elevation of 3500′ near its western margin while the rest of the plateau is about 2260′ above sea level. Its western part rises as a steep escarpment above the general level of the plateau. Thus the Ranchi plateau may be divided into two divisions viz. the 'Western Highland' and the 'Main Plateau' (Fig. 13.1).

Northwards the plateau overlooks the valley of the Damodar down a steep escarpment which has been dissected by its tributaries. The descent towards east and south is not as steep as that on the north but is marked by waterfalls made by rivers flowing down the plateau. However, the gentlest descent is towards south but even that is well defined by deeply dissected river valleys.

A major portion of Ranchi plateau is formed of Archaean rocks among which the important rock types are granite gneisses. To the south of it lies the region of Dharwarian formations. The former has intrusive relation with the latter showing that granite gneiss is a comparatively younger formation. At the foot of the northern escarpment lie the detached patches of Gondwana formations.

All these features are affected by later diastrophic forces and they are the guiding factors in the formation of the present relief.

*This paper was published in *The Indian Geographical Journal* Vol. XXXVI, No. 4, 1961.

The Orogenic Movements

The orogenic movements are confined to the sedimentary deposits of Dharwars and Gondwanas. The movement which folded the Dharwarian formations is of the pre-cambrian age and is called Archean Orogenesis. Though the ranges have been eroded away altogether in their folded form as such, and generally form low lands in the landscape, the dip of the strata in the residual hills, reflects the character of their past form. In Gangpur and adjoining parts of the southern Ranchi, the famous Gangpur anticlinorium is located. Between this anticlinorium and the granitic mass of Ranchi, there is a zone which has a synclinal structure. Further west, we meet with the alternate bands of granite-gneiss and mica-schist, the former of which forms part of an anticline and the latter that of a syncline. Further eastwards, to the north of Singhbhum anticlinorium, a syncline is located in which Dalma Traps have been deposited (Fig. 13.1). The syncline is again succeeded by an anticline to the north. The last two features dominate southeastern Ranchi.

A marked difference in topography is found in the regions of two major formations viz.—granite-gneisses and Dharwars.

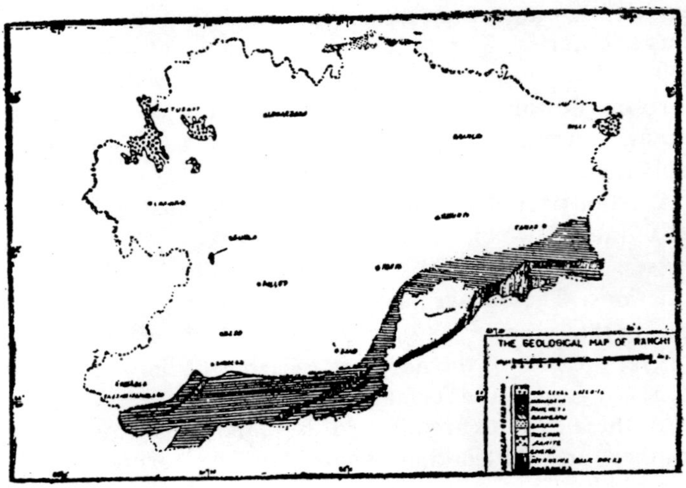

FIG. 13.1 : *The Geological Map of Ranchi*

Being homogeneous in structure, we do not find in granite-gneiss, any structural guidance in the arrangement of hills and valleys. Here the watersheds are the highlands and their extent is decided by the size of the rivers, while in the regions of Dharwars, the hills and valleys normally run along the strike direction. Besides, being residual hills, their shape and extent is decided by the outcrops of comparatively hard formations.

As usual in other places, orogenic movements are accompanied by differential movements and igneous activity. The evidence of such faulting is provided by crushed conglomerates on both the sides of anticlinoriums. The crushed zone of the north enters into southern Ranchi.

The outcrops of Dalma Trap prove the occurrence of the volcanic activity during Archaean orogenesis. The rocks of this volcanic flow are found in an elongated belt extending in southern Ranchi, Manbhum and southern Singhbhum.

Folds are also evident in the Gondwana deposits to the north of Ranchi Plateau. But the formations are subject more to faulting than to folding. In the north and south Karanpura coal-fields, the folds extend from east to west; hence it is understood that the stress exerted on the strata was in a north-south direction.

Two periods of movement have been recognised here, one which deformed pre-Mahadeva strata, that is after Panchets and the second which folded the later deposits. The direction of movement in both cases was the same.

Epeirogenic Movements

The most prominent and vigorous movement to which this area has been subjected is epeirogenic, which is expressed by the occurrence of fractures and faults accompanied by uplift. None of the geological formations has escaped this force. These movements have occurred from the time of the Archaean orogenesis up to the Tertiary period.

As mentioned earlier faulting and displacement occurred in southern Ranchi during the Archaean orogenesis. But the structure, there is so complicated and confusing that it is practically impossible to detect the amount of throw. Besides the zone of crushed conglomerates of Dharwar series, there are a number of other faults which have been traced in southern Ranchi. For example, a distinct fault has been traced along the boundary between granite mass and the Dharwars, particularly in Manbhum. Similarly, there is a well-marked fault north of Dalma Trap mainly extending in Manbhum but entering for some distance in Ranchi plateau area.

Coming to the north, along the northern margin of the Ranchi plateau runs the boundary-fault of Gondwana and that of the Damodar valley. The fault is supposed to have been formed at the time of actual deposition. The fault runs along the whole upper Damodar valley but the region adjacent to Ranchi plateau is that of the north and south Karanpura and Rampur coal-fields.

Thus, some of these faults are definitely of early ages and epeirogenic activity recurred along such weak zones in latter periods also. But some of these faults might have been formed during the Cretaceous and the Tertiary block-movement which were probably the last to affect this region. About these movements Dr. Dunn says "subsequent to the out-pouring of the Deccan Trap, there was prolonged erosion and uplift to the extent of about 1,000′ in the western Chota Nagpur. . . the remnants of the latter were left as a small plateau in the west. This peneplain was being formed throughout a great part of the Tertiary times".[2] This is the movement which made the 'Western Highland' a distinct topographical feature of the region. It retains its prominence even now in the whole of Chota Nagpur plateau.

The main plateau of Ranchi was rejuvenated subsequent to the 'Western Highland'. "Next in the latter part of the Tertiary period, uplift was renewed to the extent of about 1,000′ in the western Chota Nagpur, but in the form of a block-movement with a warping along a northsouth line which is now represented by the eastern edge of Chota Nagpur plateau. At the same time a trough developed along the valley of the Damodar, presumably due to faulting, cutting off the small Hazaribagh plateau to the North". Dr. Dunn has mentioned that later there was a third period of movement but probably Ranchi plateau and Damodar valley have taken their present form as a result of this block-movement.

The fact that Ranchi plateau has been subject to some recent epeirogenic movement is clearly expressed in its shape present topography and drainage. Besides, there are geomorphic features to prove that rising was not continuous but there must have been at least two distinct periods of uplift with a long break of standstill, while the region was sculptured by exogenic forces.

Granitic mass of Ranchi came into existence in the first era of the geological history but its present elevation ranges from 2250′—3250′. The exogenic forces working on these formations throughout the later three eras could not have left it so high. Its history is of degradation, never alternated by aggradation. Moreover, the younger deposits of the area are for the most part Gondwanas, which have their base either on the granitic formation or on the Dharwars. The thickness of Gondwanas, as estimated by the geologists is considerable but still these deposits form the lowlands.

Moreover, similar granitic masses of the south in Manbhum, Singhbhum, Gangpur and Bonai, also do not attain such elevation as the Ranchi plateau. They are only about 1000′ high on an average. But this fact does not lead to any significant topographic difference. Thus it can be concluded that the land-mass of Ranchi must have heen peneplained several times in its history and its present elevation is the product of a recent epeirogenic movement.

The 'Western Highland' which was rejuvenated earlier, is even now much higher than the main plateau and there is a distinct escarpment which separates the two (Fig. 13.2). But only the small table-lands comprising the 'Western Highland' attain the elevation of 3000′. The main river valley, i.e., of the North Koel and the Sankh, have the same average elevation as that of the main plateau. It shows two facts, firstly the rivers are unable to change these tablelands into sharp-crested ridges which might have been possible in the period in which they had carved out their valleys into gorges. Naturally the surface rocks are more resistant to erosion than the lower ones. And secondly, this part of the plateau must have been elevated to its present elevation before the rise of Ranchi plateau proper, otherwise the rivers would have been flowing on a higher level and might have narrower and deeper valleys with overlapping spurs. The longitudinal profiles might have been steeper and more irregular than they are at present. Now the waterfalls, the most prominent features of the river valleys are found either near their sources or across their tributaries. Only a few minor falls are found in the main rivers. Also, the common uplift must have been reflected in the watersheds of the rivers. Because here the rivers have to work more in order to extend their channels by headward erosion. Hence, the watersheds of the rivers must have been broader, higher and unaffected by river action. But actually the watershed between the North Koel and the Sankh is less than 250′ wide and important tributaries forming these rivers rise at a few hundred yards distance from each other on two slopes of the narrow elongated tablelands.

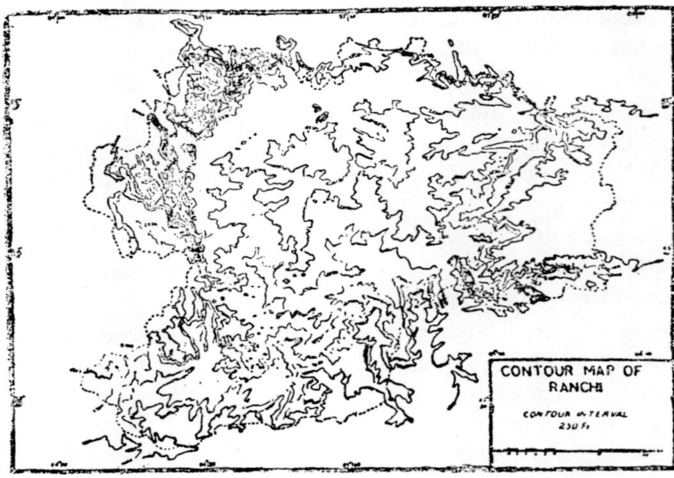

FIG. 13.2 : *Contour Map of Ranchi*

In this connection it is necessary to mention that both the 'Western Highland' and the main plateau have independent drainage systems and the watershed still runs along the escarpment overlooking the main plateau. There is not a single river of the main plateau which has extended its valley up to the interior of the 'Western Highland'. All of them either drain the scarp face or the narrow marginal zone along the escarpment.

All this discussion brings us to the conclusion that the river system of the 'Western Highland' is of older date prior to the upliftment of the main plateau of Ranchi as they show most of the characteristics of old age, while the rivers of the main plateau developed lately after its upliftment.

The Geomorphic Expression of the Upliftment on the Marginal Areas

The Western Highland

The marginal escarpments overlooking the surrounding region were originally formed by the upliftment of Cretaceous period but they vary in their geomorphic characteristics and so they need detailed treatment. The escarpment to the east overlooking the main plateau and the one to the north overlooking the Auranga and Damodar river valleys are intersected by rivers which divide them into separate blocks, while the escarpment to the south is not so well-marked and distinct but is still steeper than the southern escarpment of the main plateau.

Escarpment to the East

There can be two possibilities of the formation of this well-marked and steep escarpment. Firstly, the 'Western Highland' was covered by the Deccan Trap which served as a resistant cap-rock for the underlying granite. Thus, there was differential weathering of the 'Western Highland' and that of the main plateau, giving rise to this escarpment. But such is not the case, as investigated by Dr. S.C. Chatterji,[3] in his field survey, basalt is found even on the main plateau.

Secondly, as mentioned earlier the escarpment marks the line of a previous uplift, prior to the rejuvenation of the main plateau. But such a fault-scarp, in course of time could not have remained so distinct and steep. Thus, the first factor also seems to have given its helping hand in making the escarpment a prominent topographical feature. It is, therefore, considered that the upper part of the escarpment was formed by uplift and has become high and well-marked, in due course, by the differential erosion of granite and cap-rock basalt. So this is probably a composite scarp, the upper part being formed by uplift and the lower by erosion.

The Escarpment to the North

The escarpment to the north is of the same nature. Being part of the boundary fault of the Gondwanas it will be more convenient to discuss it with the northern escarpment of the main plateau. But it will not be out of place to mention that the escarpment in this section is highest in the whole of the Ranchi plateau. The descent on an average is between 1850′-1500′, for example, from Khamarpat (3504′) to the valley of Deonad river (1750′) or from Bulbul (3349′) to the valley of Auranga river (1750′). Probably it is due to the combination of two factors: first, the uplift of the 'Western Highland' in relation to the rift-valley of Damodar was larger as compared to the main plateau, and secondly the granite-gneiss was protected by a cap-rock of basalt on the 'Western Highland' while the Gondwanas of the Damodar valley suffered intensive erosion.

Escarpment to the South

As mentioned in the previous paragraph, the southern escarpment of the 'Western Highland' is neither so prominent nor is so high. The area is highly dissected by the river Sankh and its tributaries. Instead of well-marked tablelands, there are only small buttes and elongated hills. At present, the rocks on both the sides of the escarpment are the same, so there is no difference in the topography, either on its top or at its bottom. It is possible that the basaltic cover was removed much earlier here than in the north. And subsequently the rivers have been actively engaged in brushing off the escarpment. Hence the amount and the line of uplift is not easy to detect.

The Main Plateau

As mentioned earlier the northern escarpment of the main plateau runs along the southern boundary of the Damodar valley. This valley is one of the branches of the main Narbada-Son rift. Even between its two boundaries, the southern boundary fault is more marked. In case of Ranchi, only those portions of Damodar valley which include Ramgarh, North Karanpura coal-fields lie adjacent to it in the north. If we observe only the southern boundary of these fields it will be evident that they are abnormally straight and are practically in one straight line, extending westwards, one after another. (Fig. 13.1). The boundary of each is interrupted by the granitic out-crops. Naturally it may be assumed that this straight boundary was formed by faulting before the deposition of Gondwana beds. After the deposition the subsidence was renewed and the deposits sunk much deeper. In the case of Raniganj coal-field the sinking was about 9,000′ in the vicinity of Panchet hill.[4] The evidence of faulting of Gondwanas along the southern boundary is found in all the coal-fields of the Damodar valley. About the Ramgarh coal-field, C.S. Fox writes "the southern boundary of the field is cut-off by a strong fault".[5] A similar statement is made by him for the southern boundaries of north and south Karanpura coal-fields. Describing the structure of the South Karanpura coal-field and pointing to where the southern boundary fault is exposed, Jowet Albert is of the opinion—"The southern boundary fault is exceedingly well-exposed in the river east of Huburwa Hill, at one point a section in the bank above 25 ft high, shows grits dipping northwards at a high angle, faulted against hornblende schist".[6]

This statement proves that the southern boundary of these Gondwana deposits is a weak zone and persisted throughout the geological history of the region. Even at present the northern escarpment of Ranchi plateau runs along the southern boundary of those Gondwana deposits. As soon as this boundary is crossed, the highland is met with. But as we move from west to east, the escarpment is found more and more receding southwards away from the Gondwana deposits. Up to Rochap, a hill between Batuka and Nalkhari Nadi—two southern tributaries of Damodar, the escarpment runs adjacent to the southern boundary of Gondwana. Eastwards though it has receded southwards but two prominent spurs run up to the southern boundary of South Karanpura coal-field. Further east, the escarpment swings southeastwards, terminating in Dondag Pahar an eastwards extending spur of Ranchi plateau (Fig. 13.1 and 13.2). Below the escarpment, in this section, the land slopes gently up to Damodar. Probably because being lower in elevation and also being comparatively nearer to the sea, this part would have been attacked first by the rivers, and naturally levelled before the adjacent part in the west.

From the above, it is therefore concluded, that during the late Tertiary block-movement, the same line of weakness was utilised, and the Ranchi plateau was uplifted along the faulted zone, making the southern boundary of the Damodar valley, for it is the law of nature to utilize the same zones of weakness again and again, for the expression of any earth movement.

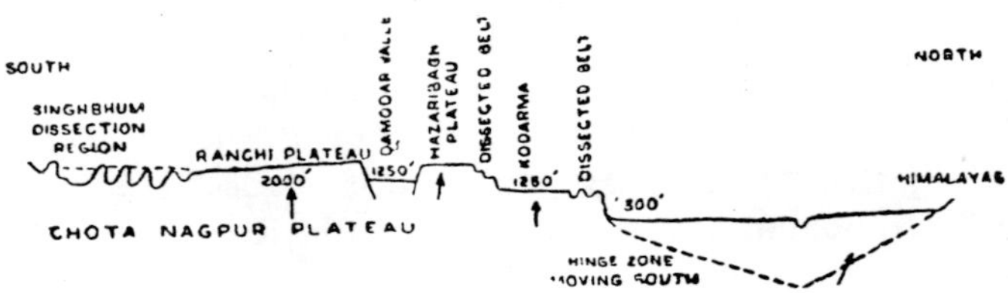

FIG. 13.3 : *Section Across Bihar*

Here the faulting has taken place at the junction of the two geological formations, so there is no question of there being similar rocks on both sides of the fault, which is an important indication of faulting. As the rocks on the down-throw side are Gondwanas, which are more susceptible to erosion than granite-gneiss, they must have been eroded more easily, hence the lower part of this fault-scarp may have been formed by differential erosion. So this fault again can be classed as a composite scarp. Thornbury says that such scarps are common in nature.[7]

Being guided by the steep slopes, the rivers which descend from the escarpment have comparatively straight courses and are naturally parallel to each other. Such are the southern tributaries of the Damodar river which descend from the Ranchi plateau. Main rivers from west to east are the following: Dhudki nadi, Dhangarha nadi Chirokhara nadi, Dainkata nadi, Saphi, Batuka nadi, Nalkhari nadi and Dhondhab nadi. In the extreme west, Gala nadi which is a tributary of the westward flowing river Auranga, also drains the scarp-face. These rivers in their turn have developed innumerable tributaries, which are actively engaged in lowering this escarpment.

Along the zone where these rivers flow through the escarpment, their gradient is steep, hence their force is utilized in cutting their beds and thereby forming narrow and deep gorges. Such gorges are found in all the above-mentioned rivers and in some of their tributaries. These gorges are alternated by long, narrow and sharp-crested ridges, which present a topography somewhat similar to the mountains. The predominance of relief in this zone is clear evidence of the work done by these rivers subsequent to the formation of the escarpment.

Another effect has been the existence of rocky outcrops in the river beds and on river banks throughout this zone. Moreover, the deposits are found just above their confluence with the Domodar only, and not in the section where they flow through the escarpment or near their sources. By observing the sources of these rivers it becomes evident that all of them have encroached into the flat tops of the plateau and are extending their shallow valleys, the flat tops of the plateau, however, continue to exist in between them. These must have been at one time part of the plateau, though they are now isolated by their headward erosion.

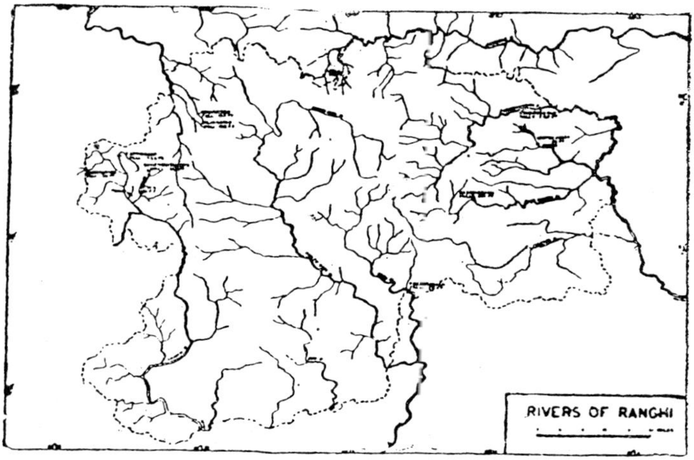

FIG. 13.4 : *Rivers of Ranchi*

Another fact which is very striking is that the valleys of the above mentioned rivers must have been interrupted by a sudden drop at the fault-scarp thereby ereating the waterfalls.

But now we find only one fall of 70 ft in the courses of Dhangarha river,[8] preceded by a gorge about 1½ miles long (Fig. 13.4). All the other falls seem to have been eliminated by the active erosion and the rivers have comparatively smoothened their valleys in longitudinal profile. Existence of this single waterfall along the escarpment is a unique feature. It has persisted for a longer period probably because of two factors. First, the rocks may be more resistant than the rocks in the other river valleys to the east; and secondly, in comparison to the other tributaries of the Damodar, which drain the scarp face this river is situated near the source of the Damodar. Owing to its location in the interior, far from the sea, the waterfall has remained unaffected by the normal process of elimination while the irregularities of the rivers to the east have been brushed off.*

All these facts lead us to the conclusion that the escarpment is in a mature stage and a time will come when these rivers will extend their valleys by the headward erosion and river capture and ultimately by their active erosion, the conspicuous nature of the escarpment will vanish.

We have emphasized the view that Ranchi plateau has been formed by block-upliftment and the Damodar valley by sinking. The existence of three hills Batuka 2425′, Aswa 2450′ and the Lugu Pahar 3200′, in this valley might however, lead to a different conclusion that the valley has been formed by erosion and not by faulting. Support for the latter view is found by geomorphologists in the argument that in epeirogenic movements, uplift always does not take place in one direction, but the reverse movement is also evident. In the case of this rift valley, the second process, being localized to the valley, led to the upliftment of the valley, instead of its further sinking; and these hills were thrown up almost to the height of the upthrow blocks on either side of it. This becomes clear by the erosion which is going on vigorously on all the sides of the hills, causing steep slopes.

Another feature may also be mentioned here, that the southern block of the rift valley, namely Ranchi plateau which being uplifted more vertically than its northern counter-

*The logic of this argument is not quite clear (Editor).

part, the Hazaribagh plateau, shows signs of tilting as well. About Hazaribagh plateau, Dr. Dunn says that "Along the northern side of the plateau the movement was more of the nature of a tilt, the surface sloping more gradually down towards Kodarma."[9] On Ranchi plateau the fact is confirmed by the shorter length of the rivers in the north as compared to those flowing to the south or to the east. Besides this, the highest parts of the plateau are found adjacent to the northern escarpment and southwards the height decreases slowly (Fig. 13.2). As described above, the valleys in the north are narrow and deep. Their youthfulness is further indicated by vertical corrosion and by headward extension of their channels. In contrast to them, the rivers flowing to the south and east, that is on the back side or on the gentle slope of the tilted block, have not only open and broader valleys (except where they descend from the plateau) but they have made their flood plains in which they meander from one side to the other. Sand deposits are also found in their valleys. However, it will be true to say that the tilt was not very acute as the rivers flowing towards the south and east do not rise from the edge of the tilted block, but take their rise from a central ridge (Fig. 13.4). South Koel adopts a circuitous route across this ridge. In this connection a doubt may arise owing to the non-existence of any evidence of tilting in the 'Western Highland' where the peculiar drainage pattern is not found. Not only the Sankh and its tributaries extend for northwards but North Koel also rises from the extreme southern part of the 'Highland'. They have open and well-developed valleys, except in the upper sections where the youthful conditions are found and the waterfalls are met with. Also the highest point of the plateau is reached in Neturhat plateau which is located in its centre.

The Eastern Escarpment

The topography of the eastern margin of the plateau (Fig. 13.2) indicates that formerly the plateau had extended more towards the east but it receded later to its present extent, mainly by the erosive action of Subarnarekha and its tributaries. At present Subarnarekha marks the eastern margin of the plateau and for a long distance the district boundary also runs along it. Beyond the valley of Subarnarekha in the east, large mesas are found which are separated from the main plateau. The northern mesa is very irregular in shape. Its prominent hills are Durgha Pahar 1874 ft in the north and Sangrampur Pahar 1801 ft, Uparhanga 2037 ft, and Bhaski 2380 ft in the south. The descent from the last one southwards is very steep and in places by precipitous escarpment. The southern mesa is called Baghmundi. It is flat topped and somewhat rounded in shape. From it rises the Chamtu hill 2390′. On the other side of the valley and analogous to these mesas, are spurs which extend from Ranchi plateau up to Subarnarekha. From north to south they are known as Dondag Pahar 1804 ft, Barkibur Pahar 1986 ft, and Chandgarhi Pahar 2081 ft. In addition to these, there are very small rounded buttes on both the banks of Subarnarekha, ranging in height between 1500′ and 2000′. This topography shows that at one time the plateau extended far eastwards and these mesas are merely its remnants. However, owing to the tilting of the plateau, these high and extensive mesas and buttes are found only in the northern part. Southwards not only the main plateau swings southwestwards but in place of mesas only small buttes are met with, and thus an extensive gently sloping plain is left on both the banks of the Subarnarekha. Its valley is bounded on the south by a high elongated ridge called Dalma to the east of Subarnarekha and Deshwar Pahar to the west of it (Fig. 13.2). The uplift is also expressed by high waterfall not only in Subarnarekha, but in its main tributaries which descend from the plateau (Fig. 13.4). All of them are preceded by deep and narrow gorges with deposits in their beds. This denotes that the line of uplift was much eastwards and Subarnarekha has carved out its valley through the plateau. The highland has receded much westwards due to the constant operation of exogenic forces, and the waterfalls have kept pace with the recession.

Moreover, contrary to the north or to the west, here the uplift took place in one type of rock—granite and gneiss. That may explain the absence of precipitous and steep escarpment. The descent from the plateau though well marked and considerable, is gentle and gradual, and for this the Subarnarekha and its tributaries are responsible. The presence of extensive stony wastes along this slope, give some idea of vigorous degradation which is going on along this escarpment.

Coming to the southern escarpment we find that the gentlest descent from the plateau is on this side (Fig. 13.2). It may be due to several factors. The gneisses and granites came into contact with Dharwars of the south, which while varying in resistance, form lowlands in general. However, the resistant rocks have formed hills and ridges, and they rise even higher than the level of the plateau proper. Naturally they obscure the prominence of movement. The obscurity has also been accentuated by the fact that here two movements have affected the area, the diastrophic in the plateau and orogenic in the south and not only Dharwars but the granitic mass has been degraded so vigorously that the land in a broad zone has lost the undulating nature of the plateau and has been converted into innumerable hills and valleys.

This southern descent though gradual is not uniform. Abrupt descents are alternated by gentle slopes. Roughly, there are three steep slopes which are well-marked on the spurs between the main rivers which are however, graded and also open. Nevertheless, the presence of these steps in the river is marked by the rocky outcrops in the water channel, or by the narrowness of the valley parallel to the steep descents on the spurs. Along each of these steep slopes three separate sets of rivers take their rise which accentuate the prominence of the alternate steep slopes. In between them on gentle slopes, the rivers do not flow along the general southern descent of the plateau but take an easterly or westerly turn. This is most probably because in this zone strike of the country rock becomes more important than the general slope. Further, the guidance of country rock can also be seen in the direction of the flow of the small rivers in the southern marginal areas where Dharwar outcrops are found. The tributaries join the master streams either from N.E. or S.W.

The presence of waterfalls also strengthens the statement. When river Sankh descends from the main plateau, it has a very narrow and straight valley. It is a perfect gorge and is headed by Pheruaghagh falls. Similarly, when Karo descends from the third steep slope forms the Pheruaghagh falls and runs through its narrow gorge. Granite being a homogeneous rock, these steeps cannot be formed by erosion and most probably mark the intermittent uplifts to which this area was subjected.

The first steep descent is from the main watershed southwards that is from 2500 ft to 2250 ft on the spurs and to 2000 ft in valleys. The second descent is from 2050 ft to 1950 ft on the spurs, that is from 2000 ft to 1900 ft. It is sufficiently steep in the southeastern section of the plateau.

Thus it may be concluded that diastrophic forces working in the geological past have left their imprints on Ranchi plateau, and to some extent they are responsible for the present height of the plateau as well as for the present base-level of erosion for the rivers of the area.

REFERENCES

1. Ranchi, a district in the State of Bihar, lies in the extreme southwestern part of it between 22°21′-23°43′ N. and 84°—85° 54′ E.
2. Dunn, 1942 : The Economic Geology and Mineral Resources of Bihar Province, *M.G.S.I.* Vol. 78, p. 12.
3. S.C. Chatterji, 1945: Some aspects of geomorphology of the Ranchi plateau, *Cal. Geo. Review*, Vol. VII. Nos. 2 & 3, p. 33.
4. C.S. Fox, 1934 : Lower Gondwana coal-fields in India, *M.G.S.I.* Vol. 59, p. 83.
5. Ibid., p. 134.
6. Albert Jowet, 1929: On the Geological structure of Karanpura coal-field, *M.G.S.I.*, Vol. 52, p.39.
7. Thornbury : *Principles of Geomorphology*, p. 249.
8. The river drains the western part of the escarpment.
9. Dunn, 1942: The economic geology and mineral resources of Bihar Province, *M.G.S.I.*, Vol. 78, p. 12.

FOURTEEN

O.P. Singh

TERTIARY UPLIFTS
A Corollary in the Evolution of Landforms in Palamau Upland*

PALAMAU Upland (lat. 23° 39′ 40″ N—24° 33′ 50″ N and long. 83° 20′ E—85° 56′ 10″ E), the northwestern part of South Bihar is a distinct Geographical unit of Chota Nagpur Highlands. This 'broken and divided country'[1] of Palamau District, Bihar is surrounded by the Kanhar River in the southwest, the Hazaribagh Plateau and hills of Aurangabad in the east, the Husainabad Plain in the northeast, the river Son in the northwest and the Western Higher Plateau, locally known as 'Pat' in the south. It is characterised by rugged terrain, poor soil, scrub forest and poor agriculture. After feeding the data of mean annual rainfall and temperature in Peltier's scale of morphogenetic region and processes, the study area exhibits moderate morphogenetic region[2] characterised by moderate chemical weathering,[3] minimum wind action,[4] but maximum pluvial erosion.

Before weaving out the subject matter, it may be pointed out that as an integral part of the Chota Nagpur Highlands both structurally and physiographically, it has much in common with the study of the Chota Nagpur Highlands because this aspect can hardly be treated in isolation for the study area only. The present surface configuration of the Highlands in general and the area under study in particular, is a function of different phases of uplifts, tectonic and the process and

*This paper has been presented to All India Social Scientist Convention, Allahabad, Feb. 1975.

magnitude of erosion in consonance with the resistivity of rocks.

The Highland consist of four erosion surfaces caused by three phases of uplifts. Firstly, the Western Higher Plateau surface with an average elevation of 1000 metres A.M.S.L., merges into the plateaux of Madhya Pradesh. The flat topped plateaux, locally known as 'Pats' are characterised by level surface and the accordance of their summit levels shows that they are parts of one large plateau. The second erosion surface is known as the Ranchi Plateau and Upper Hazaribagh plateau. The uplifted peneplain is 600 metres A.M.S.L. The third having an elevation of 300 metres A.M.S.L. covers a wide area known as Lower Hazaribagh plateau which girdles the plateau of 600 metres A.M.S.L. on all sides—from Palamau via Hazaribagh, Santhal Parganas, Manbhum into Singhbhum districts. The last erosion surface is an undulating surface formed by river valleys and lying between 150 metres—300 metres A.M.S. L.

Although several workers have tried to weave out the different aspects of denudation chronology and the geological history of the area, the credit for a systematic approach goes to Dunn[5] which formed the basis for further workers like Chatterjee,[6] Auden,[7] Singh,[8] and Ahmad.[9] According to Dunn, "Subsequent to outpouring of the Deccan Trap, there was prolonged erosion and uplift to the extent of about 300 metres in the Western Plateau ('Pat' Region)" and in the later part of the Tertiary Periods, "Uplift was renewed to the extent of about 300 metres in the Western Chota Nagpur......Along the northern side of the plateau, the movement was more of the nature of a tilt". Then, there was "an uplift of the whole Chota Nagpur plateau by about 150-300 metres A.M.S.L. in Pleistocene Time".[10]

This 'Uplift Theory' of Dunn is supported by Singh and Ahmad but contradicted by Chatterjee who proposed a "Differential Erosion Theory" for geomorphological evolution of the area. Although the 'Pat' Region has different geomorphological features yet for understanding the evolution of the study area, it is very essential to throw light on it. About the 'Pat' Region, Chatterjee has two points: (1) either the gneissic plateau was higher in the west where it has overlain by lava flows or (2) the difference in level marks the extent of differential erosion since the deposition of lavas. The first view is condemned by the followers of Dunn that it could be higher surface only due to uplifts or a more resistant rock. If the pre-lava height of Western Plateau was due to uplift, it is obvious that the uplift was of the same peneplained surface as the Ranchi gneiss plateau, hence, they must have almost same height. The second view of Chatterjee is more improbable because gneiss is not as soft rock as to have led to the difference in two surfaces of about 480 metres A.M.S.L.

The difference of the level between the 'Pat' Region and surrounding country specially the Ranchi Gneiss plateau might be attributed to the development of hard lateritic pans in the relatively drier western part. This has rather preserved the underlying basalt from rapid erosion and can be held responsible for the formation of the bold scarps in the region as well.[11]

On the basis of the field study and an examination of various topographic features, it may be pointed out that the area is apparently subjected to successive uplifts which more or less coincides with the three important phases of the upheaval of the Himalayas. As the Himalayas recorded three uplifts during Oligocene, Mid-Miocene and Pleistocene times, the Peninsular India in general and the Chota Nagpur in particular have recorded three uplifts. There may be controversy about the fact whether the uplift was in the nature of faulting or warping but there is no controversy that the uplifts did occur.[12] The youthful topographic features observed in the study area such as high gradient of streams, gorges, canyon-like features, numerous waterfalls and rapids, deep-cutting beds and interrupted courses of the North Koel and Auranga River and river capture do speak the upheaval of the study area.

Here, an attempt has also been made to consider the 'pediplain concept' for interpretation of the various erosion surfaces in the area. Towards the early Tertiary if we accept the concept of Penck and King, was uplifted with reference to the surrounding base level by about 300 metres or more. Thus resulted a tableland out of the former pediplain and near the function of two surfaces which was also the line of uplifts, there came into existence scarps which were about 300 metres or more high with the reference to the lower surface. The erosion and parallel retreat of scarps drove them inland so that the area of the newly uplifted plateau was being gradually reduced. As this process continued; the second uplift by about 300 metres or more towards the middle or later part of Tertiary periods occurred practically along the same line where the first uplift had taken place. This resulted in the formation of second line of scarps near at the site of original scarps. But by this time the original scarp had migrated inland. The result of the second uplift was that while the outer plateau was raised by 300 metres or more with the reference to the original pediplain, the inner plateau which had been raised earlier and had migrated inland, gained additional height of 300 metres and became about 600 metres altitude. The third uplift took place in pleistocene along the line of first uplift by about another 300 metres and brought into existence of the outermost line of the scarps. But by this time the earlier two lines of scarps and the plateaux above them had reduced further inland. This final uplift raised the first plateau to about 900 metres, the second one to about 600 metres and the outermost to about 300 metres A.M.S.L. The surface of 150-300 metres A.M.S.L. of the river basins is a result of stream erosion.

The landscape of the study area expresses itself in different surface forms which range from granite-gneiss domes projecting over peneplained surfaces to peaks, hills, plateaux, spurs, and scarps highly dissected and diversified by streams. Its story consists of many chapters involving erosion and deposition, folding, faulting, warping, uplifting and tilting. Volcanic outburst has also interrupted its history. The imprints of a few cycles have been preserved but others appear to have been obliterated over the vast interval of time. Thus, it may be concluded that the present morphology of the region is conditioned by the interplay of the former surviving surfaces elevated during the Tertiary Era with the forms resulting from the recurring erosion cycles.

REFERENCES

1. R.P. Singh, 1970 : Topography and towns in the Chota Nagpur highlands, *The Magadh University Journal*, Vol. III, pp. 97-101.
2. O.P. Singh, 1976 : A quantitative approach to the classification of morpho-units: A case study, *National Geographer*, Vol. XI, pp. 79-87.
3. V.N.P. Sinha, and O.P. Singh, 1976 : Weathering in Palamau Upland, *Indian Geographical Studies, Research Bulletin* No. VI, pp. 52-54.
4. O.P. Singh, 1977 : Palamau upland of Chota Nagpur highland: A study in landforms and settlements, Ph. D. Thesis (Unpublished), Deptt. of Geography, B.H.U., p. 26.
5. J.A. Dunn, 1935 : Late tertiary uplifts in Singhbhum, *Proc. As. Soc. Bengal*, Vol. 29, pp. 285-288 and also "Post-Mesozoic movements in the northern parts of the peninsular, *Memos. G.S.I.*, Vol. 73, pp. 123-42.
6. S.C. Chatterjee, 1946 : Physiographic evolution of Chota Nagpur, *Calcutta Geographical Review*, Vol. 8, pp. 39-42,
7. J.B. Auden, 1954 : Erosional pattern and fracture zones in Peninsular India, *Geographical Magazine*, Vol. XCI, pp. 89-101.
8. R.P. Singh, 1969 : *Geomorphological Evolution of Chota Nagpur Highlands*, N.G.S.I. publication, B.H.U , pp. 1-129.
9. E. Ahmad, 1958 : Geomorphic outline of Chota Nagpur, *Geographical Outlook*, Vol. II, pp. 16-22.
10. J.A. Dunn, 1941 : The economic geology and mineral resources of Bihar Province, *Memos. G.S.I.* Vol. 78, p. 12.
11. O.P. Singh, 1977 : Evolution of physical landscape in Palamau upland, *The Deccan Geographer*, Vol. XV, p. 262.
12. O.P. Singh, 1974 : Sinuosity index of rivers as a key to the geomorphological character of a Region: A case study, *The Geographical Viewpoint*, Vol. V, p. 53.

FIFTEEN

J. G. Mulay and *V. V. Peshwa*

GEOMORPHOLOGICAL STUDIES OF DECCAN TRAP AREA AROUND KHANDALA (MAHARASHTRA) WITH SPECIAL REFERENCE TO AERIAL PHOTO INTERPRETATION

INTRODUCTION

THE AREA of present research project, bounded by latitudes 18° 40′ to 18°50′ N and longitudes 73°15′ to 73°30′ E, lies around Khandala, Poona district of Maharashtra State and covers parts of Survey of India toposheets No. 47 F/5 and 47 F/6 on 1 : 63,360 scale. The Panchromatic black and white aerial photographs on 1 : 25,000 and 1 : 80,000 scale were made use of during the studies.

The Western Ghat scarp which runs almost northsouth, divides the area into two sectors—Konkan area in the west and plateau area above Ghat in the east. Konkan sector of present area of investigation is the part of a very narrow coastal strip about 80 kilometres in width, of western Maharashtra. The continental divide between Konkan area and the plateau area above the Ghat, runs more or less in northeast-southwest direction north of Lonavala and northwest-southeast direction south of Lonavala.

Almost alternate horizontal sheets of compact and amygdaloidal basalts belonging to Deccan Traps, which are intruded by number of basaltic or doleritic dykes, are exposed all through the area. The occurrence of gabbroic intrusion near Khopoli, in this area, has also been reported by Mulay and Peshwa (1980a). Eventhough the area is intensely affected by fractures ranging in length from few metres to as much as

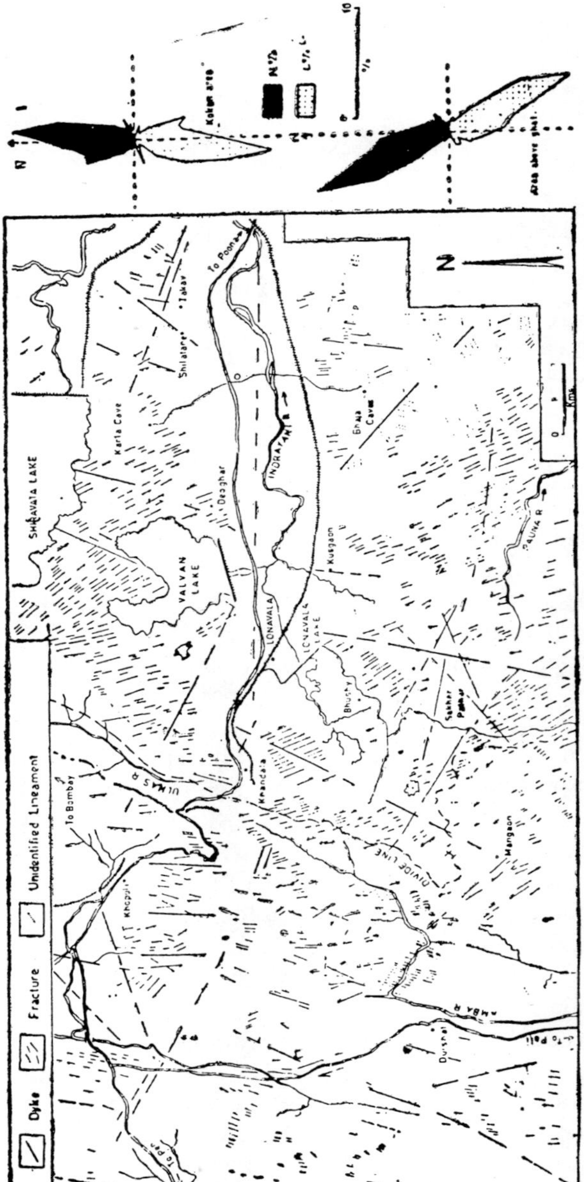

FIG. 15.1 : *Lineament Map of Khandala area based upon Aerial Photo Interpretation and Field Checks*

5 to 10 kilometres, no vertical or strike slip displacement was encountered (Fig. 15.1).

METHODOLOGY

The altimetric and hypsometric analyses were carried out on topographic sheets. The drainage map prepared from aerial photographs on 1 : 25,000 scale was used for drainage analysis. The geomorphological map is based upon the interpretation of aerial photographs on 1 : 80,000 scale. The interesting features seen on aerial photographs and Landsat imageries were checked in the field.

The investigations, thus, based upon the study of topographic sheets, aerial photographs and Landsat imageries, helped to evaluate the geomorphological evolution of the area. The compilation and comparison of the data obtained through multidisciplinary approach provided an excellent opportunity to understand the advantages and disadvantages of the aerial photographs in the geomorphological studies.

STUDY BASED ON TOPOGRAPHIC SHEETS

The altimetric frequency curve (highest point in grid square method) for Khandala area revealed the existence of several planar surfaces, which are major primary landform having large areal extent and negligible relief. (Rama Rao and Vaidyanadhan 1974). The planar surface (Fig. 15.2) between 200′ (60.96 m) to 300′ (91.44 m) is represented by the Patalganga, Utra and Amba river basins in the Konkan area and most of the part, is under cultivation. The most prominent planar surface lies at 2050′ (624.84 m) above sea level in the area and is due to Indrayani, Pauna and Shirvata river basins. The planar surface at 3000′ (914.4 m) height corresponds with the top of the Karla Cave hill range.

In order to study the relief aspects of the area, hypsometric curves following Strahler's suggestions (1952) were constructed for the representative second, third, fourth and fifth order basins. According to Strahler (1952) the hypsometric integral above 60 per cent, in between 60 per cent and 35 per cent and below 35 per cent is indicative of youthful, mature or equilibrium and old stages of aeration respectively. The

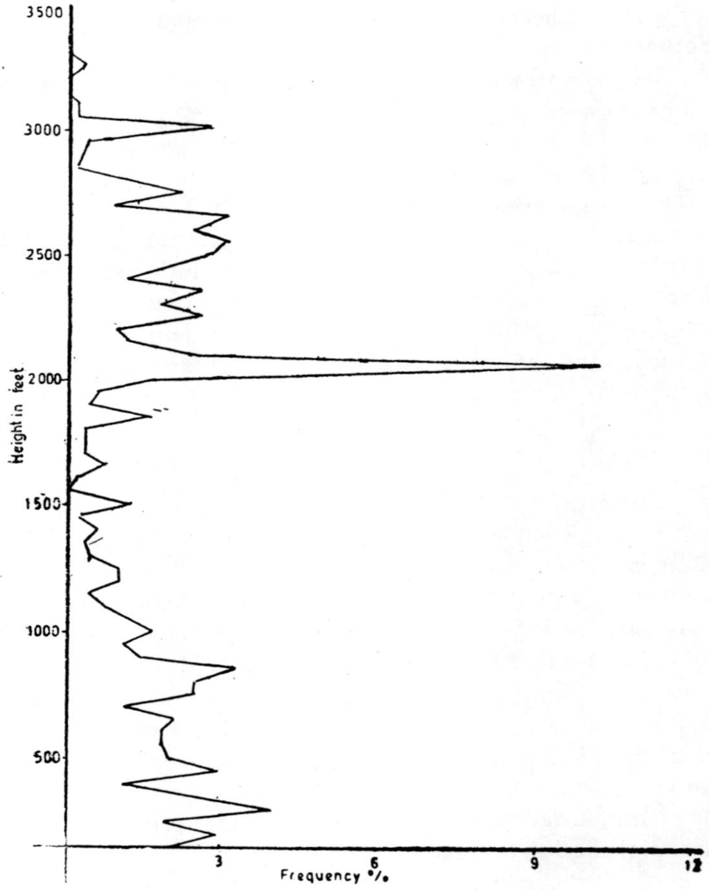

FIG. 15.2 : *Altitude—Frequency Curve for Khandala Area*

values of hypsometric integral in the present study (Fig. 15.3) show that the low order basins in both the sectors of Khandala area are in mature stage whereas the area in general has reached the monadnock stage.

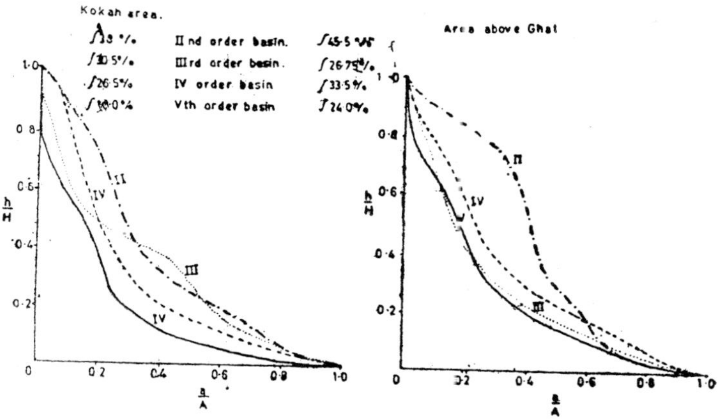

FIG 15.3 : *Hypsometric Curve for Representative Basins within Khandala area.*

STUDY BASED ON AERIAL PHOTO INTERPRETATION

The basins of all the rivers of both the sectors of Khandala area were quantitatively studied on aerial photographs (Fig. 15.4) for their linear and areal aspects and certain dimensionless parameters like bifurcation ratios and drainage density were computed.

Horton's (1945) method of stream ordering was adopted during present study. Bifurcation ratios (Table 15.1) have special significance in the study of drainage basins. According to Strahler (1964) the bifurcation ratios lie in the range of 3 and 5 for the drainage basins where there is no structural control on the drainage pattern and that the higher ratios, greater than 5, are indicative of strong structural control on drainage. It is interesting to note that the bifurcation ratios for first and second order streams are more than five, in both the sectors of Khandala area and suggest structural control on the streams of that order. The drainage of the adjoining areas is also controlled by fractures and has been reported by

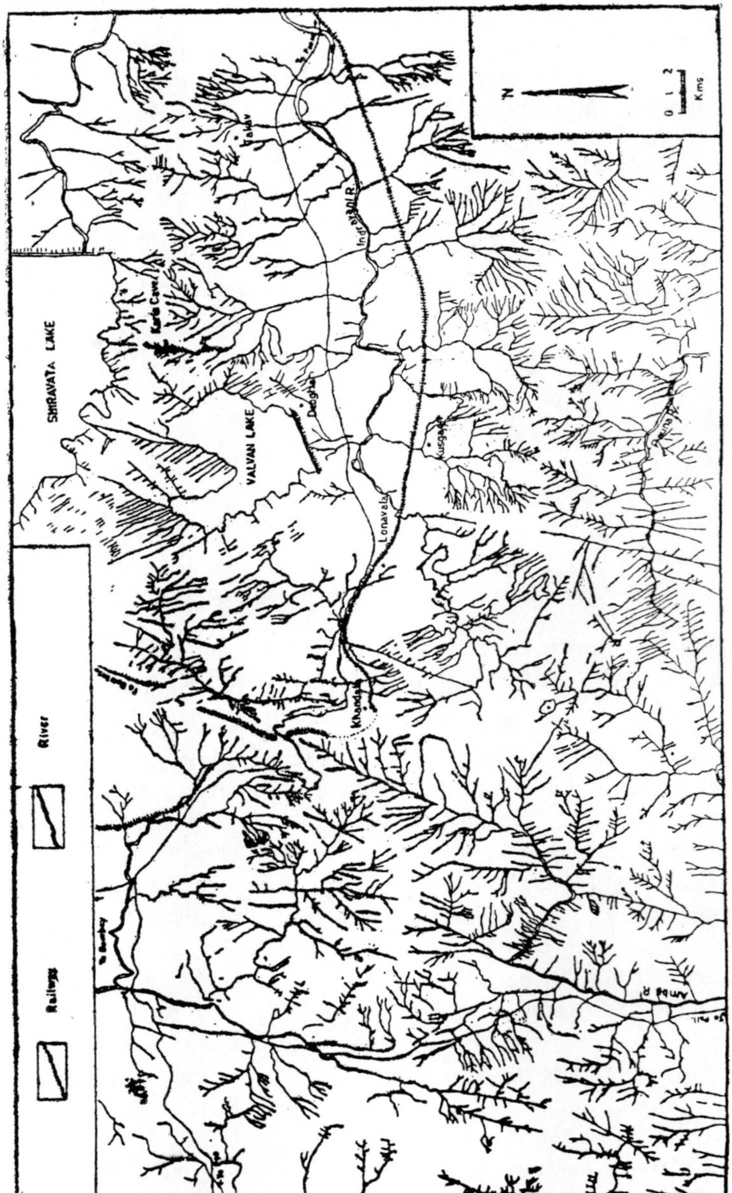

FIG. 15.4 : *Drainage Map of Khandala area Based upon Aerial Photo Interpretation*

number of workers (Kanegaonkar, 1977; Ghodke, 1978; Powar et al,, 1979a, b).

According to Horton's (1945) law of stream number, the number of streams of each order form an inverse geometric sequence with order number. The plot of logarithm of number of streams against their corresponding order fall, therefore, on a straight line. The slope of the plot gives the regression coefficient (b). Since regression coefficient is identical with the logarithm of bifurcation ratios, the bifurcation ratio could be calculated from regression coefficient. The bifurcation ratio, thus, obtained from the regression coefficient (Fig. 15.5) matches favourably with the average bifurcation ratio (Table 15.1).

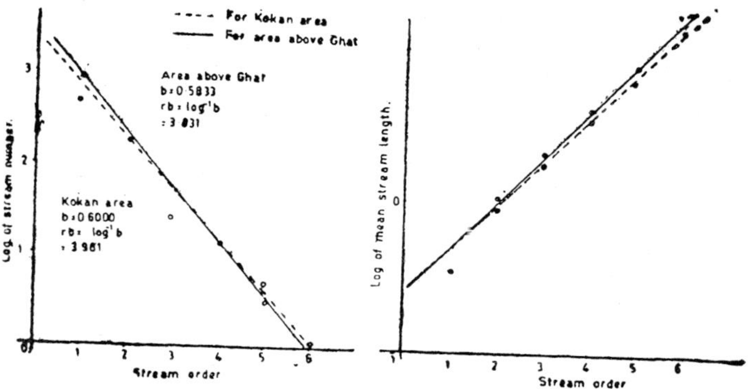

FIG. 15.5 : *Plots-log of Stream Number and log of mean stream length against Streams Order in Khandala area*

The average stream length of any order of stream is always less than the successive higher order and more than immediate lower order. According to Horton's law of stream length, a plot of logarithm of mean stream length against the stream order should yield a set of points lying essentially along a straight line. In present case the plots (Fig. 15.5) satisfy Horton's law of stream length.

The other important parameters of drainage basins are drainage density and stream frequency, both of which are related to the texture of drainage net and express areal aspect of drainage basin. Strahler (1964) observed that the drainage

TABLE 15.1 : **Bifurcation Ratios, Stream Length etc., for Khandala area**

Order of Stream	Konkan Area					Area Above Ghat				
	Number of Streams	Bifur-cation Ratio	Total Stream length	Average Stream length	Length Ratio	Number of Streams	Bifur-cation Ratio	Total Stream length	Average Stream length	Length Ratio
I	910		308.6	0.339		986		313.47	0.317	
		5.38			3.188		5.35			2.908
II	169		182.8	1.081		184		169.75	0.922	
		3.59			1.944		3.28			1.907
III	47		98.8	2.102		56		98.55	1.759	
		3.61			1.9062		3.73			1.895
IV	13		52.1	4.007		15		48.97	3.2646	
		4.33			2.0489		3.00			1.931
V	3		24.63	8.210		5		31.52	6.304	
		3.0			1.260		5.00			3.020
VI	1		10.35	10.35		1		19.04	19.04	
	Average bifurcation ratio			3.982		Average bifurcation ratio			4.072	

density increases with decrease in the size of individual drainage unit.

The average drainage density of the Khandala area is 5.12 and in certain portions of Khandala area the values are still higher, for e.g. in Pauna valley 7.17, in Amba valley 7.46 and in Valvan lake, Rajmachi area, 8.14. Drainage density is expressed in terms of miles per square mile. During present study also the same unit, miles per square mile, for drainage density is used for the sake of easy comparison.

According to Strahler (1964), in general the low drainage density is due to highly permeable or highly resistant subsoil material, dense vegetation cover and low relief whereas high drainage density is favoured in the region of impermeable or weak subsoil material, poor vegetation and high relief.

In Khandala area the conditions which govern the drainage density are of mixed type. The topography of the area is in general rugged but the vegetation is comparatively dense. The subsurface material is resistant, impermeable but fractured. The interpretation of drainage density is therefore rather difficult. However, these values of drainage densities when compared with drainage densities reported by other workers in Deccan Trap regions, compare favourably with those where there is structural control on the drainage pattern and are higher than those where there is no structural control. (Subramanian and Subramanyan, 1978, Powar *et al.*, 1979a and 1979b).

The higher values of drainage densities in this area are therefore indicative of presence of fracture zones and their control on drainage pattern.

STUDY SUPPLEMENTED BY FIELD CHECKS

Aerial photo interpretation on 1 : 80,000 scale for the whole area (Fig. 15.6) and on 1 : 25,000 scale for the Indrayani valley was carried out in order to unravel the presence of geomorphic features in this area.

Most common landforms of Deccan Trap region are mesas and buttes. On aerial photographs these features show light tone due to grass cover on the top surrounded by comparatively dark tone areas on all sides due to the steep escarpments which are usually free from any soil cover. Isolated buttes,

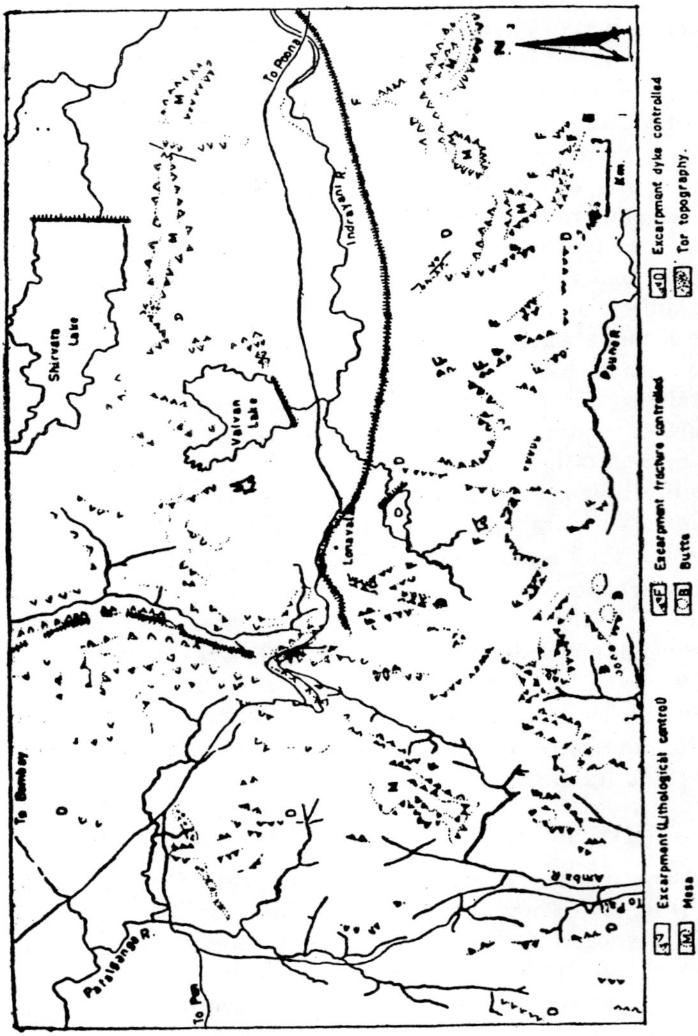

FIG. 15.6 : *Photogeomorphological Map of Khandala area*

strings of buttes and pinnacles were recorded in the area east of Mangaon in Konkan sector.

Besides this, a number of linear ridges, linear valleys and steep escarpments were marked on aerial photographs and shown in Fig. 15.6 with possible genesis.

Deccan Trap basalts in absence of structural control favours dendratic drainage pattern for the reason of their horizontal disposition and more or less uniform characters. So also the slopes of the hills, under normal circumstances are directly correlatable to the type of basalt flows, i.e., steep slopes to compact basalt flows and gentle slopes to amygdaloidal basalt flows.

However, during the study of aerial photographs and subsequent field work it was observed that the drainage pattern of Khandala area, at many places, is parallel, slopes of the hills are steep irrespective of type of basalt flows due to erosion along fracture zones and dykes. On aerial photographs a number of long narrow linear valleys, negative lineaments, are seen. The maximum number of lineaments from Konkan area follow N-N 10°E direction and in the plateau area above Ghat the trend of majority of the lineaments is N 20°-30°W (Fig. 15.1). Mulay and Peshwa (1980c) showed, on the basis of photo characters of the lineaments and field checks, the most of the lineaments of Khandala area are due to fractures and some are due to dykes (Fig. 15.1). The dykes of Khandala area also produce three types of topography-positive, negative and neutral. The valleys controlled by dykes near Khandala, Shirvala lake, Mandad and Chinchavalli village are very long, deep and steep. The hill ridges produced by dykes north of village Takave, Mankavalli are not very high but long. Mulay and Peshwa (1980b) classified the dykes of Khandala area into four group on the basis of their field and photo characters.

Though the rivers of Konkan area are considered westerly flowing, most of the rivers, especially in their source region, have approximately northsouth courses. Thus Ulhas, Amba, Utra rivers and their tributary streams have very long NNE-SSW valleys. The hills of Konkan sector also show approximately northsouth orientation.

The rose diagram based upon azimuth-frequency and

azimuth-length analysis of photo lineaments and the stream segments traced from topographic sheets compare favourably (Mulay and Peshwa, 1978).

Thus all these observations amount to the conclusion that besides lithology, the fractures and the dykes have good influence on the drainage pattern and in general on geomorphic grain of this area. Even the development of Western Ghat scarp, as stated by Das and Ray (1977), could be attributed to the easy erosion along weak zones.

Aerial photo interpretation on 1 : 25,000 scale of Indrayani valley discovered a number of destructional and constructional landforms associated with the present course of river Indrayani. The important observations are (1) partly abandoned meander channel still having connection with main course but used only during high discharge of water, (2) Completely ditached meander scroll and now totally filled by river deposits. The meander scroll supports vegetation, viz., Fluggia, Tamarix, Heteropogon etc., whereas the surrounding area is under cultivation. The meander scroll, therefore, appears as dark toned semicircular segment on aerial photographs. The samples of the sediments collected from a trench taken through meander scroll were subjected to mechanical analysis, and the results confirmed its fluvial nature. In the trench on meander scroll it was also observed that there are a few iron concretions and bedded crust developed on the gravel. Development of such crust suggest fairly high antiquity of the fill, at least few thousand years old.

Though the deposition of moderately compacted gravels have taken place in this area, the river channel is rocky. Along with this, it is interesting to note that no alluvial deposits were seen along any other river of Khandala area.

DISCUSSION

No doubt elevation of the places could be found out on aerial photographs and various photogrammetric methods have been devised for this purpose but these procedures are too technical for beginners and the height determination is not as readily possible as it is on topographic maps. It is, therefore, not advisable, particularly when topographic sheets are available, to study relief aspects of the drainage basins, like hypsometric and altimetric analysis, on aerial photograph. The use of topographic sheets in such studies saves the time. For the determination of heights of particular spots, especially those which lie in between the range of contour interval of topographic map, the aerial photographs certainly offer easier solution. In forested areas due to thick cover of vegetation the geomorphic features could not be recorded on aerial photographs.

However for the reason of scale and aim to make topographic sheets generalized, number of limitations are imposed indirectly upon topographic maps. Under these circumstances all the details of any aspect cannot be shown on topographic maps. As far as drainage of any area is concerned, it is not possible to show all the streams on topographic sheets, and the streams which are victimised are of lower order, usually first order. Tanner (1968) stressed the need of identifying first order streams on aerial photographs or in the field rather than on ordinary topographic maps. During present study also, even if the factor of scale is kept aside, the number of first order streams recorded on aerial photographs is almost double of the first order streams shown on topographic sheet and the difference goes on decreasing with the order of streams increasing. Since bifurcation ratios, drainage densities and stream frequencies, the properties of drainage basin which are often used in interpreting the geology of the area, are directly based upon stream length and stream number, the results obtained from topographic sheet in this respect are very likely to mislead the interpretation. The use of aerial photograph is, therefore, strongly advocated in the study of linear and areal aspects of drainage basins.

The greatest advantage of aerial photographs is the enormous amount of information that they contain in the form of various photo recognition elements. It is for this reason the aerial photographs have become indespensable tool in various field science and as stated by Raina (1972) aerial photographs are used maximum in geology than any other field science. The ability of aerial photographs, thus, offers multichannel approach to the geomorphological problems. Thus topographic sheet might indicate the presence of certain geomorphic feature, whereas aerial photographs in addition to that throw light on the genesis of that feature. In Khandala area the genesis of escarpments, valleys and landform could be easily determined with the help of aerial photographs and field checks. Mache and Peshwa (1978) established the relation between drainage and geology of the area around Chansura, Shahadol district M.P. with the help of aerial photograph. With the help of aerial photographs and Landsat imageries Desai and Peshwa (1978) and Naik (1979) explained successfully number of drainage anomalies of the West Coast region and genesis of geomorphic features of Talikota area (Karnataka).

The vertical exaggeration which may come through stereoscopic viewing is often advantageous in locating and studying the geomorphological features of low relief. Topographic sheets of Khandala area do not account for the presence of number of small cascades, pinnacles, dyke controlled hill ridges, whereas aerial photographs very promptly locate them. The signs of rejuvenation of the rivers of this area, deposition of scree material and sediments, land use pattern, for topography produced by gabbroic intrusion could directly be read on aerial photographs. In this area had aerial photographs not indicated the presence of gabbroic intrusion, meander scroll these features would have been missed even during field work.

Thus in short, aerial photographs have decided advantages in geomorphological investigations, they indicate effectively the presence of geomorphological features and simultaneously offer easy and direct explanation for their genesis. Thornbury's (1969) remarks that a geomorphologist who does not understand at least the basic principles of airphoto interpretation is greatly handicapped, are self-explanatory in this respect. About quantitative work in geomorphology Thornbury (1969) stated "No one questions the need for more quantitative work in this field of geomorphology. We should be careful that geomorphology does not become so immersed in mathematics, physics and chemistry that it ceases to be the study of land forms". The authors strongly feel that more and more applications of aerial photographs and Landsat imageries in the geomorphological studies would reduce considerably the unwanted quantitative work in geomorphology.

CONCLUSIONS

The geomorphological studies of Khandala area with the help of topographic sheets, aerial photographs and field checks revealed the following points :

(1) The lower order streams are in mature stage whereas higher order streams have reached old stage. The area, in general, shows monadnock stage.

(2) The bifurcation ratios, drainage densities, drainage pattern, comparison between lineaments and stream segments, and field checks favour the conclusion that the drainage of Khandala area is controlled by fractures. The development of Western Ghat scarp could also be attributed to easy erosion along fracture zones.

(3) In the study of relief aspect of the area, topographic sheets are more convenient than aerial photographs.

(4) In the study of minor geomorphological features like first order streams, cascades, etc. aerial photographs are definitely much more advantageous than topographic sheets. Along with locating the geomorphological feature aerial photographs throw light on their genesis also.

(5) Though the advantages of aerial photographs are many in geomorphology, they do not outright eliminate the need of either field work or the use of topographic sheets.

SUGGESTED READINGS

Das, S.R., and Ray, A.K., 1977 : Fracture pattern within Deccan Trap and West Coast of India, *G.S.I Misc. Publi.* No. 31, pp. 75-88.

Desai, A.G., and Peshwa, V.V., 1978 : Drainage anomalies in West Coast region of Maharashtra and Goa, India: A study based on Landsat imagery and aerial photo interpretation, *Symp. on Morphology and Evolution of Landforms*, Delhi, pp. 255-259.

Ghodke, S.S., 1978 : Geology of the area around Panvel, Kolaba district, Maharashtra, Ph. D. thesis (Unpublished) Poona. Uni.

Horton, R.E., 1945 : Erosional development of streams and their drainage basins, hydrophysical approach to quantitative morphology, *Bull. Geol. Soc. Am.* Vol., 56, pp. 275-370.

Kanegaonkar, N.B., 1977 : Geology of the Deccan Trap country around Pune, Maharashtra, Ph.D. thesis (unpublished) Poona Uni.

Mache, R.N. and Peshwa, V.V., 1978 : Photogeological interpretation and controls on drainage in the Gondwanas and Bijawars of Son valley, Shahadol district (M.P.), *Symp. on Morphology and Evolution of Landforms*, Delhi, pp. 250-254.

Mulay, J.G., and Peshwa, V.V., 1978: Fracture control on drainage pattern of the area south of Lonavala and Khandala (Maharashtra), *Symp. Morphology and Evolution of Landforms*, Delhi, pp. 222-225.

Mulay, J.G., and Peshwa, V.V., 1980a : The occurrence of gabbroic intrusion in Deccan Trap basalts near Khopoli in the Konkan area (Maharashtra), *Cur. Sci,*. Vol. 49, No. 7, pp. 356-357.

Mulay, J.G., and Peshwa, V.V., 1980b : Remote sensing of dykes in Deccan Trap area around Khandala and Khopoli (Maharashtra State), *Paper presented at National Space Sci. Symp.* Varanasi.

Mulay, J.G., and Peshwa, V.V., 1980c : Study of photolineaments and their significance in the Deccan Trap area around Khandala (Maharashtra), Paper submitted to Seminar on Application of Photo Interpretation and Remote Sensing techniques for *Natural Resources Surveys and Environmental Analysis*, Dehra Dun.

Naik, S.D., 1979 : Photogeological and geomorphological studies of Talikota area, Bijapur District, Karnataka state, India, Ph. D. Thesis (unpublished), Poona Uni.

Powar, K.B., Sukhtankar, R.K., and Patil, D.N., 1979a : Geomorphology and Tactonics of the West Coast of India between Palghar and Revas, Maharashtra, *Tech. Report.* No. 2, Deptt. of Geol., Uni. Poona.

Powar, K.B., Sukhtankar, R.K., and Sawant, P.T., 1979b : Geomorphology and Tectonics of the West Coast of India between Srivardhan and Vijayadurg, Maharashtra, *Tech. Report*, No. 3, Dept. of Geol, Uni. Poona.

Raina, B.N., 1972 : Application of aerial photographs in geological, geomorphological and tectonic study, Appreciation seminar on Use of Aerial Photo Interpretation in *Survey and Mapping of Natural Resources*, IPI, Dehra Dun, India.

Rama Rao, K.L.V., and Vaidhyanadhan, R., 1974 : The planer surface of Keonjhar region, Eastern India, *Indian Jour. Earth Sci.*, I, pp. 84-95.

Strahler, A.N., 1952 : Dynamic bases for geomorphology, *Bull. Geol. Sci. Am.*, Vol. 63, pp. 923-938.

Strahler, A.N., 1964 : Quantitative geomorphology of drainage basins and channel networks, *Handbook of Applied Hydrology*, Ed. Ven Te Chow, Mc Graw Hill Co., New York.

Subramanian, S.K., and Subramanyan, V., 1978 : Airphoto analysis of the drainage in the area around Nagbhir, Chandrapur district, Maharashtra, *Symp. on Morphology and Evolutions of Landforms*, Delhi, pp. 55-74.

Tanner, W.F., 1978 : Geomorphology-Statistical analysis, The *Encyclopedia of Geomorphology*, Ed. Fairbridge R.W., Dowden, Hutchinson, Stroudsburg.

Thornbury, W.D., 1969 : *Principles of Geomorphology*, John Wiley and Sons. Inc. New York.

SIXTEEN

Anil Kumar and *Ram Niwas Pandey*

QUANTITATIVE ANALYSIS OF RELIEF OF THE HAZARIBAGH PLATEAU REGION

INTRODUCTION

THE HAZARIBAGH Plateau Region consists of surfaces of varying elevations and reveals configuration of varying altitude, dimension and magnitude. The undulating surfaces, elevated plateaus, dissected scarps, east to west running through western rugged terrain, residual hills and valleys provide wide scope for geomorphological analysis. The relief needs interpretation, examination and analysis of its shape, slope, altitude and existing expressions.

In order to visualize the relief properties of the Hazaribagh Plateau Region in proper perspective, various methods have been employed. Only important morphometric attributes pertinent to characteristics of relief elements have been selected for the purpose of elucidation of the landscape. Their quantitative interpretation has proved itself as one of the important quantitative techniques in relief analysis by providing the scale of standardised measurement and comparison of the different morphological characteristics developed in the region. Arithmetical and Graphical methods mainly deal with relationship of area to the surface; height area relationship and their Cartographical representation portray the total character of the land surface. The statistical indices computed for different topographic elements like altitude and relief, etc., are similarly helpful in the study and comparison of erosion surfaces and their relief characteristics.

Area-Altitude Relationship

The measurement of the interrelationships of area and altitude of the Hazaribagh Plateau Region is attempted through hypsometry. The measurements of area between successive contours have been done with the help of polar planimeter. Rotameter is applied in measuring the length of contours at selected interval of class categories. The absolute height of each grid of one sq. kilometre is obtained for determining the frequency of altitude in the region. The data for each calculation are presented graphically.

The Hypsometric Curve

The Hypsometric Curve 'is an ogive or cumulative-frequency curve, which represents the absolute or relative area of land above or below each contour'.[1] Fig. 16.1 for Hazaribagh Plateau Region reveals the vertical distribution of land and helps in detecting its surfaces. The area under study is dominated by 304 metres surface and there is a gradual decline in the area above this altitude. Second surface, though in limited extent is 609 metres surface. Both these surfaces dominating the curve are marked by steep slope of the line between altitude 152 to 304 metres and again 533 to 609 metres, representing the scarps. Steepness of the curve at different altitude also provides a general impression of slope of the region. But the conclusion drawn from graph is extremely generalised.

The area-height curve represents an absolute area of land between adjacent contours. The curve reveals frequency distribution of area at 76.2 metres interval. The plot of the data distinguishes three peaks at 609, 380 and 152 metres. These breaks indicate undulating surfaces of upper Hazaribagh, lower Hazaribagh plateau surfaces besides eroded valleys and foothill areas of northern fringe. About 35 per cent of the total area of Hazaribagh Plateau Region has an elevation of 380.9 metres. Steep rises from 228 metres though fails to signify any slope but obviously represents the lower scarps, which is abrupt than the upper scarps including the upper plateau surface. Ruggedness and eastwest rise in the mean elevation of the lower plateau, all together are responsible for decrease in area as the altitude increases from 380 metres.

The slow decline of area above 685 metres is characteristic of the erosional surfaces with residual hills of various magnitude and dimension.

The comparison of these two curves clearly reveals the preference of area-height curve over the hypsometric curve.

Clinographic Curve

The clinographic curve of Hazaribagh Plateau Region (Fig. 16.1) fails to demonstrate average slope between contours. It is unsatisfactory also for the present area as the mid-altitude contours of 304-380 and 533-609 metres are longer than the base. In spite of these shortcomings

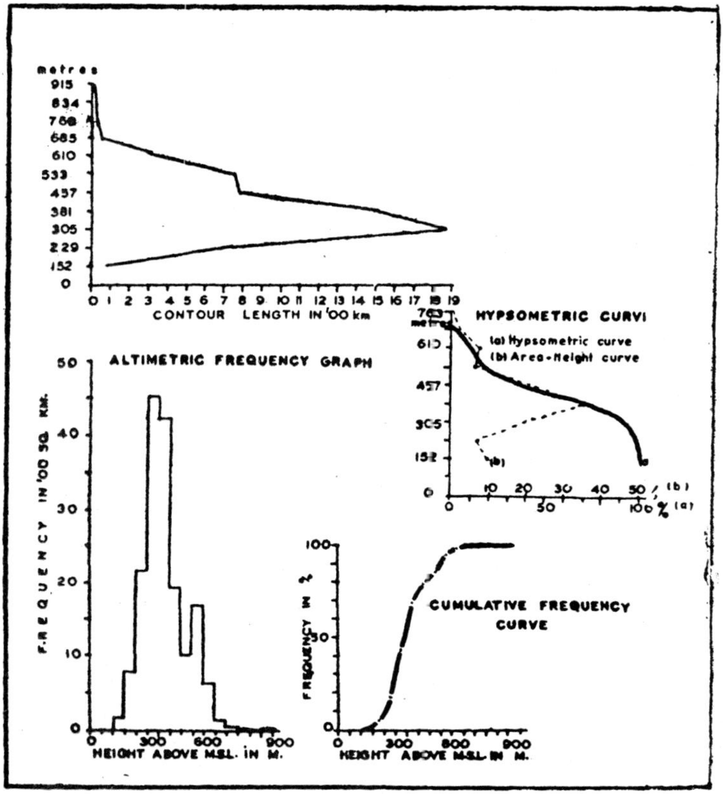

FIG. 16.1 : *Clinographic Curve*

certain relief characters can be deducted from the study of curve. Surface above 685 metres seems dissected and dominated by hilly terrain as curve takes a steep gradient. Again contour lengths of 457 and 533 metres are almost same. This again characterises the dominance of surface at particular height. This is again supported by the maxima of curves. The curve certainly provides some information regarding the space, relief and elevation of the surface of the Hazaribagh Plateau Region.

Altimetric Frequency Graph

Altimetric analysis demonstrates the numerical frequency of certain altitude levels. A statistical count of the number of absolute height in each altitude ranges per sq. km when plotted on graph or curve reveals the existence of erosion surfaces at different levels (Baulig, 1939).[2] This method is more valuable in the areas of multiple cycle than uplifted or structural platforms. The counting has been done from 17,300 grid of one sq. km. Altimetric frequency graph and curve (Fig.16.1) both reveal two surfaces at 365.7 metres and 609.5 metres above M.S.L. Its major portion is constituted of surfaces at 365.7 metres level representing the lower Hazaribagh plateau spaning over the northern and eastern section of the region. The area above 609.5 metres is less conspicuous and represents erosional surfaces of the upper Hazaribagh plateau dotted with numerous residual hills. The abruptness of curve at 152.4 metres level is indicative of the plateau character of the region. Cumulative frequency curve also indicates that up to the elevation of 457.1 metres there is rapid gain in the cumulative area with every increase in elevation. This cumulative gain is minimum after 609.5 metres (Fig.16.1).

All the hypsometric curves drawn for the area under study are concerned with the relationships of area and altitude. In spite of their respective differing usefulness; all, demonstrate the existence of two erosion surfaces.

The lower surface dominates in areal extent while upper surface is of limited extent. The broad recognition of relief personality is though suggested through graph but morphological analysis of relief requires more than the mere recognition of altitudinal surfaces. It requires careful examination of topographical maps and real consideration of slope. The statistical indices computed for different topographic elements like average relief, relative relief, drainage texture, etc. have been selected in the study of relief characteristics is one hand and in the comparison of several geomorphic surfaces at different altitudes on the other.

The broad erosion surfaces thus recognised by the altimetric method adopted in the analysis of relief, are also revealed by the superimposed profiles and long profiles of the important rivers of the Hazaribagh Plateau Region. The purpose of a profile is to show with precision, the form of the land (Miller, 1953).[3] They present the variation of altitudes of the surfaces and the magnitude of the summit levels, etc. within the region besides profiles established then as erosion surface.

The significance of valley shape in transverse and longitudinal profiles is a first step in careful analysis of relief. The Thalweg profiles of the Barakar, the Mohane and the Kunur and some of the important tributaries reflect two major breaks in slope in their courses. The study of their long profile provides an interpretation of the surface history because they are the erosional curve of the river course from source to mouth at any stage of its evaluation. The breaks in the long profiles of the Barakar, the Mohane and their tributaries suggest interruption in the progress of the fluvial cycle. Such breaks are in places subdued and in places marked by sharp knick points. The long profile of the Barakar shows two conspicuous breaks-in-slope marked by a waterfall, known as 'Tiger Fall' in its upper reaches when it descends from the upper Hazaribagh Plateau surface. There is another minor break in slope in the lower course of the Barakar. The thalweg of the Mohane also reflects sharp break in slope one at 600 metres and another at 300 metres where it descends from upper and lower plateau surfaces respectively. The break in slope in the Kunur marks the point where it plunges into the Damodar trough. The valley cross profiles of the Barakar, the Mohane and the Kunur, also trace out their different surfaces which have been carved out during the recurrence of denudation cycles. They have their sources over undulating 600 metres surface. Their cross valley sections became deep where it descends towards the lower 300 metres surface. Valley becomes open over the lower surface. Towards their down stream section cross valley section again takes a deep V-shaped form while crossing the lower scarp. The profiles while confirming the two erosional surfaces also elaborate the morphology of the area besides throwing light on their evolution.

The superimposed and projected profiles (Fig. 16.2) illustrate the morphology of the Hazaribagh Plateau Region. The profiles reveal interesting facts in terms of the variation of altitude of the erosional surfaces, the magnitude of the summit levels and progressive advancement of relief forms within the area under study.

The superimposed profiles give a panoramic view of the morphology of the area. They outline higher plateau surfaces over 600 metres and lower surfaces of about 300 metres and even 150 metres surfaces. The superimposed profiles show not only the surfaces (of some morphological unity, as an erosion surface) by the general uniformity of level of various profiles but also reflect depth of the valleys and amplitude of relief. Each erosion surface seems to be separated by scarps which are vividly expressed (Fig. 16.2). The descending plateau presents a dissected terrain above the river basin.

The crowding of profiles seems to blur some of the morphological aspects of the erosion surfaces at various elevations in superimposed profiles. Projected profiles relieve this said obscurity and outlines the summit details of 600 metres and 300 metres surfaces. Only surfaces lying south of the upper Hazaribagh plateau is lost in this map. The method is useful in detecting and demonstrating dissected erosional surfaces. The undulating terrain traversed by river valleys define themselves as step like erosion surfaces belonging to Post-Tertiary landscape cycles and subcycles. Most of the tributaries of the Barakar flow and descend the upper plateau flowing parallel to the scarps. This reflects the step-like character of scarps above the 300 metres surface. The composite profile provides a fair idea of the surfaces of the Hazaribagh Plateau Region. The isolated Parasnath Hill in the southeast (1367 metres) and Jhumra Pahar (669 metres) in the southeast corner of the upper Hazaribagh Plateau stand as remnants of some former surface. The span of various surfaces are clearly outlined in

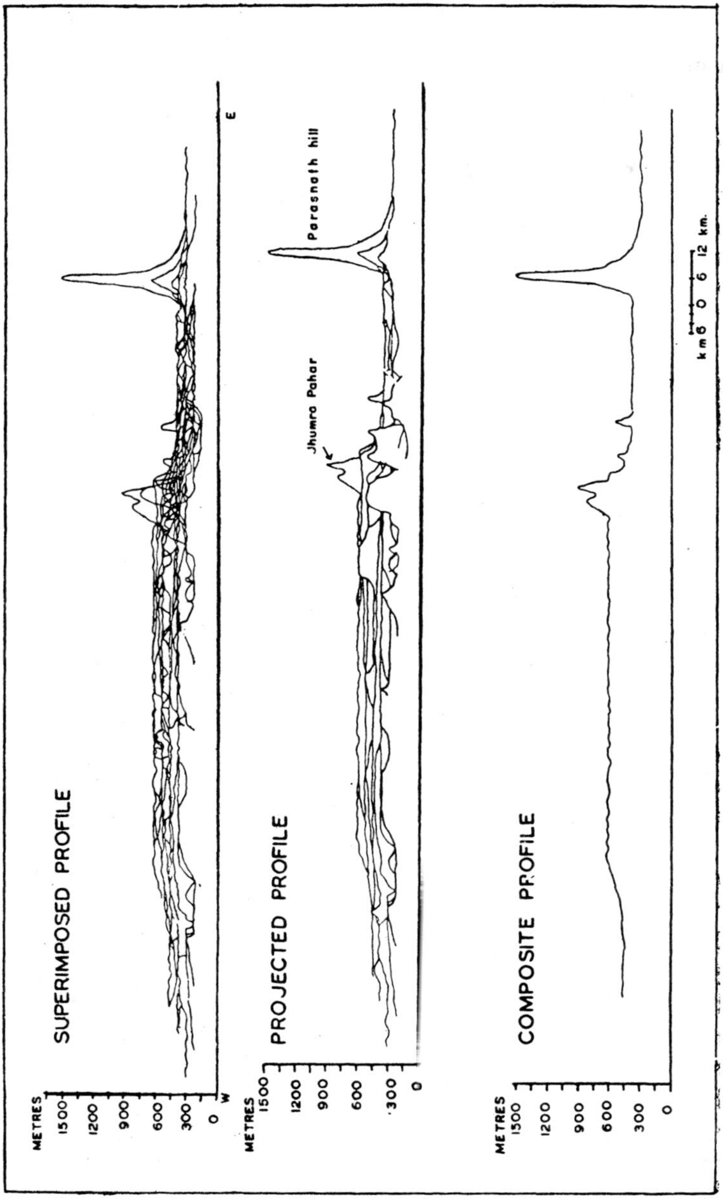

FIG 16.2 : *Profiles*

this map. The upper Hazaribagh Plateau reveals a gradual decrease in its height in the west while scarp marks its eastern margin.

The analysis of profiles indicates that there are three levels of summit height and levels of plain. The lower elevation of the plain below 150 metres is actually of the surfaces adjacent to the northern foothills in the valleys of Sakri and Lilajan. It seems possible that Tertiary upliftment and subsequent change of base level introduced the present cycle which is responsible for most of the present configuration of the Hazaribagh Plateau Region. The upper and the lower Hazaribagh plateau surfaces represent parts of a once continuous erosion surface that has been heavily dissected and differentially uplifted in Tertiary period. The development of new drainage system on these uplifted surfaces has dissected it deeply. The pre-Tertiary surfaces are best preserved on ridges, crests and summits. The initial form of the uplifted surfaces can be reconstructed by drawing generalised contours tangent to the existing contours on the interfluves (Miller, 1953).[4] The difference in the altitude and dimension is related to the Post-Tertiary erosion which has reduced them below the general level of the initial surface besides reducing initial extent of the surfaces. The generalised contours of 609.6 metres outlines the extent of the former central Hazaribagh plateau which is divided and dissected by Post-Tertiary streams (Fig. 16.3). Some of the older river systems traversing the old erosion surface also gained new energy after the uplift. The subsequent values of contours outline the probably position of uplift and early scarps. The generalised contours of 305 metres and 380 metres indicate the northern limit of the Hazaribah Plateau Region. In south the contour of 457 metres delimits the northern margin the Damodar trough. The significance of this map lies in demarcating the extent of former surfaces which have been dissected in relation to the systems that have been rejuvenated or newly formed their drainage system and created the present configuration with the progress of the denudation cycles.

The morphometric attributes, such as intensity of relief, relative relief, dissection index and drainage texture pertinent to the characteristics of landform elements conveniently present a significant zonal synthesis interms of cartographic maps.

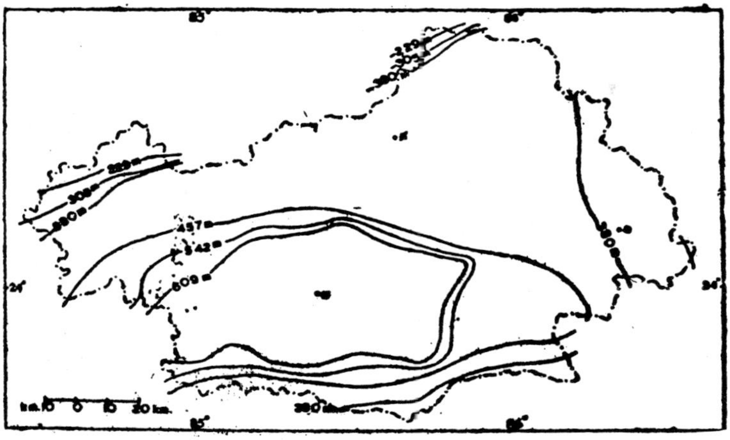

FIG. 16.3 : *Hazaribagh Plateau Region, Generalised Contours*

Intensity of Relief

The topographical map though depicts almost precise information about the relief of the region but fails to convey many features of landscape, particularly amount and intensity of the mean height, surface levels, breaks and edges. The exact height of a point above sea level is far more some purposes much less important than the relation of its altitude to that of surrounding areas (Monkhouse, 1963).[5] The Hazaribagh Plateau Region is an area of complicated relief and to make important deductions of its relief, it seems quite objective to determine the intensity of relief of the entire area of the study. The intensity of relief has been determined by finding out the average relief of each unit area of the Hazaribagh Plateau Region. The isopleth map has been prepared to show the intensity of relief per sq. km (Fig. 16.4). The map distinctly defines six categories of intensity of relief ranging from 150 metres per sq. km to above 750 metres per sq. km. The isopleth thus behaves like modified contours omitting valleys produced over the Hazaribagh Plateau Region by subsequent erosion. The modified configuration

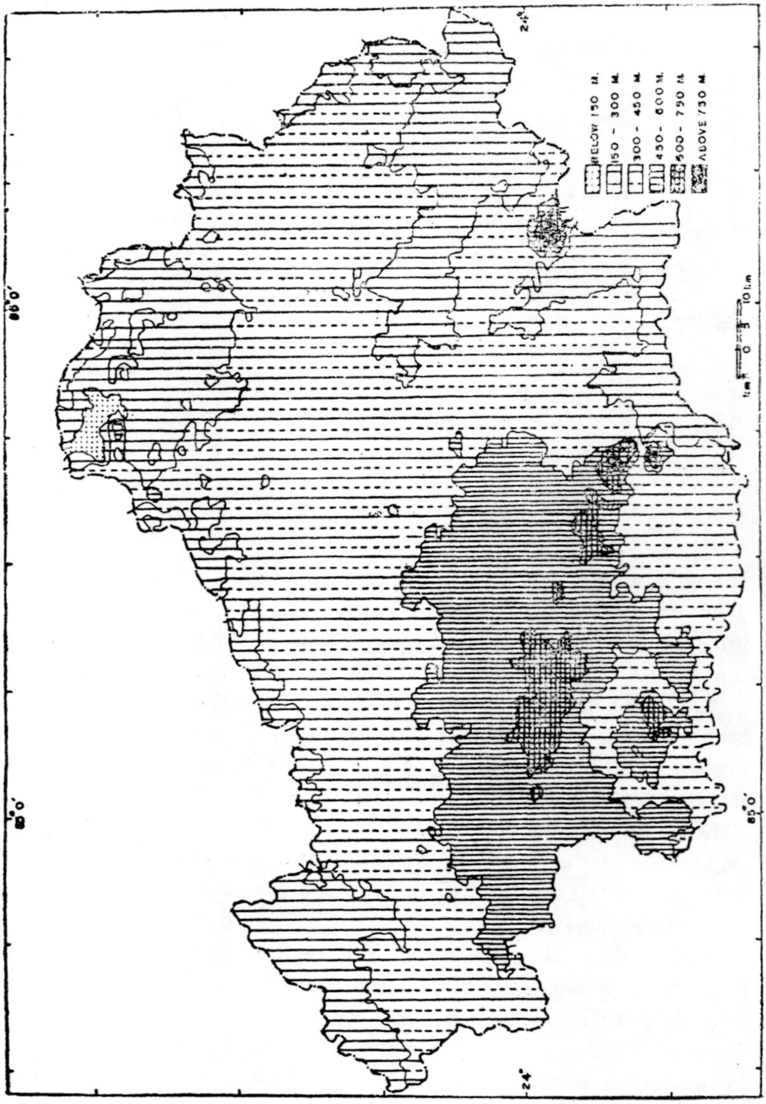

FIG. 16.4 : *Hazaribagh Plateau Region—Intensity of Relief*

map expresses the extent of surfaces of average relief above mean sea level per sq. km. It also defines the trend and distribution of the area in different categories of intensity of relief. An assessment of the intensity of relief of the area has been made in the following Table 16.1.

TABLE 16.1 : **Intensity of Relief**

Categories of intensity of Relief in metres per sq. km.	Total No. of Units in sq. km.	Percentage to the total area	Major categories
Below 150	84	00.49	Low
150-300	3959	22.38	23.37%
300-450	9548	55.19	Moderate 55.19%
450-600	3229	18.66	Moderately high 18.66%
600-760	437	02.53	High
Above 760	43	00.25	2.78%
Total	17,300	100.00	

The above Table 16.1 shows in brief the intensity of relief under different categories and the total number of units in sq. km. It further reveals that areas of low intensity relief occupying about 23.37 per cent of the total area of the region extend over the northern margin of the area marking the valley of the major river system such as the Lilajan, the Mohane and the Sakri. In the southern and southeastern part of the area under study low intensity of relief occupies the lower valleys of the Damodar and the Barakar rivers.

The moderate intensity of relief dominates the area and occupies more than half of the total area of the Hazaribagh Plateau Region. It extends over the lower Hazaribagh plateau and occupies the western portion of the Damodar valley. Few isolated high hills break its spread.

The moderately high intensity of relief occupies the fringe and scarp of the upper Hazaribagh plateau and its western dissected terrain. It accounts for about 18.66 per cent of the total area. The isopleth lines of 450 metres and 600 metres demarcate it.

The area of high intensity of relief are bounded by the isopleth lines of 600 metres and 760 metres. They comprise the upper Hazaribagh plateau and its dissected parts. From east to west Parasnath hills, Lugo Pahar, Dhamdharwa hill and Mahudi hill also reveal high intensity of relief. They comprise only 2.78 per cent of the total area.

The intensity of relief thus helps to distinguish the various surfaces minus their dissection, through its generalised configuration of average relief. It provides a good base for further detailed study of the relief.

Relative Relief

The relative relief represents actual variation of altitude in a unit area with respect to its local base level. While making the analysis of relief it is often important to relate the altitude of the height and lowest points in any particular area, that is, to ascertain the amplitude of available relief (Glock, 1932).[6] Then relative relief does not take into account the dynamic potential of the terrain but as it is closely associated with slopes it is more expressive and also useful in understanding morphogenesis. The relative relief enumerates that the steeper the slope the higher is the surface above its base.

The relative relief map of the area (Fig. 16.5) gives a clear conception of the nature and amount of local relief of the Hazaribagh Plateau Region. The map distinguishes ten categories of relative relief in metre/feet per sq. km and outlines their trend in the area. Isopleth of 15 metres relative relief demarcates the undulating surfaces of the upper and lower Hazaribagh Plateau besides the broad river valleys of the region. The western dissected terrain finds expression of its local relief through 15 to 122 metres isopleth. The prominent Pre-Cambrian hills, dominating outer fringe of Padma Ghat, namely Mahudi Pahar and Mahabar hills clearly find expression in this map. The steepness of their ridges are conveyed through the closeness of 30.5 metres to 61 metres isopleth lines. Isopleth marks the relative relief of the Parasnath hill, Lugo Pahar and Mahudi hills.

The frequency of ten categories of relative relief of the

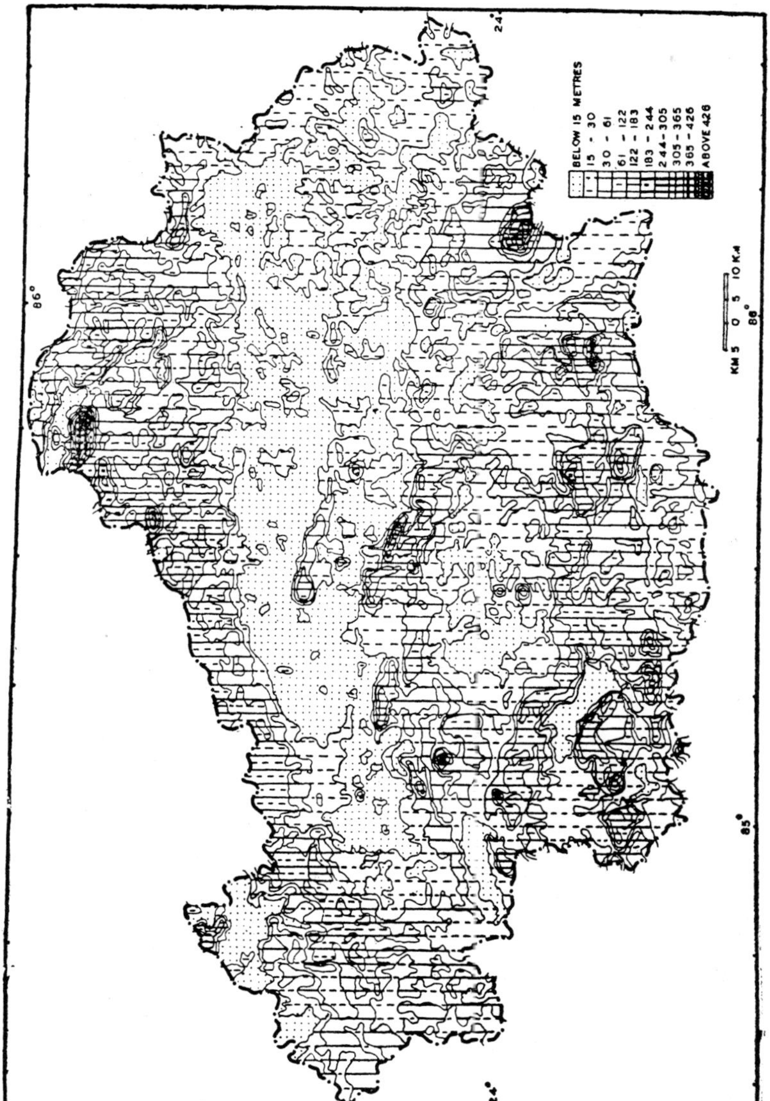

FIG. 16.5 : *Hazaribagh Plateau Region—Relative Relief*

area under study is tabulated in Table 16.2. The analysis of the table distinguishes major categories of relative relief, *viz.* low, moderate and high representing 61.0 per cent, 36.7 per cent, 2.3 per cent of the total area of the Hazaribagh Plateau Region respectively.

TABLE 16.2 : **Relative Relief**

Categories of Relative Relief	Total area in sq.km	Per cent to the total area	Major categories
Below 15 m (0—50′)	5458	31.5	Low 61.0
15—30.5 m (50′—100′)	5095	29.5	
30.5—61.0 m (100′—200′)	3591	20.8	Moderate 36.7
61.0—122.0 m (200′—400′)	2098	12.1	
122.0—183.0 m (400′—600′)	666	3.8	
183.0—244.0 m (600′—800′)	222	1.2	
244.0—305.0 m (800′—1000′)	81	0.5	High 2.3
305.0—365,8 m (1000′—1200′)	45	0.3	
365.8—426.7 m (1200′—1400′)	16	0.1	
Above 426.7 m Above 1400′	28	0.2	
Total	17,300	100%	

The area of high relative relief representing only 2.3 per cent of the total area are demarcated by 183 metres isopleth. This forms a conspicuous unit along the higher interfluvial tracts and hills. Tik-Tikwa and Muhawar hills in the northern valley of the Sakri, Dubaur hills north of Kodarma and Lilajan hills in west outline the areas of high relative relief in the lower scarps of the plateau region. Mahabar hill Jamiro hills and Barsot hill form elongated quartzite ridges along the northern edge of the upper plateau. The area is traversed by the Barakar and its tributaries such as the Kewta and Barsoti Nadi. Their tributaries have cut remarkable valleys into and parallel to the central scarp. The escarpment ends into Sundralli and Mahuda peaks near the Mohane basin. Isolated on the central plateau are several hills with high relative relief in the east and like Chendwar in the east and Kasiatu, Hesatu and Hudwa in the west.

The most conspicuous areas of high relative relief are located in the southern and southeastern parts of the Hazaribagh Plateau Region. Parasnath lying between the Barakar and the Jamunia is the area of very high relative relief. Besides it three conspicuous area of high relative relief are located in the southern part of the Damodar valley. In the western section, Mahudi hills with its 243.84 metres bold scarp; Lugu hill and Jilinga in the east reveal the high relative relief. These hills are dissected by southern flowing rivers Bahera, Bokaro and Kunur, etc. The subsequent undercutting of southern scarps by their channels has produced deep valleys of the steep slope.

The area of moderate relative relief varies between 30.5 to 183 metres and occupies about 36.7 per cent of the total area. The interfluvial ridges of dissected terrain in the west, and the scarps of the upper and lower plateau are conspicuous for their moderate relief. The Dabaur Ghats below Kodarma, interfluvial parts lying between the Sakri and Barakar basins in the north, Lilajan and Amanat river system in the west are highly dissected by streams. Padma Ghat in the north and Nayakhap Ghat in the south of upper plateau surface are the areas of moderate relief. The scarps in these areas descend step by step without revealing distinct breaks. In the southern part area of moderate relief extend southwards from the edge of the upper plateaux. These are the fluvial section of the south flowing river system. The hills with high relative relief are surrounded by the isopleth lines of moderate relative relief. The lower and dissected parts of the Parasnath, Lugu and Mahudi hills are the areas of moderate relief. An elongated zone of moderate relief finds its expression in the Kodarma plain extending roughly in west-east direction. The area is represented by Tilaiya hills and smaller hills extending eastwards upto the confluence of the Barakar and the Barsoti Nadi.

The low relative relief dominates the Hazaribagh Plateau Region and occupies 61.0 per cent of its total area. The area of low relative relief has its maximum expansion over the lower Hazaribagh Plateau along the broad valley of the Barakar. This continuity is punctuated by few isolated hills and ridges acting as secondary or Tertiary divides between the tributaries of the Barakar. The values of relative relief increases towards the southeastern section, otherwise whole area is more uniform and plain-like.

The upper Hazaribagh plateau and the Barkagaon plain in the western part of the Damodar basin, broad valley of Jamunia in the east, are few conspicuous areas of low relative relief in the southern half of the area under study. In the northern parts of the Hazaribagh Plateau Region, the areas of low relative relief are confined to the Lilajan basin in the west and Sakri basin in the east. These areas are marked with slow channel degradation and broad convexity of valley side slopes and subdued interfluves.

Dissection Index

The dissection index which is the ratio between relative relief and absolute relief, is used for the better understanding of the Hazaribagh Plateau Region. Isopleth map has been drawn to show the index of degree of dissection per sq. km (Fig. 16.6). The isopleths clearly distinguishes these broad zones of high dissection, medium dissection and low dissection. The Table 16.3 below gives the percentage of area of each class of index of dissection.

Table 16.3 summarizes the frequency of index of dissection in the Hazaribagh Plateau Region. It reveals that three major categories of low, moderate and high index of dissection registers 71.06, 24.23 and 4.71 per cent of the area respectively. The greater part of the area is dominently an area of low index of dissection. It conveys the nature of its terrain. High dissection marked by isopleths of 0.30 and above. They crowd together where hills and ridges of resistant rocks have been deeply eroded by numerous streams. They follow the trend of ridges of resistant rocks of Mahudi and Mahabar

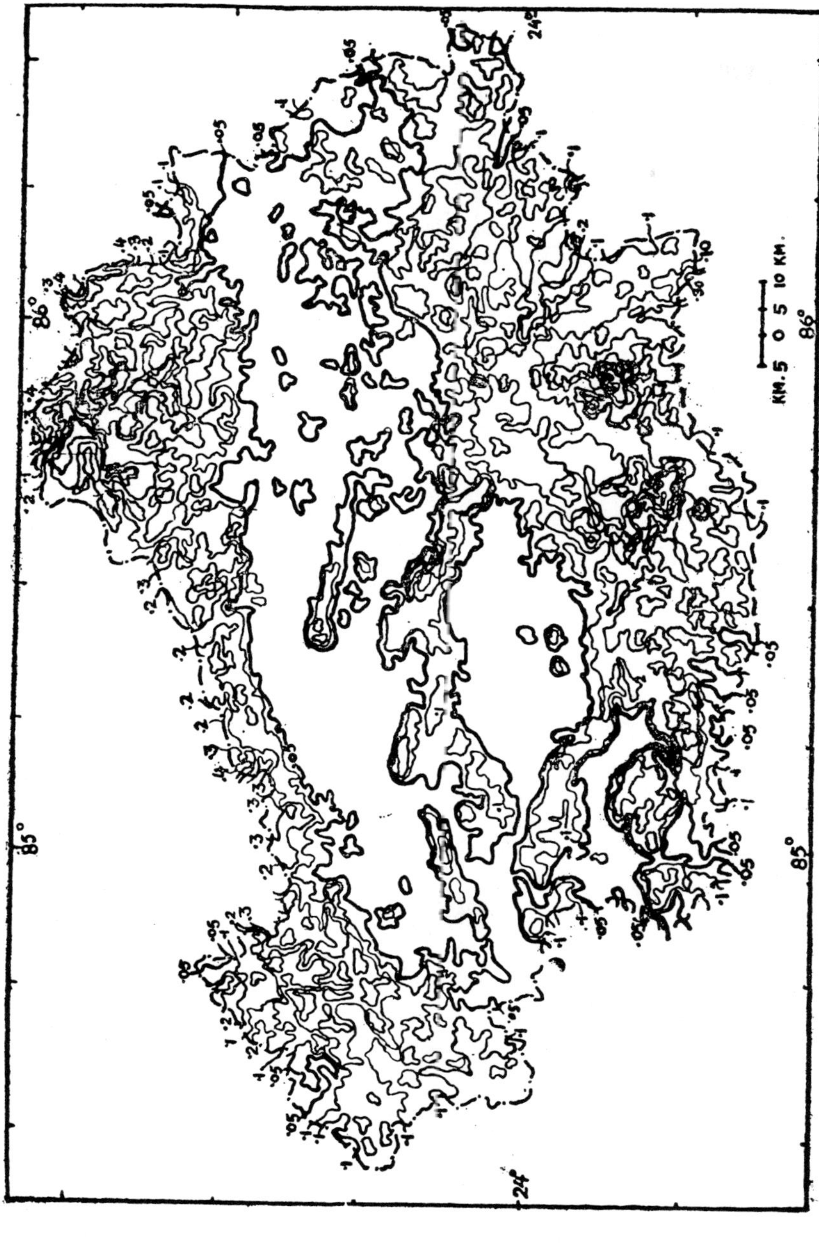

FIG. 16.6 : *Hazaribagh Plateau Region—Dissection Index*

hills near the lower parts of the Padma Ghat. Similarly in the northern scarp zones, high index of dissection are confined over Muhawar hill and Ghoranji hill in the Sakri basis, Dabaur hill north of Kodarma, Bendi hills, Lohawar Pahar north of Danea, Kolwa Pahar southeast of Hunterganj, and over the interfluvial areas of upper Lilajan basin. In the Damadar Valley such areas are confined to Satpahari, Mahudi hills, Lugu, Jilinga and Parasnath hills from west to east. The crowding of isopleths reveals that index of dissection rapidly increases towards the crest of hills and ridges. About 4.71 per cent of the area of the Hazaribagh Plateau Region indicates the segments of high index of dissection.

TABLE 16.3 : **Index of Dissection**

Index of Dissection per sq. km.	Area in sq. km.	Percentage of the total area	Major categories of index of dissection
below 0.05	7651	44.22	Low
0.05-0.10	4645	26.84	71.06%
0.10-0.20	2951	17.03	Moderate
0.20-0.30	1245	7.20	24.23%
0.30-0.40	529	3.05	
0.40-0.50	177	1.03	High
0.50-0.60	52	0.30	4.71%
Above 0.60	60	0.34	
Total	17,300	100.00	

The area of moderate index of dissection clearly follows the trends of moderate relative relief and covers 24.23 per cent of the total area of the Hazaribagh Pleteau Region. It forms a narrow and irregular zone all along the scarps of the lower and upper Hazaribagh Plateau Region. In the upper Lilajan catchment area and in the Sakri basin moderate dissection is more prominently outlined. Moderate index follows the trend of interfluvial sections of the Barakar and the Jamunia, the Kunur and the Barakar in the eastern and southeastern parts of the area under study. Some of the prominent ridges of the Barakar basin, like, Tilaiya hills are clearly outlined by 0.10 isopleths.

Areas with low index of dissection represents the old erosional surfaces and broad river valleys of the area under study and includes highest percentage of the Hazaribagh Plateau Region. In this respect it spans over the areas of low relative relief. The undulating erosional surface of the lower Hazaribagh Plateau Region is well demarcated by isopleth of 0.95 and 0.10. The trend of this surface is punctuated by moderate dissection in the lower section of the Barakar basin where dissection is prominent over residual ridges and hills. Second prominent patch of low index of dissection is confined over the undulating upper Hazaribagh Plateau surface. It extends westward to Chatra plain.

The Damodar basin is punctuated by low index of dissection where streams descending from the upper Hazaribagh Plateau have excavated broad valleys. In most of the areas like Barkagaon and Mandu, area of low index or dissection follows the trend of scarps.

Distributional Pattern of Relative Relief and Index of Dissection

An analysis of relative relief values and index of dissection reveal that a cross-correlation exists between them. Their relation reveals the nature of terrain and their development in the Hazaribagh Plateau Region. A two-way classification of the units in sq. km of the relative relief and index of dissection is depicted in the following Table 16.4.

The Table 16.4 shows that maximum frequency of units falls in the category of low relative relief and index of dissection. Frequency tend to decrease with increase in values. High relative relief has higher values of dissection index. This interrelationship is illustrated in the statistical plot map (Fig. 16.7), which indicates that in the areas of low relative relief maximum values of index of dissection is below 0.10 and seldom approaches the moderate values. With increase in relative relief, the values of index of dissection similarly shift towards moderate to higher values. Plot also reveals that frequency similarly decreases as the value increases. This relationship suggests the mode of relief development in the area. The fluvial erosion has dissected the plateau rims to

Table 16.4 : Relative Relief Index of Dissection

Relative relief in M/ feet/in per sq. km.	Index of Dissection								Total units sq. km.	%
	.05	.05-.10	.10-.20	.20-.30	.30-.40	.40-.50	.50-.60	>.60		
Less than 50	4703	661	64	29	1	—	—	—	5458	31.5
50-100	2588	2306	177	14	5	3	1	1	5095	29.5
100-200	345	1561	1538	134	10	1	1	1	3591	20.8
200-400	10	108	1078	677	206	14	5	—	2098	12.0
400-600	3	7	77	312	175	73	14	5	666	3.8
600-800	1	2	5	60	100	29	13	12	222	1.2
800-1000	1	—	1	14	25	26	4	10	81	0.5
1000-1200	—	—	—	1	4	16	7	17	45	0.3
1200-1400	—	—	—	4	2	3	4	3	16	0.1
Above 1400	—	—	1	—	1	12	3	11	28	0.2
Total	7651	4645	2951	1245	529	177	52	60	17,300	100.00
Per cent	44.22	26.84	17.03	7.20	3.05	1.02	0.30	0.34	100%	

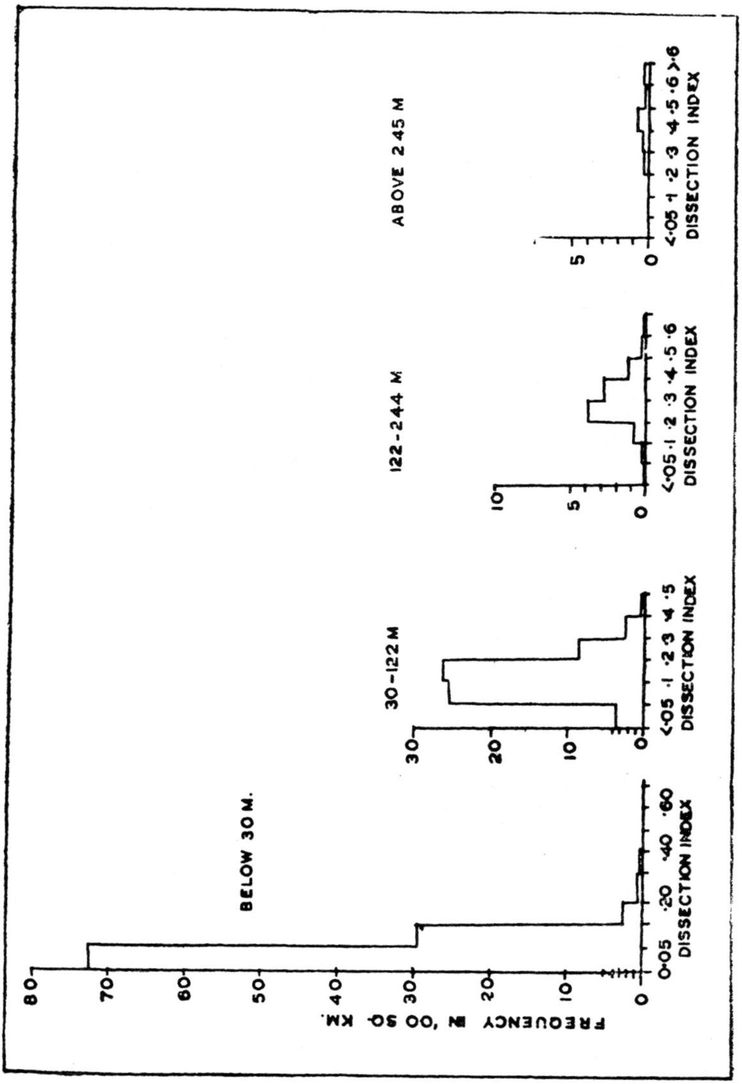

FIG. 16.7 : *Statistical Plots, Relative Relief vs. Dissection Index*

produce step like scarps of moderate relief where dissection is more moderate than intense. High relative relief is confined to the areas of exposed and resistant rocks.

The study of hypsometry has established the magnitude of altitude and their relation with the area. The study of these graphs has revealed surfaces at various altitudes. The facts thus gained have been confined to the profiles. Profiles also distinguished them as erosional surfaces which are represented by scarps of recent origin. Some of the remnants of earlier erosion cycles have been clearly outlined in these maps.

The intensity of relief map has simply traced out the different surfaces and provided a simple physiographic outline of the Hazaribagh Plateau Region. The analysis of relative relief further brings out the characteristics of the amplitude of the available relief and their areal distribution. Some of the obscure elements of the physiography have been revealed in this map. The study of the index of dissection map and its comparison with the relative relief map has been very useful as they together emphasize major relief features, their extent, continuity and inter-relationship. The index of dissection map to a great extent proved useful in removing the obscurity of relative relief map.

Review of these maps and diagrams reveals various surfaces of different magnitudes. The undulating surfaces and basins of low relative relief and index of dissection are inherited surfaces while moderate high values of the attributes represent the uplift of the old erosional surface. The displaced surfaces have evolved scarps along their margins. Tectonics of the area, relative relief, reflecting degree of dissection and frequency of drainage in response to the lithology have produced terrain of unequal dimension and varying elevation.

REFERENCES

1. J.I. Clark, 1966 : Morphometry from maps, in Durry, G.H. (ed) *Essay in Geomorphology*, Heineman, London, p. 240.
2. H. Baulig, 1939 : Duex methodes d'analyse morphologigue appliquies a' la hante Belgigue, *Bull. Sec. Belge Etudes Geog*, 9 pp. 165-84, in Durry, G.H. *Essays in Geomorphology* (Ed) Heineman, London.

3. A.A. Miller, 1953 : *The Skin of the Earth*, Mathuen, London, p. 50.
4. Ibid., p. 50
5. P.J. Monkhouse and H.R. Wilkinson 1963 : *Maps and Diagrams*, Mathuen, London, p. 67.
6. W.S. Glock, 1932 : Available Relief as a factor of control in the profile of a Landform, *Jur. of Geol.* Vol. 49, p. 74.

SEVENTEEN

A. Kumar and *R. N. Pandey*

SALIENT CHARACTERISTICS OF SLOPE PROFILES, HAZARIBAGH PLATEAU

THE HAZARIBAGH Plateau Region, located in humid sub-tropical climate, depicts its morphological character by the appearance, dimension and magnitude of slope. It is an ancient landmass where weathering, fluvial erosion and earth's movements have played a significant role in shaping its present landscape. The plateau surface expresses subdued topography with gentle slope over its planation surface. The slopes of the Hazaribagh Plateau Region seem to have been influenced by the structure and the fluvial erosion system.

Slope are the fundamental types of landscape feature so they need careful study (King, 1967). The areal distribution of the average slope of the Hazaribagh Plateau Region depicts the salient features of its surface in relation to slope, structure, dissection and relief. But the visible slope profiles of some surveyed residual hills provide a basis to identify the phases of slope development. The survey of the visible slope form has been carried out with the help of Abney level, tape and ranging rods by selecting three important sites in the three accessible localities of the area: (i) Profiles have been obtained by traversing the hill slopes of Kurmi, Koya and Lamki hills located near the Parasnath hill station. (ii) Near Surajkund, the northern slope of Doma hill and the western slope of Chadri Pahar have been measured. The northern and southern side slopes of Poraiya hill, near Barhi-Kodarma Road, have been measured. The measured data of slope profiles have been plotted over graph and slope angles have also been noted above the profile units. A slightly generalised intercept profile has been derived from the measured profile. This involves the recognition of curved and rectilinear portions of the profile as a whole. Following the method suggested by Young (1971), the slope element and segments, minimum and maximum segments of these profiles are recognised for the detailed analysis of hill slope form characteristics (Fig. 17.1).

The slope form of Kurmi, Koya and Lamki has developed over weathered and rugri covers granite gneiss hill. While the Poraiya, the Doma and the Chadri Pahar are quartzite hills. The general character of these profiles can be established simply on the basis of ratio of curved segment in profiles. This requires the recognition of profiles simply on the basis of its convex concave appearance. The slope forms of these profiles clearly reflect the curved units and amount of curvature of the measured sections of profiles. The length of the curved units and their terminal angles have been calculated from the measured data and profiles. The measured value expresses the relative length of convex and concave units of profiles as a percentage of the total length of the profiles (Table 17.1).

TABLE 17.1: **Percentage of the total Length of the Profiles**

Curved units of slope	Slope Profiles									
	Kurmi		Koya		Lamki		Poraiya		Doma	Chadri
	W	E	E	W	N	S	N	S	N	W
LX %	46.6	60.0	50.0	27.7	28.7	62.5	30.1	66.6	63.6	66.6
LV %	66.6	48.0	52.5	79.5	82.2	55.5	90.4	33.3	54.5	71.1

The measured value in Table 17.1 expresses the general tendency of hill profiles in the sense whether concave or convex units taken together are increasing or decreasing the steepness of profiles (Kumar, 1975-76). It is obvious that difference between LX and LV values are high in Lamki and both profiles of the Poraiya hill. Out of these, three profiles are conspicuously concave. Concave units are occupying maximum length in six profiles. Southern slope of Poraiya out of three other profiles showing higher values of convex length, is apparently

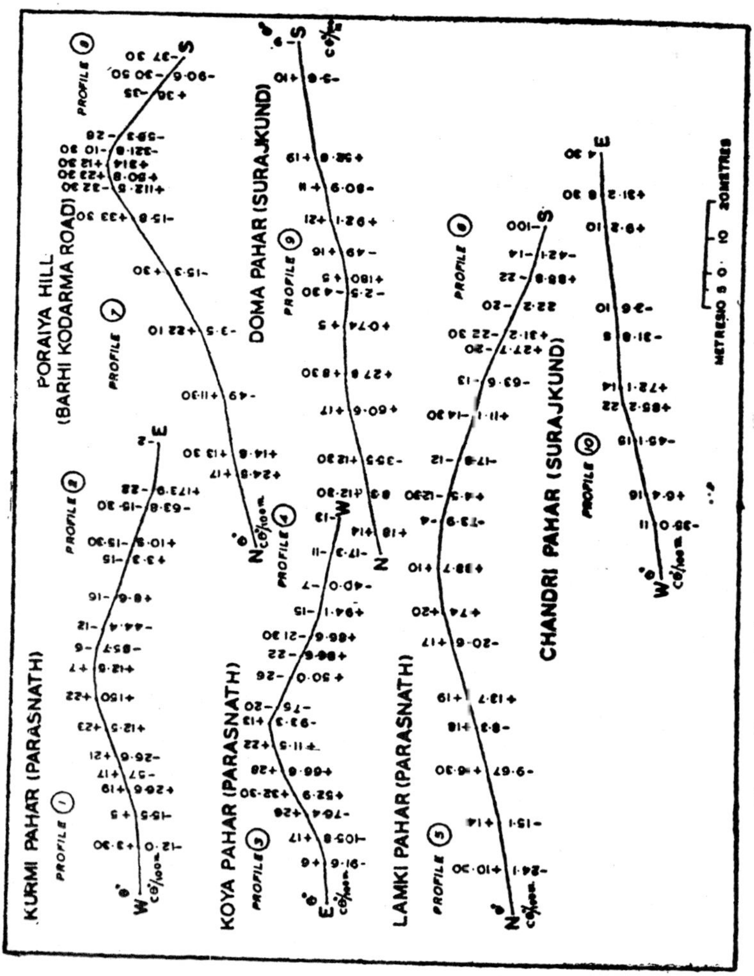

FIG. 17.1 : *Slope Profiles of the Selected Residual Hills*

convex in form. The dominating values of convex or concave element of profiles suggest the flattening or steepening tendency of profiles. But measurement of relative length of these profiles also includes the upper and lower segments of the curved units. It means that some of the maximum and minimum profile units are included twice in calculation.

The evolution of angular change in profiles also involve the change in curvature of profiles. The total amount of angular change of each group of profiles units (AX% and AV%) as percentage of the total angular change involved in the curvature of profiles is included in the Table 17.2.

TABLE 17.2: **Percentage of the total angular change involved in the curvature of profile**

Angular change of curvature units	Slope Profiles									
	Kurmi		Koya		Lamki		Poraiya		Doma	Cnadri
	W	E	E	W	N	S	N	S	N	W
AX%	41.2	55.6	49.0	22.0	25.9	53.0	25.3	67.0	54.0	48.2
AV%	58.8	44.4	51.0	78.0	74.1	47.0	74.7	33.0	46.0	51.8

Table 17.2 illustrates increased difference between the angular changes of convex and concave units in the western slope of Koya hill, northern slope of Lamki Pahar, northern and southern slope of Poraiya hill. This difference is less in the profiles of Kurmi, Doma and Chadri Pahar. The measurement of the amount of convexity or concavity is derived and listed in Table 17.3 with the help of Table 17.1 and 17.2. The value of general profile form is then established by dividing CX and CV. The values of curve ratio (CR) for the profiles are listed in Table 17.3.

It can be noted from the Table 17.3 that western profiles of Kurmi hill, Koya hill and Chadri Pahar have the apparent straight tendency as their curvature ratio (CV) seldom exceeds 1. Here straight tendency is again marked in those profiles where curvature ratio, though less than 1, is 0.99.

The visual impression gained from looking at Fig. 17.1 is given numerical expression.

TABLE 17.3: **The Value of Curve Ratio for the Profiles**

Curvature & curvature ratio of Profiles	Kurmi W	Kurmi E	Koya E	Koya W	Lamki N	Lamki S	Poraiya N	Poraiya S	Doma N	Chadri W
LX/AX=CX	1.13	1.07	1.02	1.03	1.10	1.17	1.18	0.99	1.17	1.38
LV/AV=CV	1.13	1.08	1.02	1.01	1.11	1.13	1.21	1.00	1.18	1.37
CX/CV=CR	1.00	0.99	1.00	1.01	0.99	0.99	0.97	0.99	0.99	1.01

Almost all profiles has a near straight slope with a tendency towards concavity. These hillocks present a conical or dome shaped appearance and since they are side of the ridge or hills may be concave (Chibber, 1953). The basal concavity of profiles are well advance, as a result convexity is only limited to the narrow crestal section. The degree of back wearing is prominent in the basal section of the slope, as waste slope forms a prominent section of the slopes. The lithology also seems to have influenced the steepness of the profiles near the crestal sections particularly in the profiles of Doma and Chadri hills composed of weathered quartzite. Even the Poraiya hills composed of mica-schist is conspicuously concave as the waste matter is actively removed by the stream flowing few metres away from the basal section of the slope.

Convex and concave segments include those units which are rectilinear sections or maximum segments of profiles. Its absence or presence determines the complexity of profiles and nature of slope development. The relative length of maximum units tend to decrease as profile approaches towards a perfect convex-concave form. Generally rectilinear segment occupies 20 per cent of the total profile lengths. But in the case of these profiles rectilinear units are occupying 8 to 28 per cent of the total length of profiles. Only in the southern profile of Poraiya hill ratio is about 40 per cent. In profiles where this ratio is more than 20 per cent it tends to increase the general steepness of profiles and their basal section similarly reveals conditions of slope replacement through active denudation. The shorter length of the maximum units in the profile of Kurmi, Koya, Doma and Chadri are an indication to the process of slope flattening (Carson, 1972). Each profile in absence of free face at crestal section, reveals a shorter crestal slope but longer and well developed basal concavity. It seems that crestal section are controlled by the basal concave form and rectilinear slope units. The analysis of rectilinear sections further reveals that over Poraiya rectilinear slope is steeper with an average of 30° to 35° of slope. It is more relevant when we find that here rectilinear slope units are not only steeper but also occupy maximum length of the profile. Over other hills the west looking slopes are steeper than the east looking profiles. The rectilinear slope angle varies between 16° to22°

TABLE 17.4 : **The Rectilinear Slope Angle**

Profiles	Slope Segment in per cent		
	Convex	Rectilinear	Concave
Kurmi W	37.5	12.5	40 5
,, E	37.5	25.0	37.5
Koya E	41.8	8.0	50.12
,, W	35.0	8.5	55.5
Lumki N	25.0	25.0	50.0
,, S	54.14	25.27	18.18
Poraiya N	55.5	11.11	33.3
,, S	55.0	30.00	15.0

Angle Characteristics of Profiles

Profiles are composed of different units having different angles of slope. The association of slope angle with the curved units of profiles requires analysis of slope angle and their frequency in profiles. It is possible to establish the relative steepness of profiles with frequency classification of slope units. The plotted values of the total slope units of these slope profiles reveal a steepness of slope profiles. The angle frequency graph (Fig. 17.2) expresses bidomal character of angle concentration over the profiles.

It discloses that 10° to 22° of slope, with frequency of over 58.9 per cent of the total slope units, represents steep section of profiles. In frequency graph second frequency slope angle group is less than 10° of slope. It covers 22.76 per cent of the total length of the profiles. The steepest profile unit occupies

17.95 per cent of the total length. The high frequency of profile units with low to moderately slope angle indicates the mode of slope development. Steeper angle are less stable in slope than the gentle slope angles. These are the areas of slope replacement, replacement of map angle by lower angle slope units. This is obvious in the profiles where basal concavity is prominent form of the profiles. The specific concentration of gentle and steep slope units over profiles further discloses that lower angle values are more frequent in the basal portion of profiles. These slopes are rather process-controlled than structure.

The frequency of Rectilinear and Convex segments of profiles gradually increases often 8° of slope. Slope units with less than 16° of slope are confined to the basal concave section of profiles. High frequency of moderate to low slope units further reveal the relative period of slope development. Slope seems to have achieved late maturity of their stage of development.

Curvature Frequency

Profile curvature is the rate of change of angle with distance down the true slope, expressed in degree per 100 metres (Young, 1972). The convex or concave curvature of the slope profiles and their frequency is analysed to confirm the form characteristics of profiles. Curvature of slope units in respect of upper and down slope units has been calculated using the equations:

$$C = 100 \frac{Q_a - Q_b}{0.5\,(D_a - D_b)} = 0°/100\text{ m}$$

For convenience seven broad classes of profile have been recognised.

Steeply Convex———C=Above +100°/100 m
Moderately Convex—C=+10° to +100°/100 m
Slightly Convex———C=+1° to +10°/100 m
Near Zero Curvature—C=+1° to−1°/100 m
Slightly Concave——C=−1° to−10/°100 m
Moderately Concave−C=−10° to−100°/100 m
Steeply Concave——C=Above −100°/100 m

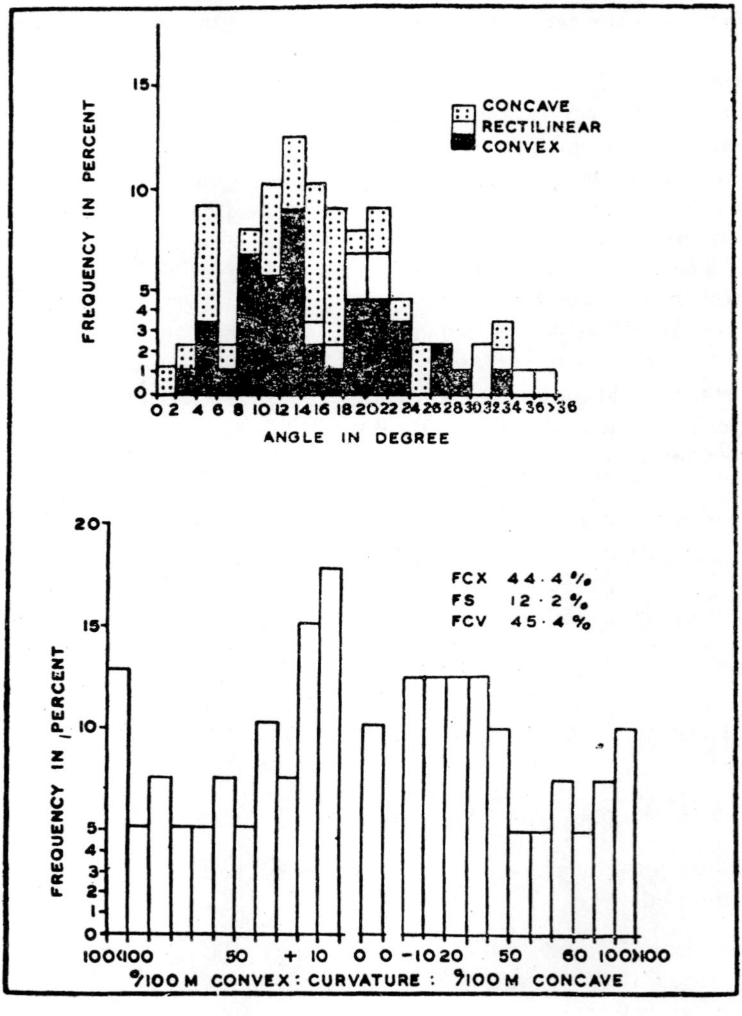

FIG. 17.2 : *Frequency Distribution of Angle and Curvature in the Selected Residual Hill-Slope Profiles.*

Curvature frequency graph (Fig. 17.2) illustrates the high frequency value to $\pm 10°$ to$\pm$ 100°/100 metres group, where they occupy 69.3 per cent and 77.5 per cent of convex and concave units respectively. Curvature frequency rapidly declines with increase in curvature value. Slope units of near zero curvature, representing rectilinear segments of profiles occupy only 10.2 per cent of the total portion of profiles. Under slightly curvature positive curvatures predominates 17.9 per cent against 12.5 per cent negative curvature. Distribution points out that due to restricted convexity near crest, positive curvatures are available in the crest in the middle portion of profiles, as in profiles of Doma and Chadri Pahar. Negative curvature has developed in the basal sections of almost every profile. Slope units with $-10/100$ m of negative curvatures are available in the basal sections of profiles, where due to horizontal extension of slope units, a gradual ducrease in carvature value develops.

Length and Angle of Slope Units

Each curved units of profile are related to the relative length of profile units which is one of the chief characteristics of slope. The correlation of length of slope units and amount of slope angle summarises that units with steeper angles are relatively shorter in their lengths.

In each profile length of units with more than 20° of slope is less than 10 metres. Exception are the profiles of Poraiya hills whose northern slope is apparently dominated by longer and steeper concave form while southern slope has longer and steeper mean convex segment. The two profiles has moderately steeper convex and steeper concave profile curvature (Profile No. 7 and 8 in Table 17.5).

Table 17.5 shows average length, slope angle and curvature of convex and concave segments of each profile. The average length of units within convex in between 8.3 to 13.8 metres while unit length is 8.0 to 15.7 in concave segments. In profile No. 5 and 10 concave units with 15 metres of length are associated with 14.15° and 11.3° of slope. In convex segments longer slope units, as in profile 5.9 and 10 reveal an average of 15°, 11.13° and 9.20° of slope respectively.

The analysis of the slope profiles of these residual hills thus concludes that they have reached a developed maturity

TABLE 17.5 : Mean Length Angle and Curvature of Profile Segments

Profile No.	Convex Segment				Concave Segment			
	No. of Units	Mean Length of units in Metre	Mean Angle	Mean curvature 0/100 m	No. of units	Mean length of units	Mean angle	Mean curvature 0/100 m
1	3	9.3	16.20	50.40	4	10.0	11.7	79.3
2	3	8.5	11.20	61.26	3	11.6	13.0	64.4
3	3	8.3	21.00	76.46	4	9.5	20.2	91.23
4	3	10.3	14.20	59.40	5	8.0	20.18	44.10
5	2	13.5	15.00	41.45	4	15.0	14.15	15.15
6	5	10.4	14.24	35.25	2	9.5	12.00	40.30
7	5	10.0	19.50	86.36	3	21.6	21.13	99.40
8	3	9.6	24.30	158.49	—	—	—	—
9	8	13.3	11.13	49.52	4	8.7	9.7	58.34
10	5	13.8	9.20	40.58	4	15.7	11.3	27.40

where structure is not a controlling factor. The basal section suggests their relative stability. These residual landforms are subjected to complex attack with a variety of erosion processes. Water is an important agent for removal of material from the slope and there is a strong element of linearity in its attack. Down slope movement of material over the whole slope seems to be slow and intermittent. The field study of the residual hill-slope reflects the processes forming them. The formation and maintenance of slope is dependent on the environment which is the sum total of all factors that contribute towards landform development (Sparrow, 1965). The field study reflects that the removal of waste from the base of the slope is impeded so that hill slope profile will continue to become gentler with time in the Hazaribagh Plateau Region.

SUGGESTED READINGS

King, C.A.M., 1967 : *Techniques in Geomorphology*, Edward Arnold, London, p. 51.

Young, A., 1971 : Slope profile analysis—The system of best units; slope form and process, *Int. Brit. Geog. Spe. Pub.* No. 3, pp. 1-13.

Kumar, A., 1975-76 : Morphological properties of the slope profiles of the western plateau scarps, Ranchi, *Geog. Outlook*, Vol. XI, p. 87.

Chibber, H.L., 1953 : The landforms, life and radial drainage of mount Parasnath, Hazaribagh District, Bihar, *NGSI Bull.* No. 18, 41.

Carson, M.A. and Kirkby, M.J., 1972 : *Hill Slope Form and Process*, Combridge University Press, London.

Young, A., 1972 : *Slope, Oliver and Boyd*, Edinburgh, p. 137.

Sparrow, G.W.A., 1966 : Some environmental factors in the formation of slopes, *Geog. Jour.* Vol. 132, Part 3, p. 390.

EIGHTEEN

P. S. Dhinwa

STUDY OF SLOPES IN THE BERACH BASIN, RAJASTHAN

INTRODUCTION

THE STUDY of slopes is one of the most significant aspects of geomorphology. There is no exaggeration in saying that geomorphology is primarily a study of slopes. The slope is main property of landforms that affects activities of man like agriculture, transport and human settlements. As a matter of fact the study of slopes is not only significant from academic point of view but also from the point of view of human activities. In India, the problem of slope development has not been intensively studied. Few studies of Pandey (1968), Rai (1971), Subramanyan (1976), and Sharma and Padmaja (1977 and 1978) are worth mentioning. A detailed work is needed in this direction.

The present paper is an attempt to enumerate and analyse the slope, hill-side slopes, Strahler's slope height curves, slope histogram and factors and processes which affect slope and its derivatives. The longitudinal and the latitudinal extension of the Berach Basin is 73°30′ to 75°15′ E and 24°20′ to 25°15′ N (Fig. 18.1). The Berach Basin occupies an area of 8951.35 km². The Berach Basin as a case study comprises of Pre-Aravalli, Aravalli Super Group, Post-Aravalli and Vindhyan Super Group formations (Fig. 18.2). The Berach river drains through Udaipur, Chittaurgarh, and Bhilwara districts. In comparison to the Aravallis, the Vindhyans reveal very peculiar features of slope development. It is attributed to the fact that the Vindhyans are more susceptible to erosion. For instance the four slope elements, viz., waxing slope, free face, detrital slope and pediment as enunciated by King (1957) have been clearly observed in the formations of the Vindhyan Super Group while the free face is missing in the Aravalli's. Climatically the Berach Basin falls into BSh type of climate as specified by Koppen (1953) which represents to hot and dry seasons. The highest temperature 43.3°C in the Berach Basin is recorded at Chittaurgarh and the maximum value 861.00 mm of mean monthly precipitation has been observed at Nimbahera.

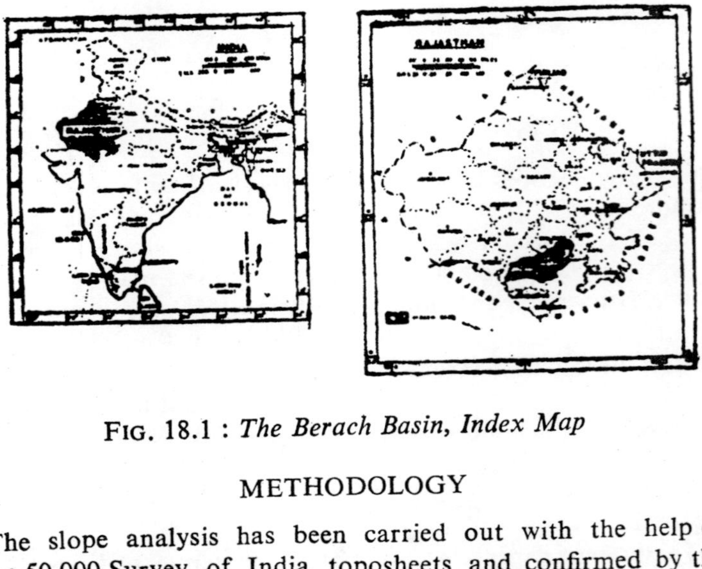

FIG. 18.1 : *The Berach Basin, Index Map*

METHODOLOGY

The slope analysis has been carried out with the help of 1 : 50,000 Survey of India toposheets and confirmed by the field observations. The study of slope development is conducted under the following two heads: regional slope, and hill-side slope development.

Regional Slope

Slope Analysis

In the present case study of the Berach Basin the maximum value of slope (24°14′) is exhibited in Chittaurgarh Hills. The Chittaurgarh Hills belong to the Vindhyan Super Group and comprises of conglomerate, shale, sandstone and limestone

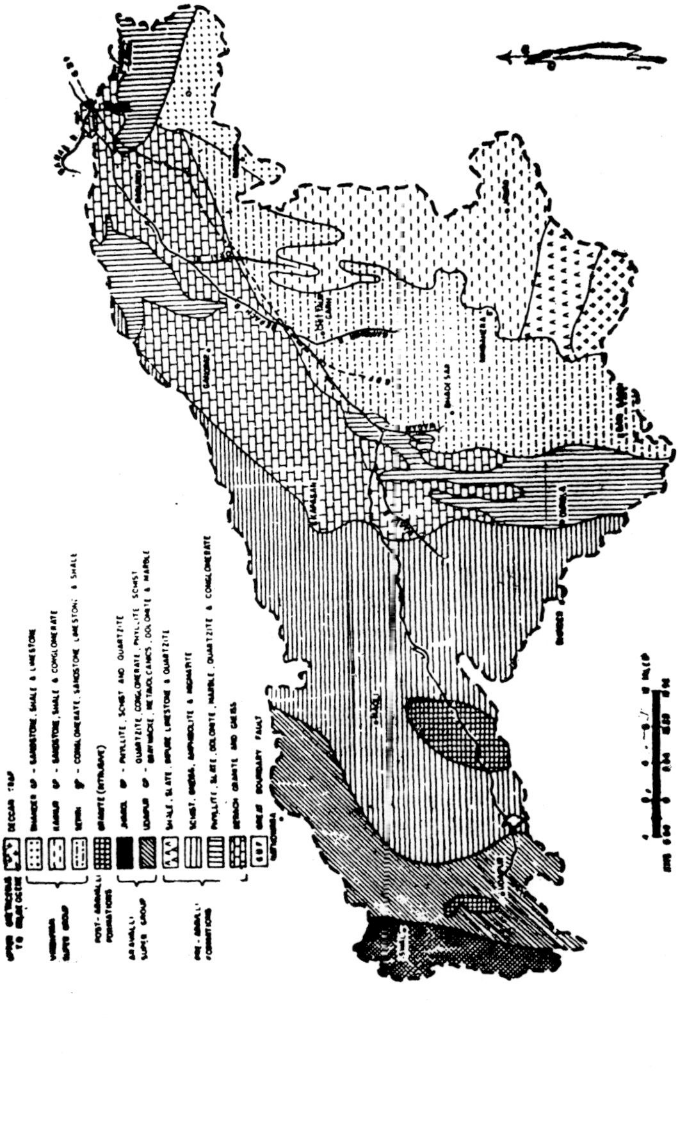

FIG. 18.2 : *The Berach Basin, Geology*

rock types. Due to the presence of sandstone at and near Chittaurgarh, steep scarps have been noticed as peculiar features of sandstone topography. On the other hand the minimum value of slope (2° 04′) has been observed at the place of confluence of the Berach and the Banas. The average slope of the Berach Basin decreases from west to east. The range of slope varies from 17° 57′ at the place of origin of the Berach river to 2° 04′ at its confluence (Fig. 18.3).

Slope Histogram

The slope histogram reveals that the maximum value of the frequency of occurrence (164) has been noticed at an angle of 0° to 5°. The next surface exhibiting the frequency of occurrence is between 5° to 10°. The third and uppermost surface having been displayed by 12 frequencies of occurrence is represented up to the angle of 24° 12′. The slope histogram also reveals that there exists a reverse relationship between the frequency of occurrence and slope, i.e., the frequency of occurrence decreases with an increase in degree of slope, thus both the variables exhibit a negative function (Fig. 18.4).

Slope-Height Curve

In this curve, the average slope for inter-contour areas is calculated by Strahler's method*. The values of slope are plotted on the horizontal axis and the contour heights on the vertical axis (Fig. 18.5). Uniform slopes appear as straight lines parallel to the horizontal axis and concave or convex appearance of the profile indicates the change in slope value. The analysis of slope-height curve indicates that the relationship between slope and height varies from place to place and

*Strahler's slope-height curve :

The area between each pair of adjacent contours is measured and their mean length computed. The mean width of each intercontour area is obtained by dividing this area by the mean length of the two contours. The tangent of the angle of slope is derived from dividing the contour interval by the mean intercontour width.

$$\tan\alpha = \frac{\text{Contour interval}}{\text{Mean inter-contour width}}$$

For the area above the highest contour the mean inter-contour width is replaced by half the length of the highest contour.

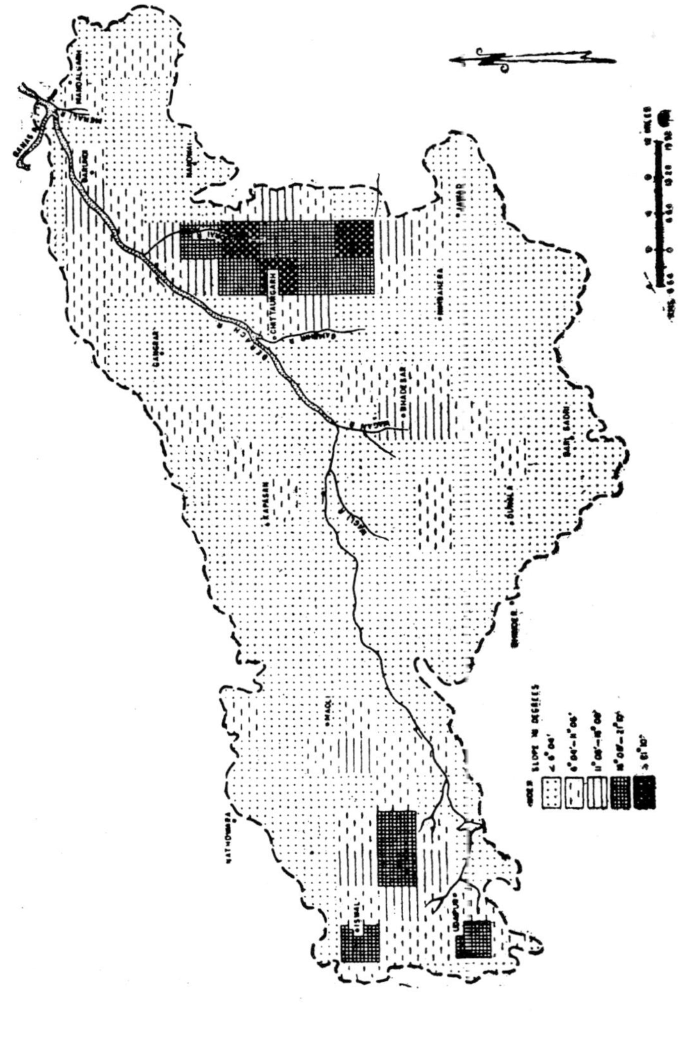

FIG. 18.3 : *The Berach Basin, Slope Map*

the nature of the curve changes with a change in topography and geological formation. For example, basins 3, 4, 7, 9, 12, 14, 15, 17, 18, and 19 reveal that in the lower part of the curve slope decreases with an increase in elevation while in the upper part of the curve the value of slope increases with an increase in altitude. These basins belong to the Pre-Aravalli and the Vindhyan Super Group formations.

On the other hand basins 2, 16 and 20 exhibit that in the lower part of the curve slope increases with an increase in elevation while in the upper part of the curve the value of slope decreases with an increase in elevation. These basins fall in Pre-Aravalli, Aravalli Super Group and the Vindhyan Super Group formations.

Apart from the above cited, basins 5 and 8 exhibit that the value of slope increases with an increase in elevation. The basins fall in the Pre-Aravalli and the Vindhyan Super Group

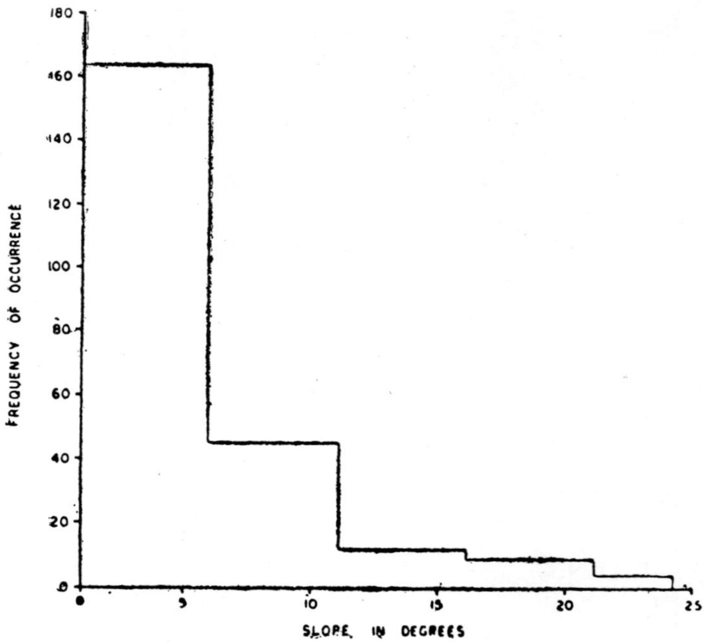

FIG. 18.4 : *The Berach Basin, Slope Histogram*

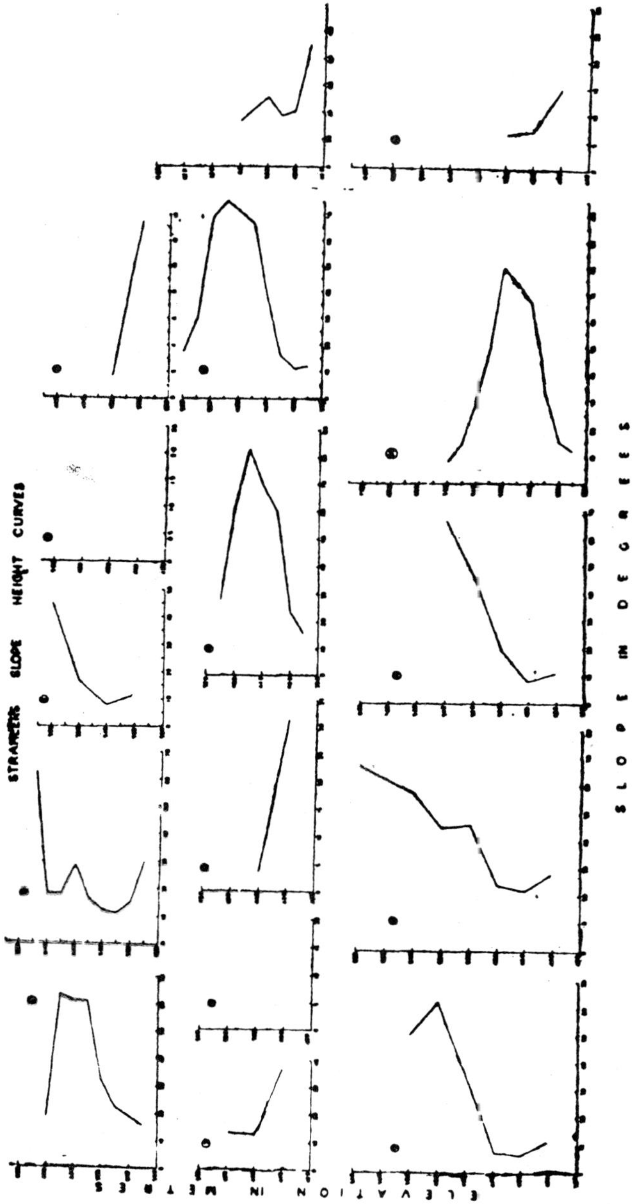

FIG. 18.5 : *The Berach River Basin, Strahler's Slope-Height Curves*

formations. While in case of basin 6, the slope decreases with an increase in the altitude which belong to the Berach granite and gneiss type of rock of the Pre-Aravalli formations.

Hillside Slope Development

The concept of hillside slope development has had a rather controversial history which continues down the present and probably much more has been written about it than any other geomorphic idea. There are three main theories of slope evolution, which are termed as slope decline, slope replacement and parallel retreat. These theories are associated respectively with the names of Davis (1909), Penck (1927) and King (1957). Besides the pioneer studies of Davis, Penck and King on the evolution of hill-side slopes, the contributions of Savigear (1952), Culling (1963), Young (1963), Birot (1949), Dresch (1949), Wood (1942), Bryan (1940), Fair (1947), Pallister (1956), and Strahler (1952-1956) are worth mentioning who have worked on this problem in different environments. In India, the problem of slope evolution has not been thoroughly studied but the studies of Pandey (1968), Kumar (1978), Rai (1971), Subramanyan (1976), Sharma and Padmaja (1977 and 1978) are worth mentioning.

Theory of Slope Decline

The exponent of the theory of slope decline was W.M. Davis. Davis' theory is a clear statement of a type of evolution, and is capable of being translated into quantitative terms, consonant with its relation to the cycle of erosion. The theory implies that slope form is time-dependent, possibly this is the feature that offers an opportunity of indirect testing by field observation. Davis holds scientific priority regarding the hypothesis that the convex upper parts of slope are produced by soil creep. He described qualitatively the mechanism of movement in creep by alternate dilatation and contraction of the soil under the influence of gravity.

Theory of Slope Replacement

The main aim of Penck's theory was to relate exogenetic forces and landforms, both of which can be observed to endogenetic processes, and so discover the nature of the latter by a process of denudation. Penck's analysis of exogenetic processes rests on his discussion of the properties of the regolith cover on slopes, and mechanisms of weathering and surface processes, among which he considered primarily was mass movement. Penck attempted to establish a relationship between the nature of slope and the tectonic history of a region. Convex slopes were formed in periods of accelerated uplift, straight slopes in periods of constant uplift, and concave slopes in periods of decreasing uplift or stability. Penck started with a straight rock slope and assumed equal weathering over the whole of it. The weathered fragments crumble and fall under the influence of gravity, all except the lowest adjacent to river level. This lowest fragment can't move as there is no gradient below it. In the theory of slope replacement the process follows a law that "the flattening of slopes always takes place from below upwards".

Theory of Parallel Retreat

L.C. King's theory of parallel retreat on slopes is part of a wider scheme of landscape evolution. He argues that they are the "natural" product of slope evolution, whether by flowing water, mass movements, or both. Full development of the elements depends on local conditions, chief among which are a strong bedrock and adequate relief. If either of these conditions is lacking the scarp may be absent, the debris slope is necessarily also missing, and a decadent convexo-concave hill-slope then results. As recognised by Wood (1942) and redefined by King (1957) the four elements of slope are as under :

1 Crest (waxing slope)
2 Scarp (free face)
3 Debris slope (constant slope), and
4. Pediment (waning slope).

SLOPE FORMING FACTORS AND PROCESSES

The hillside slope forming factors and processes vary with reference to the varying environment. Structure, climate, relief and vegetation are the basic factors which play a vital role in slope formation. The present area of the Berach Basin exhibits primary endogenetic and secondary exogenetic slopes. But the endogenetic slopes that have been produced by faulting in this area have now been obliterated, altered and modified by exogenetic processes operating on them since Mesozoic period. Hence majority of slopes are exogenetic in origin. In the present area, hillside slopes have been formed due to the combined result of weathering, mass movement, soil creep, slope wash or surface wash and sheet erosion or rill erosion.

ELEMENT OF SLOPE FORM

The analysis of various relief profiles, superimposed profiles and field observations reveals the presence of the following slope elements in this region :

1. Crest,
2. Scarp,
3. Debris slope, and
4. Pediment

Crest

It is the uppermost part of the profile. And also known as summit area of a hill or slope; the profile is usually convex. Weathering and soil creep are the main processes forming this convexity. During the field study it was observed that both Aravalli Super Group and Vindhyan Super Group formations possess rounded hill tops. The roundness is very conspicuous in the Aravalli hills. The profiles of hill tops are convex in shape and guided by the process of soil creep. In Vindhyan sandstones the crest is endowed with steep slopes in comparison to the Aravallis.

Scarp (free face)

It is the bedrock outcrop on the steepest part of the slope which is the most active element in backwearing of the slopes as a whole. Backwearing is caused by rillwash and landslides. As suggested and pointed out by Sharma and Padmaja (1978) that Vindhyan sandstones generally forms steep scarps. The author also observed in the field the similar features. In comparison to the Aravallis the Vindhyans reveal steep scarps.

Debris Slope (constant slope)
It is formed by detritus fallen from the scarp above. Its angle is determined by the angle of repose of the coarse material.

Weathering reduces it to finer particles, which are then removed by wash, flowing as rills or turbulent sheet flow. Due to gentle slopes of free face the maximum amount of debris accumulation has been observed in Udaipur hills in comparison to the Chittaurgarh hills.

Pediment (waning slope)
It is the broad concavity extending from the base of the other elements to the stream or alluvial plain. Although frequently veneered with detritus, it is essentially a rock cut feature. It is produced by surface wash and its profile may approximate to any hydraulic curve. It was observed during the field study that in the Vindhyan Super Group due to steep free face the debris rests on the pediment zone and sometimes also transported by ephemeral streams to foothill regions. While in case of Aravalli Super Group the pediment zone is present but not so remarkable because the debris rests at the debris slope and the pediment zone is occupied by the plain surface.

CONCLUSION

The present study of slopes of the Berach Basin upholds uniformitarian theory of slope retreat and not the slope decline. The Vindhyan Super Group formations are characterized by all the four slope elements while the free face is missing in the Aravallis. Where the scarp element of the slope is missing for which it can be inferred that on such slope profiles scarp face has already been consumed and hill slopes develop convexity in upper parts and concavity in their lower parts through the process of slope retreat.

SUGGESTED READINGS

Bryan, K., 1940 : The retreat, of slopes, *Annals Asso. Am. Geogr.* 30, pp. 234-268.

King, L.C., 1957 : The uniformitarian nature of hill slopes, *Trans. Geol. Soc.* Edinberg, 17, pp. 81-102.

Pandey, S., 1968 : Evolution of hillside slopes and around Jodhpur. *Nat. Geogr. Jour. India*, Vol. 14, No. 1, pp. 32-40.

Rai, R.K., 1971-72 : The study of hillside slopes in the Sonar-Bearma Basin, *Geographical Outlook*, 7, pp. 167-174.

Sharma, H.S. & Padmaja, G., 1979 : The study of slopes in the Mej Basin (Rajasthan) Geographer, Aligarh.

Sharma, H.S. et. al, 1978 : Evolution of hillside slopes of the Morel Basin (Rajasthan), *Proc. Symposium on Morphology and Evolution of Landforms*, Department of Geography, University, Delhi, Dec. 1978.

Subramanyan, V., 1976 : Hill slope analysis of the Deccan Traps around Sagar, M.P.; *Geol. Sury. Ind. Symp.* Vindhyans Bhopal, (Abstract).

NINETEEN

P. B. Saxena

A GEOGRAPHICAL STUDY OF LANDSLIDE
Its Genesis, Intensity of Soil Depletion and Zoning in the Alaknanda Valley (Garhwal Himalayas)

INTRODUCTION

ONLY when a rock has been disintegrated and decomposed by weathering processes, the weathered debris soaked into rain water may slide downwards due to gravity force. Such a sudden gravitational or downward slope movement of a weathered rock is called landsliding. It is one of the interacting processes of mass wasting. The collective term for all gravitational or down-slope movement of weathered debris is called mass wasting. It consists of two criteria: (i) The type of material in motion, (ii) The type and rate of movement (sliding, flowing and falling, creeping, sludging and slumping etc.). Generally landslides occur when hill side or valley side slope falls due to certain geological climatic and biotic factors.

The area under study consists of a part of the Alaknanda Valley (in the Garhwal Himalayas) which extends from Vasndhara to Deoprayag. The river Alaknanda rises from the Satopanth and Bhagirathi Kharak Glaciers and makes a confluence with the river Bhagirathi at Deoprayag of 723 metres from the sea level.

Geoidentical Conditions for Landsliding

Geologically the Alaknanda Valley consists of three tectonical separable major litho-stratigraphical units known as the Dudhatoli Group, the Garhwal Group and the central crystalline Group. The zone between Deoprayag and Koteshwar (Bhagirathi Valley) constitutes of moderately metamorphosed rocks particularly phyllites and quartzites forming the Dudhatoli Group, which is separated by the North Almora Trust tending almost in NW-SE direction.[1] Further the rocks of the Garhwal Group are exposed upto Vishnuprayag and are primarily quartzites, slates, schists and carbonate rocks occasionally intruded by meta volcanic. The Garhwal Group which is separated by the Main Central Thrust existing in NW-SE direction, consists of several shear and fracture zones viz., Kalliasor, Nandprayag and Belakuchi which have highest frequency of landslides. The Northern zone extending from Vishnuprayag to the village Mana and its environs along the river Alaknanda and Dhauliganga, consists of schist, gneiss and granite of the Central Crystalline Group, which rests over the Garhwal Group. It is observed from the Fig. 19.1 that all Thrust zones have a general trend of NW-SE just parallel to the Himalayan Ranges; but the Alaknanda river flows in NW-SE direction. With the result the thrust faults are lying almost perpendicular to the river Alaknanda. Similar case happens with the tributaries of the Alaknanda river joining it almost perpendicularly (ranging from 60° to 90°) susceptible for landsliding.

Geomorphologically the valley consists of three[2] major distinct landforms of high hills, hill sides (Repose slope of Talus or Scree) and River Valley formations (incised meanders and benches). High hills are generally the conical shaped peaks ranging above 1500 m. They also become overhanging scarps near the thrust faults. The high-hill landforms generally behave like watersheds of the major tributaries of the Alaknanda; consequently rills, gullies and seasonal channels have been developed to make terrain more dissected. The vegetal cover is very scant and hill tops are necked.

The hill-side landforms exist between high hills and the river valley ranging from 500 to 1500 m. The hill sides generally consist of talus or screes. They have repose slopes near thrust faults where they are fully developed. The hill-sides have thick vegetal cover but become scant due to redundancy of landslides caused by thrust zones.

The river-valley-formations consist of incised meanders and

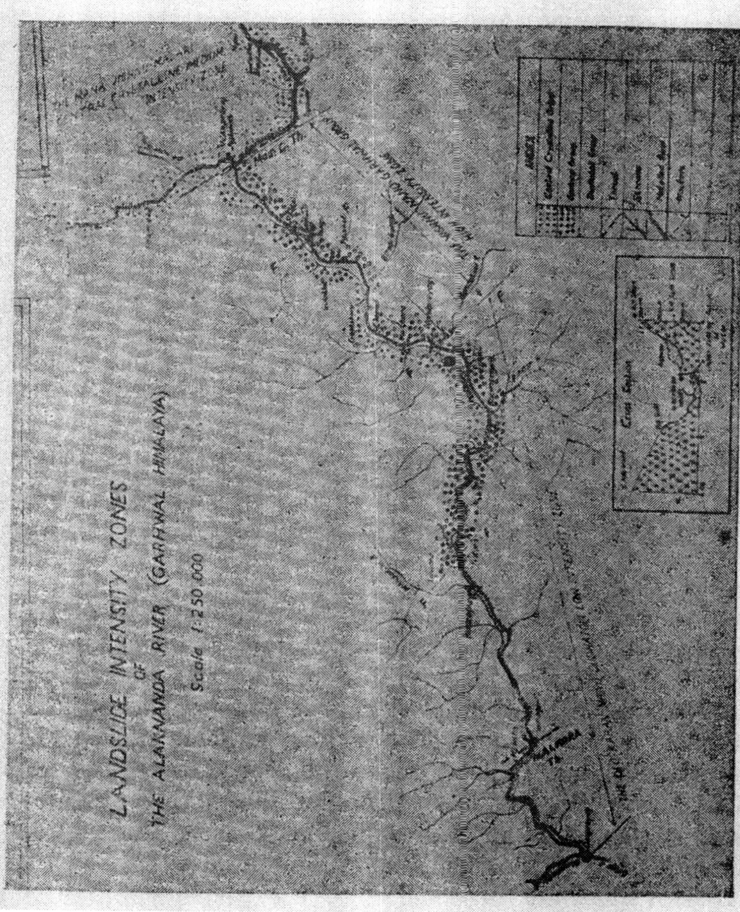

FIG. 19.1 : *Landslide Intensity Zones of the Alaknanda River (Garhwal Himalayas).*

river benches of 500 m height and below it. The incised meanders have gentle slopes with thick vegetal cover and steep slope with scant vegetal cover. The drainage system is highly related with above distinct geomorphological formations. The main tributaries of the Alaknanda like Dhauliganga, Nandakini, Pinder and Mandakini rising from high hills isolated watersheds contribute their waters to the Alaknanda forming subparallel drainage. Consequently the hill-side slopes have been transversely gullied and dissected. Several streams of the Alaknanda river contribute their waters to it through underground passage which activate the conditions for landslides. The Patal Ganga, the tributary of the Alaknanda presents conspicuous residuals of underground drainage near Belakuchi.

In addition to the above factors the temperate and subtemperate conditions of the Garhwal Himalayas activate weathering and under-cutting action of running and ground waters of the drainage. The average annual rainfall in Garhwal Himalayas, below 3000 m height is 132 cm. Above 3000 m height, snowfall occurs during winters. Thus the greater moisture content in soil and air as observed, the Garhwal Himalayas are found richly moistened causing landslides.

Genesis of Landslides

Mass wasting wherein a mass of rock or weathered debris moves down hill along discrete shear surface is defined as a slide. The general word landslide is frequently extended to include all rapid forms of low slide or fall because most of the movement in all three categories is along surfaces of separation. The slide may be subdivided (a) on the basis of type of material and (b) the nature of mass movement. The subcategories of landslide include: slump, rock slide, block slide, debris slide or debris avalanches.

Slumps which are rotational slips, are generally observed besides the concave-convex surface of rupture near the Main Central and Almora thrusts in the valley. Slumps generally occur in the thick, homogeneous cohesive meterial like clay. The surface plane along, which the slump takes place, constitutes spoon shaped structure. Its upper surface is generally found tilted backwards because the entire mass rotates as the lower portion moves outward down the slope. Slumps are mostly observed in the Mandakini and Dhauli valleys where the hillside slopes are highly moistened. The size of slumps varies from ten to hundreds of metres wide near the Central Thrust.

Rockslide

It is the small form of landslide. The movement remains relatively rapid. Rockslides occur where steeply dipping bedded strata exists parallel to the surface slope. They occur when a heavy rain or freezing and thawing provides fluid pressure or vibration breaks off obstructions and reduces the coefficient of friction on the glide plane and a detached slabor block slides down. It is generally observed in every crumpled, loose, fractured and powdered rocks strata in the Alaknanda Valley.

Debris slide

It generally includes the slip of the glacial debris or old collection of terminal morains. In the Dhauliganga Valley there a few localities of Malari, Lata, and Helang villages, which had been settled on the old glacial debris, the thickness of soil is comparatively greater and the spread of its weathered material forming fine and thick soil clearly indicates the slip of the debris of old terminal morains of the tributary glaciers. This phenomena prevails in the Alaknanda Valley.

Avalanches

The most rapid flowing sliding or falling mass movements are called avalanches. An avalanche usually begins with a free fall of a mass of rock or ice which is pulverized on impact and flows at great speed, made fluid by the pressure heated air and water entrapped in the mass. This phenomenon prevails in the Alaknanda Valley towards the present locations of the Glaciers and their environs. Avalanches are generally occurred in the glacial and pereglacial regions in the Garhwal Himalayas from 1000′ and above.

Genetic Classification of Landslides

Nature of movement	Nature of process			Type of movement		Type of material	
	With increasing ice content in rock and soil	With increasing water content in rock and soil	With increasing moisture content in rock and soil	Movement	Bed rock	Coarse grains	Fine grains
Slow			Slumping	Rotational	Rock slump	Debris slump	Earth slump
Rapid	Debris avalanches	Landslides	Debris slide Debris fall Rock slide	Trans-lational	Block slide Rock slide Rock fall	Debris slide Debris fall	Earth block slide

Factors Affecting Landslides

Slope as a Factor of Landsliding

Slope plays a dominant role to create gravity force for mass wasting processes like landsliding, sludging, slumping and soil creeping etc.[3]

The slope on the steep rock face called free face is influenced by frequent action of weathering process. The weathering of free face causes debris to form talus or scree at the base of the slope. The talus initially produces constant slope at the base of free face to form repose slope as observed along the Almora thrust and Main central thrust. The hill-side repose typed slopes of these thrusts are susceptible to landsliding. The highly crushed phyllites crumpled and powdered talcose rocks found along the thrust zone of the Garhwal Group consist of repose typed slopes near Belakuchi.[4] When the crushed phyllites and powdered talcose rock are converted into clay, may act as lubricant during the rainy season and landslides are resulted. Such types of land-slides generally occur in the Garhwal Group near Nandprayag and Karanprayag.

Landslides also occur when the base of talus (forming the toe of repose slope) exist besides the river valley and it is eroded by the under-cutting action[5] of river water during the flood period. This type of landslide occurred in Belakuchi and destroyed its settlement in the year 1970.

Excessive Human Interference

The valley is richly endowed with various natural resources like vegetation, minerals, soils and water even its geonomic condition is still poor because the said land resources are depleted by excessive erosion and frequent occurrence of mass wasting as a consequence of deforestation, overgrazing, unscientific blasting for construction of roads and dams. Deforestation which is unrationalised practice prevails on the hill side slopes to widen roads and construct new forest roads and dams. Present investigations state that the rock slips near Lata Reni and Helang villages in the Dhauliganga valley occurred in the year 1978 on account of the removal of vegetal cover for agricultural purposes.

Landslides are also caused due to[6] unscientific use of dynamites to blast the hills or hillocks to widen the highway and construct new roads. The dynamites of proper intensity are not used but they are used in a haphazard way with the result the remote hills are cracked and slide down during the rainy season as observed in the Bhapkund, Bhagyur and Pandukeshwar villages (Fig. 19.2).

The seasonal movement of floating livestock loosens the soil cover of the repose slopes some of which are made on debris of the old landslides as noticed near Chamoli and Chhingka where graziers have winter camps of the livestock. The graziers sometimes make harm to the forest by cutting main branches and twigs for fodder. They also graze their livestock on newly formed repose slope to loosen the soil cover consequently the slope is weakened and rock slips occur.

Land Depletion by Landslides

In the present study the area of the land, depleted by the landslides has been measured from aerial photograph, and the marked sites of landslides on photographs have also been confirmed in the field. Primary data for landslide investigations have been collected from 40 villages generally affected by landslides during rainy season annually. In the Garhwal Group, the depletion of agricultural and forest land caused by landslides ranges from 1 to 1.5 hectares per landslide. The density of landslides per km^2 varies from 7 to 10 observed in the villages of Maithiyana, Chamoli, Pipalkoti, Belakuchi, Pakhi and Helang, etc.

In the Central Crystalline Group, the depletion of agricultural and forest land ranges from 0.5 to 1 hectare per landslide. The density of landslide per km^2 varies from 3 to 6 in the villages of Pandukeshwar, Hanuman Chatti, Dhak, Tapovan, Reni, Jelam and Malari, etc.

In the Dudhatoli Group, the agricultural and forest area damaged by landslides is less than 0.5 hectare per landslide. The density of landslides per km^2 is lowest i.e., ranging from 1 to 3 in the villages of Kaliasor, Rudraprayag, Ratura, Ghultue, Langasu, etc.

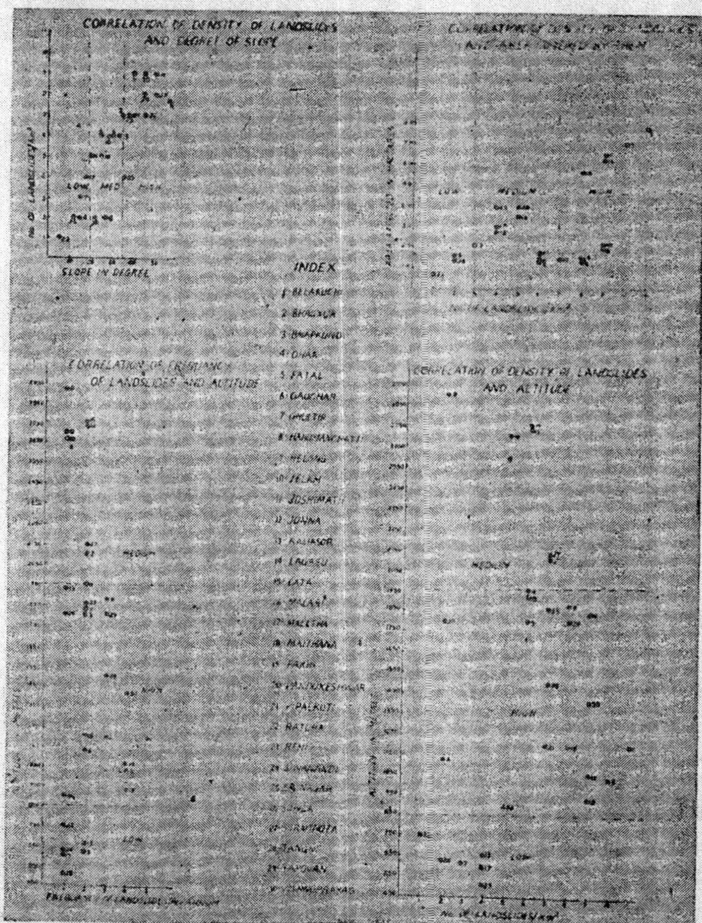

FIG. 19.2 : *Indices of Zoning of Landslides*

Zoning of Landslides

After studying the nature and degree of slope, intensity of landslides (frequency of landslides per year), density of landslides (frequency of landslides per sq. km of the selected villages) and agricultural land depletion (hectares per landslide per village) with the help of scatter diagrams aerial photographs, the author has reorganized the valley into the following landslide intensity zones (Fig. 19.2). The intensity of landslide has been evaluated by recording the frequency of landslides occurred per annum per unit area. The intensity has also been related with the density of landslides and land depletion (Fig. 19.2).

(I) The Vishnu-Nand-Garhwal Group High Intensity Zone.
(II) The Mana-Vishnu-Malari Central Crystalline Medium Intensity Zone.
(III) The Deo-Karan-Nand-Dudhasoli Low Intensity Zone.

I. *The Vishnu-Nand-Garhwal Group High Intensity Zone*—It has highest landslide density ranging from 7 to 10 landslides per sq km in the villages of Nandprayag, Mathiana, Chamoli, Pipalkoti, Pakhi, Tangni, Belakuchi, Langasu, Helang, etc. The intensity of landslides varies from 3 to 5 landslides per year in the villages of the zone. As regards the depletion of agricultural land caused by the landslides, it ranges from 1 to 1.5 hectares per slide.

From the above observations it is clear that the Zone from Nandprayag to Vishnuprayag is highly susceptible for landslide mud-flows and soil creeping. The zone consists of loose boulders, slate, phyllite and schist, which are generally crushed, faulted and fractured. Besides these the zone is also influenced by underground flow as exposed in the forms of springs, seepages, and the ground flow of 'Patalganga'. The zone is also depleted by the unscientific blasting for road excavation, deforestation over-grazing and bare-fallowing.

II. *In the Mana-Vishnu-Malari Central Crystalline Medium Intensity Zone*—Landslide density ranging from 3 to 6 landslides per sq km in the villages of Vishnuprayag, Pondukeshwar, Hanumanchatti, Dhak, Tapovan, Bhagyur, Reni, Lata, Suraitheta, Junna, Jelam, Bhapkund and Malari is observed. The intensity of landslides varies from 2 to 3 landslides per year in the villages of the zone. The depletion of agricultural land caused by the landslides ranges from 0.5 to 1 hectare per landslide. This zone consists of schist, gneiss, granite and loose boulders, causing medium intensity of landslides.

III. *The Deo-Karan-Nand Dudhatoli Low Intensity Zone*—It has lowest density of landslides ranging from 1 to 3 landslides per sq km in the settlements of Srinagar, Kaliasor, Rudraprayag, Ratura, Sivanandi, Chultir, Gauchar, Karanprayag, Sunla, and Langasu. The average intensity of landslides varies from 1 to 2 landslide per annum in the villages of the zone. The area damaged by each landslide is less than 0.5 hectare. The rocks are mainly phyllites and quartzites of Dudhatoli Group. The quartzites are hard and compact and phyllites are less crushed causing minimum number of landslides.

Preventive Measures

(i) High power explosives should not be used for road excavation.

(ii) Underground drainage system should be adopted for small channels and streamlets. There should be small drainage channel on the road side and water collected from such drainage should be joined to the underground system of drainage to minimise the roadside erosion.

(iii) Stout embankments should be constructed at the toe of slopes and on the upper side of the roads.

(iv) To protect the roads from the erosion, high embankments knitted with ironwires should be constructed.

(v) Large scale deforestation, over lapping and over-grazing and bare fallowing should be checked.

(vi) Unscientific farming on the valley slopes should also be prevented.

(vii) Afforestation and fruit gardens should be encouraged on the valley spurs and slopes.

(viii) The dry and perennial streams flowing across the roads should be allowed to drain through the cement pipes or iron pipes.

(ix) The long creepers and *Agave sisalana* should be grown on the bunds of the spur-terraces to protect them from erosion.

(x) The conservation measures for watershed management should also be brought into practice.

The above mentioned conservation measures are immediately needed to protect the High Intensity Zone which is running under huge depletion.

REFERENCES

1. A. Gansser, 1964 : *Geology of Himalayas*. Inter. Science Publishers, London, pp. 102-193.
2. R.S. Millar, 1970 : Geomorphic features in the catchment area of river Alaknanda between Joshimath and Chamoli, *Proc. of Seminar on River Valley Project*,Roorkee University.
3. L. Bloom, Arthur, *Geomorphology : A Systematic Analysis of Late Cenozoic Landforms*, pp. 163-193.
4. G. Kumar, 1970 : Alaknanda Tragedy : A geotectonic evaluation. *Proc. of Seminar on River Valley Projects*, Roorkee University.
5. Nitya Nand and Prasad, 1972 : Alaknanda Tragedy ; A geomorphological appraisal, *National Geog. Journal Varanasi*, Vol. 18 (34), pp. 206-12.
6. Personal investigations made at Nandprayag and Joshimath during field studies.

TWENTY

M. Uttara Prabhu and *G. Soundaravalli*

INSELBERG TOPOGRAPHY

INTRODUCTION

INSELBERGS, dotting parts of South India are very conspicuous. A region has been chosen to study and analyse, inselberg topography, which extends from 9°N to 11°N and 77°E to 78° 30′ E. This region covers a major part of Tamilnadu and a portion of Kerala to its west. The selected region, comprises a part of southern Coimbatore, western Ramanathapuram, western Tiruchirapalli, western Pudukottai, northern Tirunelveli and the whole of Madurai districts of Tamilnadu as well as the eastern parts of Ernakulam, Kottayam and Quilon districts of Kerala.

The region includes both mountainous terrain and the coastal plain, exhibiting climatic variations ranging between perhumid and arid conditions. The inselbergs found here present themselves in different forms like massive monoliths, weathered inselbergs, castle kopje and sugar loaf hills etc. yet they are collectively treated as inselbergs and studied accordingly since all of them are the products of the same geomorphic processes.

The region is divided into three major units, on the basis of the interactions of certain factors like relief, structure, lithology, climate, vegetation and the geomorphic evolution.

These three major units are (1) the Western ghats (2) Tamilnadu upland (3) Tamilnadu lowland (Prabhu, 1979) (Fig. 20.1).

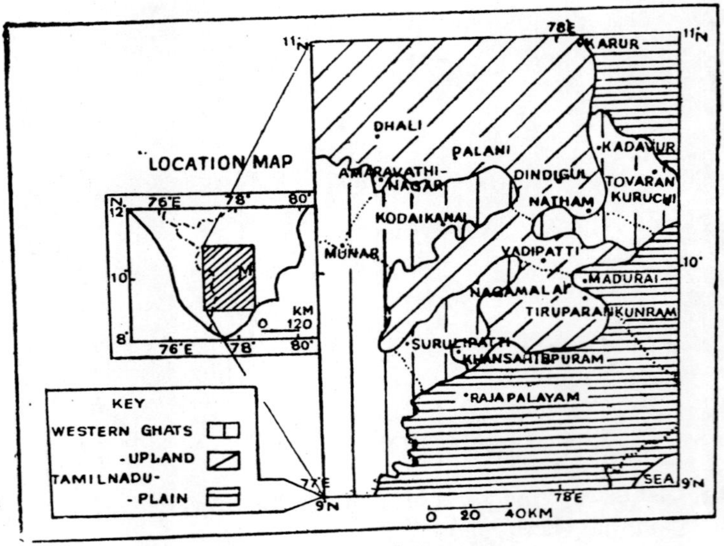

FIG. 20.1 *Land Systems*

TABLE 20.1 : **Three Major Units**

Sl. No.	Land System	% of the area	Altitude range (m)	Sub-Units
1.	Western ghats	30	900-1800	Anaimalais Palanimalai Periyar Plateau Varushanad and Andipatti hills, Sirumalais and Karandamalais.
2.	Tamilnadu upland	38	300-900	Kongunadu Plateau Cumbum and Vaigai Valley, Tiruman-galam plain.
3.	Tamilnadu lowland	32	100-300	Lower Cauvery plain Vaigai plain Southern plain, Coastal plain.

1. Steeply sloping side of an Inselberg-Khan Sahibpuram

2. Gneissic outcrop modified by weathering and water action—Thirumani ar Bednatham, Madurai

3. Rock flower by spheroidal weathering—Northern foot-hills, Anaimalais

4. Elephant skin appearance by ribbed weathering—Sanjeevi Malai, Rajapalayan

5. Compartmental weathering near Amaravati Nagar

6. Boulder field and weathered incelberg-South Eastern Kadavur

7. "Y" Joint in a Tor—Uttupuram, Vadipatti

8. A tor near Dhali, Udumal Pet

9. Rounded Blocks in a tor near Karur

10. Mattucherrukal (Tomb—Stone Topography) Aadavur

11. Overhanging rhombohedral Quartzite—Nagamalai, Madurai

12. Minor fiuting and exfoliation scar—tail end of Yanaimalai, Madurai

13. Quartzite over granite—Touaran Kuruchi

14. Core stone occurrence—Sanjeevimalai Rajapalayam

15. Sugal loaf hill—Tiruparankunram hill

MORPHOGENIC ENVIRONMENT

Stratigraphically, the whole region is covered with Precambrian rocks, igneous, sedimentary and metamorphic. Granitic gneiss, comprising ortho, para and injection gneisses, together with crystalline rocks have suffered widespread folding, faulting and intrusion. The archean metamorphics include charnockites and the associated hybrid gneisses and migmatites.

These crystalline rocks are regrouped into five major types (Tricart, 1972). Weathering plays a decisive role in the ultimate form these rocks exhibit.

The Acidic Rocks

These include granites (pink and grey younger intrusions), granitic-gneiss and acidic charnockites. The degree of weathering depends on its texture and the presence of joints. The coarse crystalline rocks, principally granites and gneisses tend to break down in two stages. Firstly the joints are weathered and piles of boulders are formed. Secondly the rocks are attacked chemically in one of their main minerals so that they lose their strength and the rest fall as a mass of granular debris composed largely of the unweathered minerals.

Acidic charnockites occur along with bands of basic metamorphic types in Palanimalais. The intermediate type associated with a few bands of garnetiferous gneiss, occurs to the west of Sivagiri, Puliangudi and in Varushanad hills. In other sections of the region under study, charnockite occurs as thin bands. The granites having a set of strong joints, occur over Sirumalai area. Granite gneiss is found in the Cumbum rift valley while to the north of Kadavur, the granitic gneiss is confined to the plain. Small bodies of pink granites are noticed in the biotite hornblende gneiss in the form of bands, lenses and islands over the Kongunadu plateau.

The Basic Types

These include basic igneous rocks (gabbro) and pyroxene gneiss. Amphibolites, hornblende-biotite and pyroxene granulite of the region come under this group. The Anaimalais, Palanimalais, Varushanad hills, Sirumalais and Periyar Plateau have these rocks but in varied proportions. Leptynites occur south of Kukkal and in the northern section of Kodaikanal.

These absorb very little quantity of water, resulting in a less regular drainage. Besides the region has very little rainfall during the dry season, which is reflected in sparse vegetation and in weathering. Pyroxene granulite contains high content of ferrumagnesium minerals. Whenever the dry season is long, these have a tendency to develop into ferrugenous plateaux or cuirasses.

The Siliceous Group

Quartzites in the area fall into this category. Since quartzites are recrystallised by metamorphic processes, these represent the most resistant rock known. Quartzites occur as their bands occasionally as thick beds extending over a few kilometres.

The Calcareous Group

Calc-gneiss, crystalline limestone etc. are grouped in this. Often these have more water in circulation and they are removable even in semi-arid area, during moist season.

Climate

The variations in tropical climate are noticeable in the region. According to Ramamurthy (1976) it can be classified into (1) Tropical maritime (monsoon), (2) Tropical interior (semi-arid) and (3) Tropical mountain type (perhumid). The average monthly temperature is above 20°C, the highest occurring in May (Coimbatore 27°C; Madurai 30°C). Towards the interior, the diurnal range is higher. Kodaikanal, a representative station for the mountain section, experienced minimum temperature range, during S.W. monsoon and maximum in the winter season.

The rainfall (more than 60%) is concentrated during October and November. Where the rainfall range is 700-800 mm, the potential evapotranspiration is uniformly high. The maximum mean annual actual evapotranspiration occurs at Madurai for a radius of 50 km. The mean annual water deficiency is more than 800 mm over a greater part of the region to the south and the east.

TABLE 20.2: **Climate and Vegetation**

Sl. No.	Area	Vegetation	Rainfall (mm)	Dry period
1.	Western Ghats	Mountain evergreen forest	1500	Leaving one month dry
2.	Periyar Plateau and Kottaimalai section	Semi evergreen	1500	1-2 months dry
3.	Higher elevation and windward sides of Anaimalais, Palanimalais and S.W. part of Periyar Plateau	North tropical deciduous.	1000-2000	4-6 months dry
4.	Eastern sides of the Western ghats, Eastern Varushanad hills, Srivillipu-thur Sivagiri hills.	Dry deciduous	600-1100	More than six months dry
5.	Lower slopes of northern Planimalais, S.W. slopes of Varushanad, Puliangudi-Puliyara Section of Periyar scarp.	Savana woodlands	600-800	6-8 months dry
6.	Tamilnadu upland section, secondary growth in deforested sections of Anaimalais and Palanimalais.	Thorny scrub	600-800	8-9 months dry
7.	Karandaimalai, Western and Southern slopes of Kadavur.	Open scrub woodlands.	600-800	more than 9 months dry
8.	Greater part of the study area.	Dry Xerophytics	500-600	8-9 months dry
9.	Coastal Section.	Saline forest	—	—
			(mud flats and lagoons)	

Vegetation

The Western ghats section is covered with intertropical forest and the eastern lowland section with Savannah. Coimbatore, Tiruchirapalli and Madurai experience 'bixeric' regime with two dry periods while Kodaikanal has 'axeric' regime without any dry season (Meher Homji, 1975).

MORPHOGENIC PROCESSES

The effectiveness of the various agents was calculated for the study region, using the methods of Peltier (1950) and Leopold *et al.*, (1964) where precipitation and temperature for 4 representative stations were taken into account.

TABLE : 20.3 : **Effectiveness of the Various Denuding Agents**

Station	Annual R.F. in inches	Temperature	Weathering Chemical	Region weathering	Wind action	Mass Movement	Fluvial Erosion
Coimbatore	24.1	76.6	Weak	Very slight weathering	Moderate	Minimum	Moderate
Tiruchirapalli	34.2	80.6	Moderate	Moderate chemical weathering	,,	Moderate	Maximum
Madurai	35.6	81.5	,,	,,	,,	,,	,,
Kodaikanal	65.8	57.0	Strong	Strong chemical weathering	Minimum	Minimum	Moderate

Coimbatore, located on the Kongunadu plateau at the south-eastern part of the Palghat Gap has moderate wind action and fluvial erosion however, the dominant process is mechanical weathering. Tiruchirapalli located on the Cauvery plain to the east has the maximum fluvial erosion and wind is partly responsible in sculpturing the landforms. Madurai, in the eastern-central part, representing the southern plain, has wind and running water as the dominant agents. Kodaikanal, representing the mountain section, is located in the upper Palanimalais where strong chemical weathering occurs. The climate as related to weathering is expressed here as a weathering statistic as per Meyers formula. Coimbatore, Tiruchirapalli and Madurai have the values ranging from 2.1 to 3.1. This covers most of the basic and acidic rock terrain which get weathered into montmorillonite and Kaolinite respectively. Part of Kodaikanal area is laterized. However, this region generally lies within the approximate limit of red weathering crusts.

INSELBERGS

The result of the above processes stands out as inselbergs of different types in the region under study. Generally they are rocky residual hills or uplands, rising in isolation, above a uniform plain. These maintain steep angle at their junction with pediment. This angle varies from 20°—35°. It appears as though the inselbergs owe their forms to the combined actions of weathering and runoff (Plates 1 and 2).

The Western ghats and the Mountain Section

Under the thick canopy of evergreen and semi evergreen forest, daily temperature variations do not penetrate deeper into the ground except at few centimetre deep (10-15 cm). Further down, the soil temperature remains constant. The bald peaks represent exfoliation domes and exfoliation scars are seen in some of the perched boulders. The weathering is deep extending as far down as 15 m.

Inselbergs are massive in appearance whenever weathered amphibolites are encountered. This may be due to the fact that these rocks are highly susceptible to spheroidal weathering. Minor forms of inselbergs are seen wherever the rocks vary in composition. On the northern foothills of Anaimalais, one such minor form exists in gneiss resembling a flower (Plate 3). When calc granulite is weathered, another minor form of monolithic dome occurs. The surface exhibits ribbed weathering, resembling elephant skin in appearance (Plate 4). These forms are common in Palanimalai and in Varushanad hills.

The outliers of these ridges as in the Periyar Plateau section, show extensive stripping giving rise to crocodile skin surface.

The mountain section on the eastern side is marked by abrupt scarp face, which separates another zone of inselbergs at the foothills. Probably in this place, scarp retreat may be closely related to the inselberg evolution. The processes, responsible for the above tendency are highly disputable. Sheet floods and lateral stream planation along with weathering play equal role in the formation of inselbergs.

Tamilnadu Upland Section

The semi-arid conditions of this section are responsible for discontinuous and yet intense chemical weathering. The rocks disintegrate more deeply. Weathering is irregular and rapid.

The upper surfaces of joint blocks tend to be more weathered than the lower surfaces. Joint spacing is responsible for compartmental weathering (Twidale, 1976) where regular weathering front remains as the residual hills. Near Amaravathinagar (Plate 5) a ridge shows the signs of block disintegration on the upper slopes.

Along a cart track, in Kadavur section, Calc-granulite appears to be weathered into white or puff coloured powdery substance but feldspathic gneiss stands out as an inselberg (Plate 6). The foothill is strewn with pebbles and angular boulders. Resistant rocks thus stand out as residual hills.

Role of jointing in accelerating weathering granite as well as in gneiss is unmistakable. Tors are well developed in these rocks eventhough there is a disputable question whether inselberg is related to tors. Near Uttupuram, in Vadipatti, angular blocks, broken along Y-joints appear as castle kopje (Plate 7).

Quartzites are the least susceptible to weathering. These ridges wherever they are, appear as the residual hills. In Cumbum-Vaigai Valley, the boulders are rounded, mainly due to spheroidal weathering.

Tamilnadu Low Land

The inselbergs are easily distinguishable from the adjacent lowland and valley section. These residual hills are bare and domal in appearance. The denudation of more fragile materials in which the resistant core is enclosed may be the cause of such forms. Since the inselbergs and the plain represent two different levels, these are sometimes considered as parts of exhumed topography. Kongunadu Plateau, being considered as the oldest planation surface initiated during Jurassic period supports the view of double planation theory of Budel, to explain the origin of these inselbergs. Dindignal and Thanthonimalai represent monolithic domes as exhumed surfaces. A minor form on Thanthonimalai is the predominance of weathering pits, aligned to the joints.

Tors dot the landscape over the area extending between Udumalpet-Karur towns. Tors sometimes appear as the weathered inselberg surface while at places, they stand out isolated like inselbergs themselves (Plate 8). Tors have rounded blocks as in Karur (Plate 9) where the rounding of the joint masses of tors could be regarded as the modification imposed subareally on the original pattern of unrounded blocks.

When gneissic rock interbanded with calcareous rocks, have steep dipping joints, the weathering is accentuated along the joints. The residual stands out steeply dipping towards the surface, giving rise to tomb stone topography, locally termed as Mattucherrukal (Plate 10). The boulders are strewn over the adjacent plain.

The granitic gneissic terrain has monolithic domes, appearing to be detached from the adjacent hillocks. These break the monotony of the plain as in Tirumangalam. Exfoliation and spheroidal weathering play a dual role in sculpturing. In Tiruparamkunram the vertical joints in the rocks enable the formation of angular tors while spheroidal boulders are seen piled up with window like opening between them. Accumulation of talus at foothill is a clear evidence to show the degree of weathering.

Quartzite outcrops mostly display orthogonal joints, resisting weathering. Yet signs of hydrations are evidenced along with joints. Prominent tors with angular blocks are common in Nagamalai section. An overhanging rhombohedral quartzite block on a tor is a minor form, the origin of which is partly due to weathering and partly due to wind action (Plate 11).

Anamalai is a monolithic dome, 3 km long, 50 m broad and 343 m high towards its head section (Plate 12). The sides of this hill display prominent flares. On the top and on the slopes, minor fluting and exfoliation scars are encountered respectively. The older exfoliation scars reflect the joint pattern.

Quartzite (Plate 13) over granite exhibit steeply sloping inselberg while the slopes are gentle, smooth and covered with thorny scrub vegetation, as in Tovarankuruchi.

Sometimes, along the slopes, inselbergs are highly weathered with indications of laterization, where core stones are exposed as in Siranjeevimalai near Rajapalayam (Plate 14).

Sugar Leaf Hills

These correspond to a less jointed mass in the midst of more densely jointed rocks. These forms are restricted to parts of inselbergs in Kandagirimalai, 16 km. from Palani and in Tiruparankunram hill (Plate 15).

It is very difficult to trace the origin and evolution of various types of inselbergs over an old peneplaned surface like the one in the study region of Tamilnadu. The role of petrological variations, structural differences, as well as climatic and biotic factors is evident. Eventhough inselbergs are scattered in all the units of land system. Tamilnadu upland and lowland section seem to be the zone of inselberg dominance. Delineation at micro-level is not possible at this stage. This needs intense and closer examination.

SUGGESTED READINGS

Leopold, L.B. et al., 1964 : *Fluvial Processes in Geomorphology* (reprint).

Meher Homji, V.M., 1975 : Ecological climatology and Vegetation. *Plant Ecology-Cartography*, 26-42.

Peltier, L.C., 1950 : The geographical cycle in periglacial regions as it is related to climatic Geomorphology. *Annls. Asso. Ame. Geog.* Vol. 40, 214-36.

Prabhu, M. Uttara, 1979 : Some geomorphic forms between 9° and 11° N latitude in South India, *U.G.C. Project*. (unpublished).

Ramamurthy, K.M., 1976 : Climate of Tamilnadu. *I.G.J. Golden Jubilee Souvenir*, 1-10.

Tricart, J., 1972 : *Landforms of the Humid Tropics Forests and Savennas*. English Edition—Longman.

Twidale, C.R., 1976 : *Analysis of Landforms*—Wiley.

TWENTY ONE

Satya Prakash and *P. B. Saxena*

MORPHOLOGY OF THE SUBMARINE TERRAIN OF THE NORTH INDIAN OCEAN AND ITS GEONOMIC POTENTIALITY

INTRODUCTION

THE INDIAN OCEAN is the most complex of all the world oceans. To the west and east of Indian Ocean lies the relatively stable continents of Africa and Australia, separated from Antarctica by the wide Southern Ocean. To the north it is surrounded by the active plate boundary of the Gulf off Aden Red Sea complex, the Himalayan region and Sumatra and Java trenches. The Indian Ocean has features which are peculiar as well as it has large assortment of features which are found in other oceans. It offers an interesting study in the field of submarine geomorphology and geonomy. Moreover, due to its geostrategy and geonomic potentiality, Indian Ocean has become an arena of power rivalry among the super world powers. In this paper, the authors have concentrated their study on the Northern part of Indian Ocean which is endowed with sedimentary geosynclines of Indus and Ganges Cones containing physio-bio-chemical treasures in the form of hydrocarbons for petroleum; manganese nodules rich in manganese, nickel, copper and cobalt.

BRIEF HISTORY OF DATA ACQUISITION AND RESEARCH IN INDIAN OCEAN

The study of geomorphology of submarine terrain can be done with the help of bathymetric and seismic data. Early studies of the Indian Ocean by early explorers and surveyors were limited to the discovery of new places and obtaining nearshore bathymetry for charting, navigation, fishing and trading. Early bathymetric data was collected by East India Company 1601 to 1862 Bombay Marine (1742 to 1832), former Indian Navy (1832-1862) and the Marine Survey of India (1874-1952). The 'naturalist' on board of surveying ships used to collect information on flora and fauna other interesting natural phenomena. The nearly cruises on which geological and geophysical work was done were those of Dana (1928-30), Snellius (1929-30), Mabaliss (1933-34), Challenger (1949-50), Albatros (1950-52), Ob (1955-57). The results were published by Wiseman, Seymour and Swell (1939), Fairbridge (1948 and 1955), Gutenburg and Richer (1954).

The first really substantial investigation of the Indian Ocean came with International Indian Ocean Expedition (IIOE-1959 to 1966) when 14 countries with their 46 research vessels participated in Marine Research. A compilation of IIOE data has been published in a digestable Atlas form—*The Geological-Geophysical Atlas of Indian Ocean*, edited by Udintsev and published by Academy of Sciences USSR (1975). As an outcome of evaluation of IIOE data, geomorphological concept of Arabian Sea and the Bay of Bengal was further enhanced by the publication of research paper on Sea Floor Spreading (Vine and Matthews, 1963), physiographic diagram of Indian Ocean (Heezen and Tharp, 1965), Thickness and Structure of Sediments (Nepro chnov and Kholopov, 1969) in Bengal deep sea floor and geosyncline (Curray and Moore, 1971) and general studies by Laughton, Matthews, Fisher, Sclater, Mckenzie, Heyvs, Luyendyk, Simpson, etc.

The accomplishments of IIOE led to specific expedition like drilling programme of Challenger (1971-72) known as Deep Sea Drilling Project (DSDP). Challenger drilled 72 holes during 7 cruises and synthesis and analysis of data has been published in Indian Ocean-Geology and Geostratigraphy by Heirtzler, Bolli, Davies, Saunders, Sclater. Apart from above Lamont Doherty Geological observatory, USA, and Scripps Institution of Oceanography have also carried out research on geomorphology and Marine resources of Indian Ocean. Many research papers have been presented in several conferences.

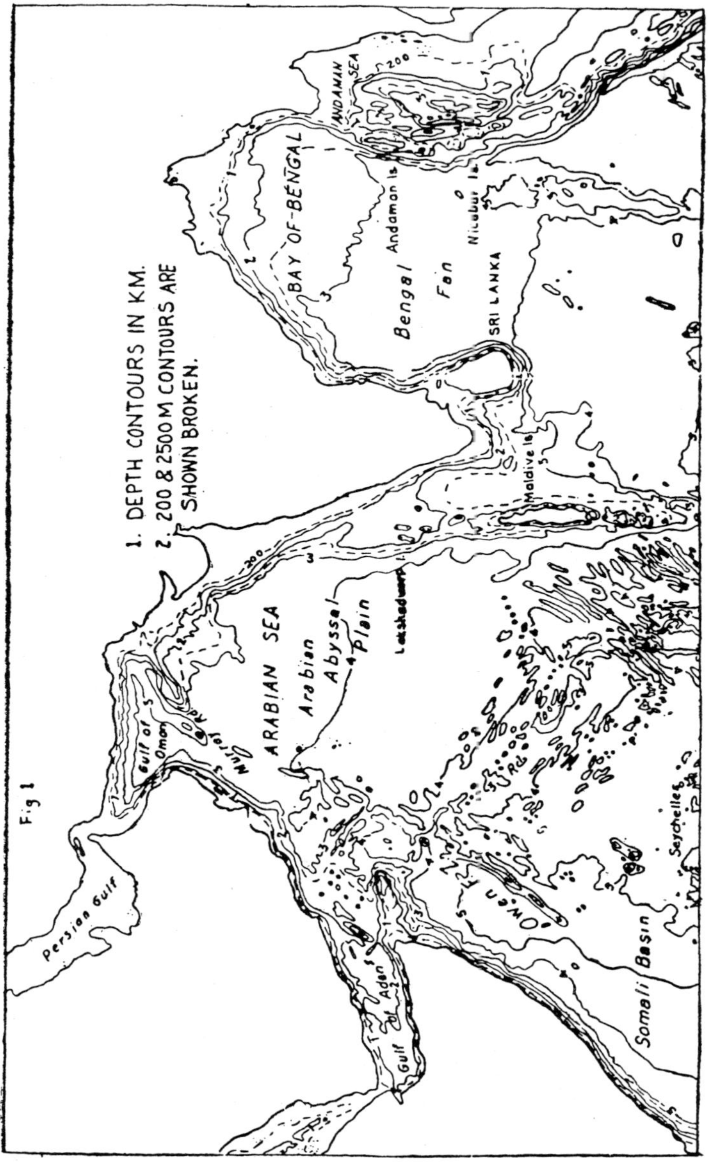

FIG. 21.1 : *Bathymetric Chart of the N. Indian Ocean*

In Indian, institutions like Oil and Natural Gas Commission (O.N.G.C.), National Institute of Oceanography, Goa (N.I.O.), Naval Physical and Oceanographic Laboratory, Cochin (N.P.O.L.), Naval Hydrographic Office, Dehra Dun (N.H.O.) Geological Survey of India, Physical Research Laboratory, Ahmedabad (P.R.L.) are actively engaged in acquisition and analysis of Bathymetric/Oceanographic/Geophysical data. The ONGC has carried out tremendous geophysical and seismic observations and drilling on the continental shelf regions. The NHO has collected vast bathymetric and oceanographic data of North Indian Ocean particularly through the programme of General Bathymetric Charts of Oceans under International Hydrographic Organisation.

THE ORIGIN AND TECTONIC SYNTHESIS OF INDIAN OCEAN

The origin and evolution of Indian Ocean can best be proved with the help of the theory of Continental Drift which is now further elaborately explained by the latest theory of Plate Tectonics. The geoscientists have been re-assembling the Jigsaw pieces and have been estimating the timing of drift, the direction of tectonic forces and the origin of major physiographic structures of Indian Ocean. However, it will be worth to review various opinions before arriving at latest and modern viewpoint. Carey (1957) developed the idea of 'Continental Blocks' that rotate around hinge points and open up 'Chasms'. The hypothesis of 'Sea Floor Spreading' was propounded by Dietz (1961), Hess (1962), Wilson (1963) and Vine and Matthews (1963) with the help of magnetic Anomaly records. Heezen and Tharp laid emphasis on Mid Oceanic Ridge Systems and Fracture Zones. Le Pichen and Heirtzler (1968) drew isochronous lines to analyse magnetic anomaly and concluded that initial separation of Pangean Continents took place in Lower Cretaceous (140 M.Y.). McElhinney (1970) with the help of paleomagnetic data, concluded that initial movements in Gondwana land began during Upper Permian but break up took place in Upper Cretaceous. Curray and Moore (1974) analysed Seismic reflection profiles, gravity profiles and modified Smith and Hallam's (1970) reconstruction by enlarging

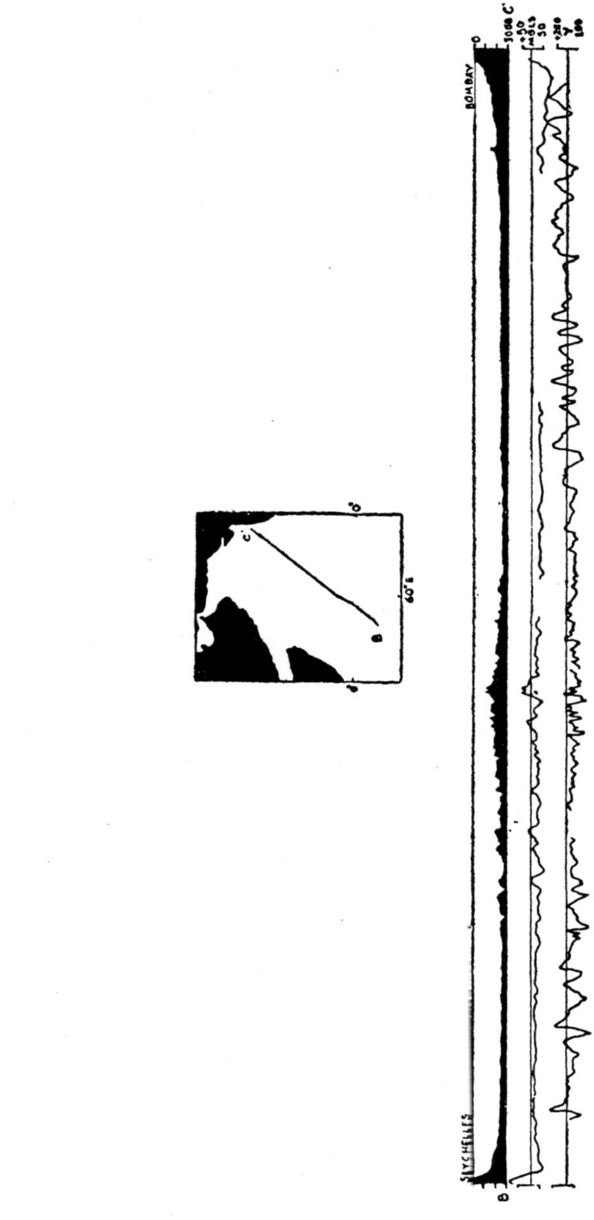

FIG. 21.1(a) : *Profiles of Rathymetry, Gravity and Magnetics*

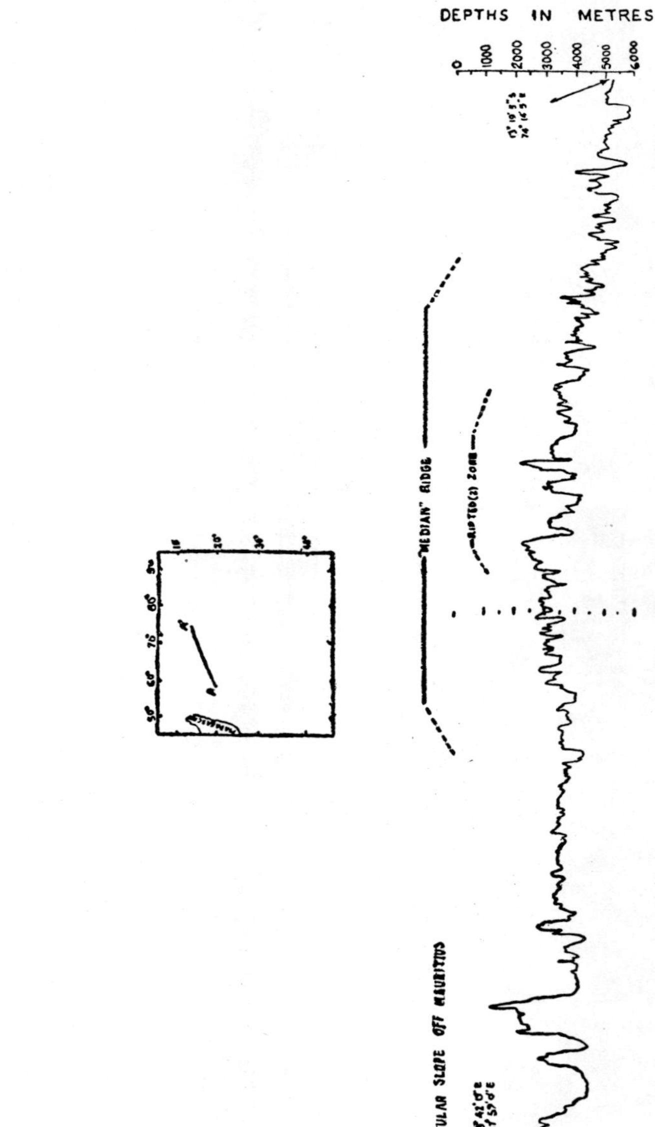

FIG. 21.1(b) : *Topographic profiles on the median ridge, Central Indian Ocean*

Gondwanic India. According to them there were 3 hot spots which joined arms to form the rift between plates and early break-up took place, in early Cretaceous. Working backwards in time and terminating in Late cretaceous, Mckenzie and Sclater (1971) identified three phases of spreading; (a) Present to Oligocene (36 m.y.) when opening of Red Sea and Gulf off Aden took place, uplift and denudation of Himalayas started; (b) Eocene to Paleocene (36 m.y. to 53 m.y.) when separation of Australia from Antarctica took place; (c) Paleocene to Late Cretaceous (53 m.y. to 70 m.y.) when northward movement of Indian Plate took place.

The DSDP results have generally supported above periods of movements with slight modifications. In early Cretaceous, Africa and S. America Split away from Madagascar, India, Australia and Antarctica. In Mid Cretaceous, Africa picks up Madagascar and India separates from Antarctica. In 533 m.y. Australia separates from Antarctica. In late Eocene and early oligocene there is a re-orientation in spreading direction, the changes Fracture Zone opens, the Ninety East Ridge moves from S.E. Indian Ridge moves, S.W. Indian Ridge is created. Major change in marine morphology occurred at the onset of rapid uplift of Himalayas.

PRINCIPAL HYPSOMORPHIC REGIONS OF NORTH INDIAN OCEAN, THEIR MORPHOLOGICAL AND STRUCTURAL CHARACTERISTICS

In order to classify and demarcate the morphological features into Hypsomorphic Regions, the authors have constructed a 'Three dimensional model' using 'Contour layer and Tint Method' with the help of bathymetric data and sounding profiles.

The authors have identified the following Hypsomorphic Regions in the various types of morphological features found in North Indian Ocean :

The Continental Margins

The Continental Margins are the outer submerged edges of the continental blocks which protrude into the sea and generally consist of continental shelf, slope and rise. These

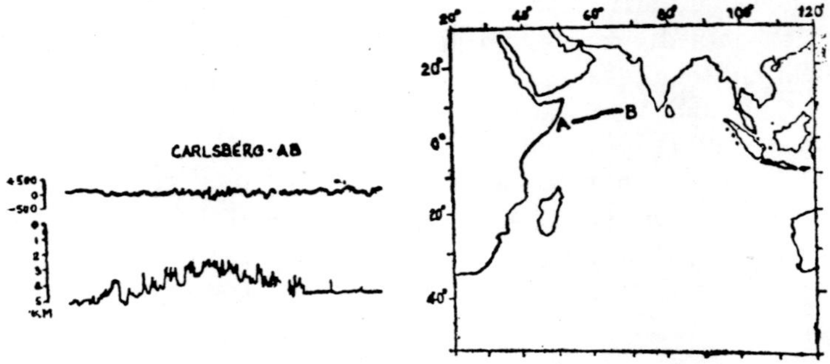

FIG. 21.1(c) : *Bathymetric and Magnetic Profiles across Carlesberg Ridge.*

are the transition zones where the continental crust thins out and changes to an oceanic crust. Some margins are therefore marked by volcanism while some are quiescent and are filled up at places with trough like structures. The continental margins of N. Indian Ocean are somewhat narrower on average except off N.W. Indian subcontinental, off Ganges delta, Gulf off Martban and Eastern side of Andaman Sea. The shape of the continental margins have to some extent been influenced by the sedimentation by rivers. The relief on the shelves is extremely subdued and consists largely of submerged features of former shoreline. The continental margins of N. Indian Ocean can be grouped into 3 main regions :

Indo-African Continental Margins

Margins off East Africa lack distinctive shelf, slope and rise sequence due to thick cover of sediments (9 to 19 km). Off Somali margins are narrower (5 to 15 Sea miles). Arabian and Iranian Continental Margins are also extremely narrow (10 to 30 Sea miles) but off Cape Monze, they gradually widen to 150 Sea miles and continue so up to Kutch. The Kathiawar margins are 75 to 100 Sea miles wide where Deccan traps are extending into Shelf which is overlain by thick Tertiary sediments. Off Cambay and Bombay, margins widen to more than 200 Sea miles and form Cambay Basin where sediments up to 2500 metres are found overlying Deccan Traps. The surface rocks consist of volcanic lava which is more acidic. More basaltic of Deccan Trap are found off West Coast of India overlying the metamorphic and igneous rocks of the Archean's Shield. Further South, we find Pleistocent and recent sediments in narrow belts. At Southern tip of Indian Peninsula the margins are 55 Sea miles wide and consist of Archean rocks with thick cover of Tertiary sediments. In Gulf off Mannar margins are narrower. Large reserves of Petroleum lie in Persian Gulf which is at present the major Petroleum producing region of this area. Oil has been discovered in Bombay High region.

Indo-Burman Continental Margins

Sri Lanka Margins are very narrow (10 to 30 Sea miles) which consists of Cretaceous and Tertiary rocks overlain by Pleistocene and Recent sediments cut off at places by numerous

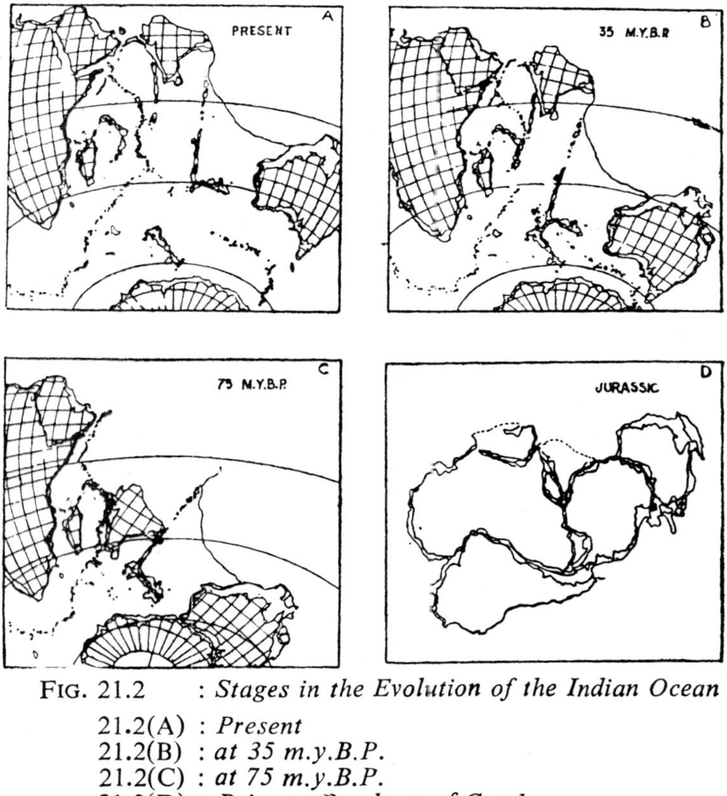

FIG. 21.2 : *Stages in the Evolution of the Indian Ocean*

21.2(A) : *Present*
21.2(B) : *at 35 m.y.B.P.*
21.2(C) : *at 75 m.y.B.P.*
21.2(D) : *Prior to Break-up of Gondwana*

canyons. On east coast of India margins are very narrow (25 to 40 Sea miles) but the rivers Krishna, Godawari and Mahanadi have deposited thick layers of recent sediments have been cut at places by numerous canyons. The margins again widen from 75 to 120 Sea miles. Off Ganges delta containing Cretaceous and Tertiary Sediments of about 6 km thick lying over basement of lava. The Arakan Coast margins again narrow down (20 to 40 Sea miles) where Tertiary sediments bordered by Cretaceous sediments are found. Large amount of Petroleum have been predicted in Deltaic Margins of Krishna, Godavari, Mahanadi and Ganges. Oil has been discovered in Godavari Delta Margins.

Thai-Burma Andaman Continental Margins

The Martban Margins are 100 to 150 Sea miles wide and run straight east to west and turn to south due to seaward extension of Burmese volcanic trend and gradually narrow down to 30 miles South of Mergue. The Malacca Strait is quite broad. The dominant structures features in Andaman Sea are related to Indonesian Arc which consists of primary double arc, the inner volcanic trend and the outer sedimantary arc on which lies Andaman and Nicobar Islands. Considerable Petroleum is expected to be explored in Margins around Andaman Sea. Oil has been discovered in Andaman Islands.

Submarine Canyons and Deep Sea Cones or Fans

Abyssal cones are found associated with Ganges and Indus rivers. The cones postdate the first collision of India and Asia and uplift of ancestral Himalayas at the end of Paleocene. These contain uppermost geosynclinal pile of sediments up to 4 km. Underlying the cones are continental rise sediments up to 12 km thick which extend into Indus and Ganges Valleys, deposited off the margin of India following its separation from Antarctica and Australia in Cretaceous. The Channels in these cones are tens of kilometers wide and hundreds of metres deep and are bounded by natural levees which terminate abruptly about 450 km downslope from the Indus and Ganges Canyons mouths.

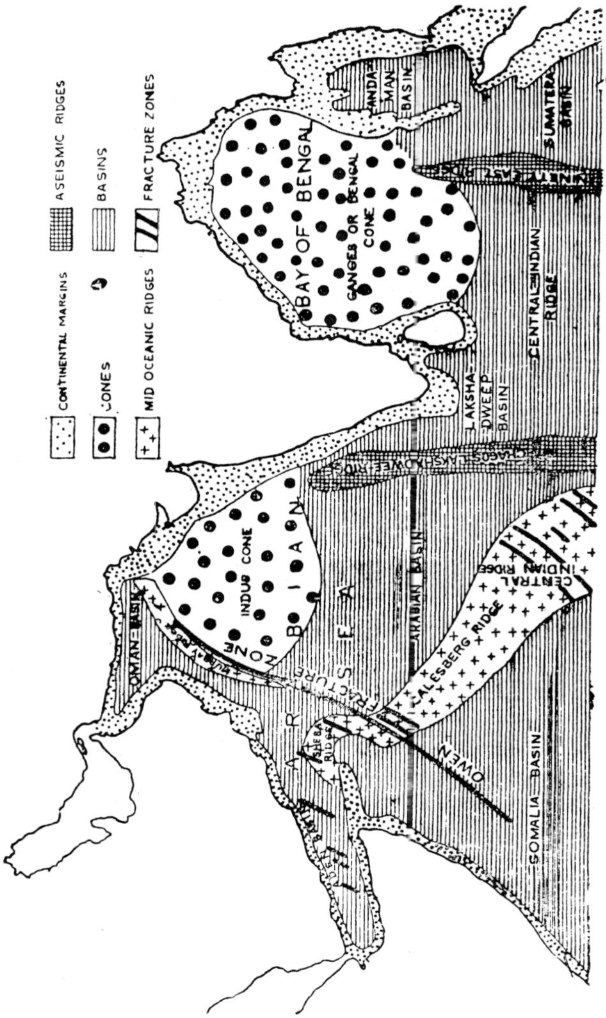

FIG. 21.3 : *Principal Hypsomorphic Regions of North Indian Ocean*

Indus Cone

It extends approximately from 25°N to 10°N, with a total length of about 900 Sea miles with a width of about 600 Sea miles. Sediments have been derived predominantly from drainage of the Indus, some of which drains through the submarine canyon—'The Swatch' and are distributed by turbidity, currents through a system of channels. Oil bearing structures expected under these thick sediments.

Ganges Cone

It extends from 20°N to 7°S, with a total length of 1600 Sea miles with a width of 600 to 800 Sea miles. Sediments have been derived from drainage of the Ganges and Brahmaputra added by Krishna, Godavari and Mahanadi. The enormous sediments enter the submarine canyons lying off these rivers and are distributed by turbidity currents through a system of channels. Huge oil bearing structures are expected to lie under these thick sediments.

Mid Oceanic Ridge System

Central Indian Ridge

The most conspicuous feature in the hypsometry of North Indian Ocean is the Central Indian Ridge that lies in the centre of Indian Ocean around 70°E longitude and which is a section of continuous global feature of World Midoceanic Ridge System encircling the Plate boundaries. It is characterised by seismic activities, numerous earthquake epicentres and fracture Zones. In addition to abundant basalt, gabbro, green schist, serpentine and peridoite also crop out on this ridge. Along the ridge crest, unconsolidated sediments are generally absent or thin patchy.

Carlesberg Ridge

The name was given to that part of Central Indian Ridge which turns to North West from Equator.

Geophysical surveys (Owen 1961-63) have proved that linear magnetic anomaly pattern arose from a combination of Sea Floor Spreading and reversals in earth's magnetic field. Owen surveys showed that the ridge did not continue into

Socotra, but was displaced northward some 300 km by Owen Fracture Zone which runs between Karachi and Somali Basin. The Sea mount chain on the ridge are essentially linear and parallel to the axis of the ridge.

Sheba Ridge

It is a North-Westward continuation of Carlsberg ridge after a displacement along Owen Fracture Zone.

Murray Ridge

It extends from Karachi towards SW for about 400 Sea miles. It seems to be a continuation of Kirthar mountain system. However, its link as continuation of Central Indian Ridge can also not be ruled out.

The mid oceanic ridges are important features for tectonic evolution of Indian Ocean but are without oil potential.

Aseismic Ridge Systems

Characteristic of Indian Ocean is the abundance of relatively shallow ridges that are free from earthquake activities and show considerable elongation in North-South direction, These meridional highs resemble tram tracks fancifully marking the drift of Indian Plate after sundering of Gondwanaland.

Chagos-Lakshadweep Ridge

It comprises of three major Groups of Coral atolls that cap shallow, steep-sided plateau. These lie on a ridge some 2700 km long that gently slopes from the base of the plateaus at 2000 metres to abyssal depths of 4000 metres. The atolls show evidence of growth on steadily subsiding platform. Beneath the coral lies a feature of volcanic origin which can be associated with Deccan Traps of India and would have originated in early or middle Cretaceous to early Eocene. The ridge is also linked in north with outer edge of Kori Bombay High, though covered with thick sediments.

Ninety East Ridge

It consists of a single ridge about 200 km wide, in places flat-topped and other places fractured and mountainous, that stretches 4800 km as northsouth. It runs along 90°E longitude and slowly disappears beneath the sediments of Bengal fan system near 9° North. Depth over the ridge from 1800 to 3000 metres. The ridge is unrelated to Andaman Arc.

Both the ridges are very important features for tectonic evolution of the Indian Ocean but are without oil potential.

The Ocean Basins

The Indian Ocean is divided into a number of isolated Ocean basins by the Oceanic ridges. These are flattest surfaces with gradient from 1:1000 to 1:7000 except for isolated peaks, hills and mid ocean ridges. Crustal downwarping has taken place beneath the Indus Cone and Ganges Cone. These sediments have been derived from erosion of the Himalayas with a thickness from 1 to 3.5 km in Arabian Basin and from 1 to 6 km in Central Indian. The Basins are most interesting areas from the point of view of geologist because of hydrocarbons and manganese nodules containing high grade minerals. The following Basins lie in North Indian Ocean.

Arabian Basin

It is more than 4000 metres deep part of Arabian Sea which is bordered in north by Indus Cone and in south by Carlsberg Ridge. In its west lies Sheba Ridge and in its east lies Chagos-Lakshadweep Ridge. It has lesser sedimentation (1 to 2 km) but occurrence of manganese nodules containing rich mineral concentrates have been identified in this Basin.

Central Indian Basin

It lies South of Sri Lanka and Ganges cones and is bordered in west by Chagos-Lakshadweep Ridge and Central Indian Ridge and in east by Ninety East Ridge. Depths are more than 4000 to 5250 metres. It has only 1 to 2 km sedimentary thickness but occurrence of manganese nodules containing rich mineral concentrates have been identified in this Basin.

Lakshadweep Basin

It is only a 2000 to 3000 metre deep Basin lying between Chagos Lakshadweep Ridge in west and south Indian continental margins in east. It has 2 to 3 km sedimentary thickness and has some petrolium potential.

Andaman Basin

It is an enclosed basin surrounded by Andaman Nicobar in west, Thai-Burman continental Margins in north and east and Sumatra in south. It separates two Tertiary Basins. It has depth from 2000 to 4000 metres with sedimentary thickness up to 1.5 km. Petroleum potential is expected in this Basin as well.

Somali Basin

It extends from Socotra Island in north to Seychelles in south and Somali in west and Calsberg Ridge in east. The basin is more than 5000 metres deep. It has good Petroleum potential due to thick (10 km) sedimentary deposits.

Oman Basin

It is a triangular Basin with depths between 2000 to 3500 metres. It is bordered in north by Makran continental Margins. Sedimentary thickness between 3 to 4 km are expected to lie here. The Basin has some petroleum potential.

Fracture Zones

The Indian Ocean is cut by several prominent fracture zones which offset the axis of Mid-Oceanic Ridge. Most important and well surveyed is the Owen Fracture Zone which lies at N.W. end of Calsberg Ridge. It runs from Somali Basin to Karachi. Several small right lateral fracture zones offset the Calsberg Ridge. These fractures destroy the continuity of Median Valley and its anomaly.

Sea Mounts

These are large isolated elevations characteristically of conical form. Nikitin Sea mount lies in Central Indian Basin and has depths of 1549 metres.

Ocean Deeps

There are few Ocean deeps in North Indian Ocean. Worth to mention is wheatby deep in Arabian Sea which has depths of 5464 metres. Few Ocean deeps are also seen in Andaman Sea.

GEONOMIC POTENTIALITY OF THE NORTH INDIAN OCEAN

The hidden physio-bio-chemic treasures of the Indian Ocean, reflect its height potentiality in the form of hydrocarbon spills (in sedimentary basins for petroleum) Iron manganese nodules (in deep sea basins) and calcium deposits. (Corals on volcanic islands of Maldives).

Petroleum Potential

For evaluating the potentiality from the view point of petroleum, the hydrocarbon deposits are to be investigated which depend upon the thickness of sedimentary formation, period of sedimentation with its source and the nature of the hypsomorphic unit.

After studying the above parameters of the North Indian Ocean it may be concluded that areas of probable hydrocarbon prospects lie in the Continental Margins, Cones and deep sea basins as detailed in the table.

Generally the thickness of the sediments tend to decrease gradually from continental margins towards the abyssal plains as depicted through isopachs in (Fig. 21.4). The highest isopach (5 km) lies in central part of Bay of Bengal. Similar trend is followed in the distribution of hydrocarbon deposits.

Mineral Potential

The organic debris drops into sea bed about 5000 metre below and provides food for sea floor animal colonies. This debris contains variety of trace metals such as cobalt, copper, nickel, zinc, barium which are in concentrated form. The peculiar formation, known as manganese nodules are of crucial importance in the further concentration of these elements. These nodules, occur at varying population, densities and morphology. It has lately become clear that colonies of animals live on the nodule surfaces and contribute to the growth of the nodules. These animals have a life style similar to that of corals because they use iron and manganese oxide in their armour. Manganese nodules have been investigated in the Arabian Basin and Central Indian Basin as given in Fig. 21.5. These nodules have rich mineral concentrates.

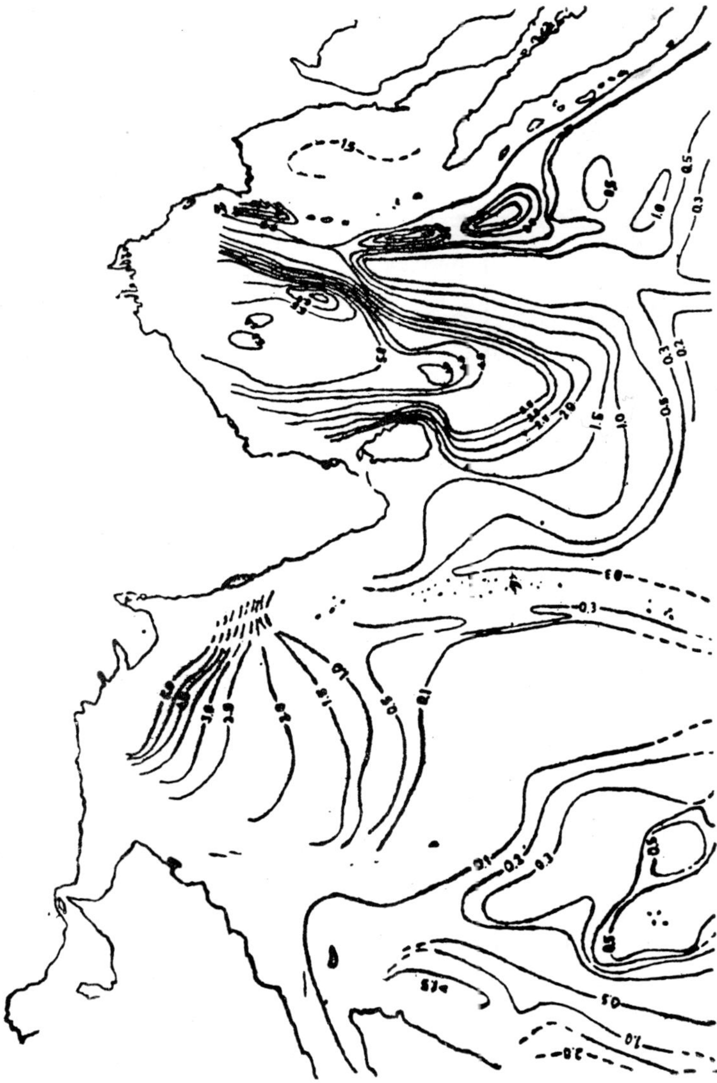

FIG. 21.4 : *Unconsolidated Sediments Thickness—North Indian Ocean, in kilometres*

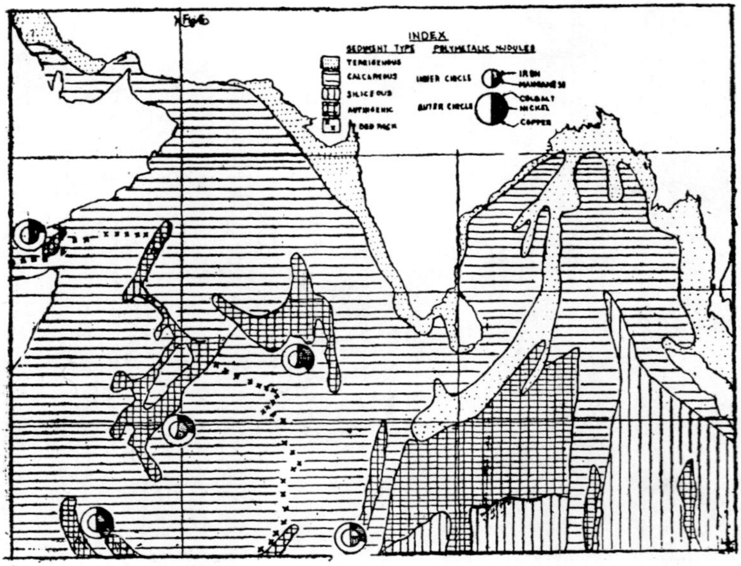

FIG. 21.5 : *Sediment type and Polymetallic Nodules*

Besides the above, the North Indian Ocean is endowed with biogenic calcareous deposits, Shell deposits, several types of fishes viz. Oil-Sardin, Mackerel, Shrimp. The annual catch varies from 25,00,000 tons to 35,00,000 tons.

CONCLUSION

Thus the North Indian Ocean is the most complex of all the World Oceans. Morphotectonically it is engraved with well defined marine configurations of Continental Margins, submarine canyons and basins which are rich in hydrocarbons and manganese nodules for producing petroleum and high grade minerals. Therefore, due to its geonomic potentiality and geostratic location, it has become an arena of power rivalry among super world powers who are desirous to take the lion's share out of future mineral products. The world powers are also desirous to settle the national jurisdiction of Exclusive Economic Zone (EEZ) according to their requirements so that one of them could play major role in international programme of conservation of Marine resources. Therefore, it is in India's interest to get actively involved in deliberations of this programme. It is in the light of these geonomic potentiality that India should try to declare Indian Ocean as a Peace Zone. The main aim of this paper is to acquaint the geoscientists, particularly the geographers, about the origin and structure of Indian Ocean, history of data acquisition, and present nature of data. The authors have also identified and classified the major hypsomorphic regions of North Indian Ocean. Some light has also been thrown upon the geonomic potentiality of this Ocean.

SUGGESTED READINGS

Bezrukov, P.L. and V.F. Kanaev, 1963 : Principal structural features of the bottom of N.W. part of Indian Ocean, *Dokl. Akad. Nauk* 153 (4) pp. 926-929.

Belousov, I.M., 1965 : Bathymetric chart of the northwestern part of Indian Ocean, *Oceanological Researches* 13, pp. 163-171.

Bergh, H.W., 1971 : Sea floor spreading in the southwest Indian Ocean, *J. Geophys. Res.* 76, pp. 6276-6282.

Booda, L., 1977 : The Oceans in our future-Futurist.

Curray, J.R. and Moore, 1971 : Growth of Bengal deep-sea fan etc. *Geol. Soc. Amer Bull.*, 82, pp. 563-572.

Deep Sea Drilling Project, 1973-74 : (DSDP) Reports.

Dietz, R.S. and Holden, J.C., 1970 : Re-construction of Pangea etc. *J. Geophys Res.* pp. 4939-4956.

Dietz, R.S., 1961 : Continent and Ocean basin evolution by sea floor spreading, *Nature*, 190, 854-857.

Fisher, R.L.; Sclater, J.G. and Mckenzie, D.P., 1971 : Evolution of central Indian ridge etc. *Geol. Soc. Amer. Bull.* 82, pp. 553-562.

Fairbridge, R.W., 1965 : The Indian Ocean and the status of Gondwana-land, *Prog. Oceanogro*, Vol. 3, pp. 83-135.

Gansser, A., 1966 : The Indian Ocean and the Himalayas, *Eclogae Geological Helvetiae*, 59, pp. 831-848.

Ghosh, A.M.N., 1959 : Possible oil bearing regions of India, 5th world petroleum congress New York.

Heezen, B.C. and M. Tharp, 1964 : Physiographic diagram of Indian Ocean, *Geol. Soc. Ame. Boulder.* Colo.

Laughton, A.S.; Mathews, D.H. and Fisher, R.L., 1971 : The structure of the Indian Ocean, *The sea* 4(2) Maxwell chap. 14.

McElhinney, M.W., 1970 : Formation of the Indian Ocean, *Nature*, 228, pp. 977-979.

Mckenzie, D. and J.G. Sclater, 1971 : The evolution of the Indian Ocean since late cretaceous, *Geophy. J. Roy. Astrot. Soc.* 25, pp. 437-528.

Neprochnov, Yu.P., 1964 : Structure and thickness of the sedimentary lay of Arabian sea, Bay of Bengal and Andaman Sea, *Trud. Inst. Okeanal* 64, pp. 214-226.

Ocean Education Project, 1974 : Manganese Nodules Mining.

Smith, A.G. and Hallam, A., 1970 : The fit of the Southern continents, *Nature*, Vol. 225, pp. 139-144.

Stewart, R.J., 1979 : Oil spills and offshore petroleum, *Technology Review*.

Udintsev, G.B., 1963 : Indian Ocean, bathymetric map, Moscow.

Wertenbaker, W., 1977 : Mining the wealth of Ocean deep, new York *Times Magazine*.

Whitemore, W.F., 1978 : Otec : Electricity from the Ocean-technology Review, 58.

Wiseman, J.D.H. and R.B. Seymour-Swell, 1937 : The Arabian floor of the sea, *Geol. Mag.* 74, pp. 219-230.

TWENTY TWO

Satya Prakash and *Jag Mohan Agrawal*

MORPHOMETRY OF NORTH-EAST ARABIAN SEA
Kori High to Bombay High

INTRODUCTION

THE NE part of Arabian Sea 'Kori High to Bombay High' has become an arena of power rivalry due to its expected petroleum potentials and has therefore aroused the curiosity of various national and international marine scientists to evaluate the origin, geology, structure, tectonics and geomorphology of this area which is now the scene of intense marine explorations and oil field developments. The earlier work on this area was carried out by foreign ships during the International Indian Ocean Expedition (1959-1965) which is contained in the Geological and Geophysical atlas of the Indian Ocean published by Academy of Sciences USSR. Hezene and Tharp prepared the physiographic diagram of Indian Ocean and described the geomorphic features of the area. A large amount of bathymetric data was collected by Indian Naval Ship *Darshak* during her *Oceanovax Cruise* 1973-74. The authors have conducted their studies based on this data in conjunction with the earlier data held in archives of Naval Hydrographic Department.

ORIGIN AND TECTONIC SYNTHESIS

The origin of the Arabian Sea can best be proved by the theory of continental drift presently known as plate tectonics theory. The geoscientists have been reassembling the jigsaw pieces and have been estimating the timing of drift, the direction of tectonic forces to assess the tectonic origin of the major physiographic structures. The initial break of Gondwanic India took place in the Early Cretaceous with separation of India from combined Australia and Antarctica in Late Cretaceous (75 M years ago). The spreading direction changed to NS and continued until Early Oligocene (35 M.Y. ago) when plate motion changed direction with India converging towards Asia in northeasterly direction.

The major tectonic and volcanic activity in the area in NW Arabian Sea was associated with the subsedence of the Lakshadweep Ridge in Palaeocene and erruption of Deccan Besalt from Palaeocene to Miocene when the continental margins of Western Coast of Indian Sub-continent started taking shape along with the faulting and subsedence of the Lakshadweep ridge. The terraces and benches on the slope area are of either stratigraphic origin, i.e., hard rock exposures or deposition of sediments in the faulted areas. The major features of continental shelf are of recent origin, i.e., during the Holocene (10 M.Y. ago) when sea level was lower. During this period oolites were deposited on the shelf in shallow waters, coral and algal reefs thrived on the outer shelf off Bombay and Kathiawar and in the Gulf off Kutch.

CLASSIFICATION OF MORPHOMETRIC FEATURES

The principal morphological features of the area have been indentified based on the critical analysis of the bathymetric data, visual inspection of the three dimensional model, Hypsographic Curves drawn along the sounding tracks and the profiles drawn by the automatic profile recorder of the computer (HP 9830) and was further supplemented by the new concept of "Submarine Terrain Evaluation' (Agrawal J.M., 1977). The morphological features of the area can be broadly classified into the following categories :

1. Continental shelf
2. Continental slope and terraces
3. Continental Rise
4. Other peculiar features like Seamounts, canyons (Fig 22.1)

The Continental Shelf

The width of the shelf is broader off Kori High and Bombay High (85 to 140 Sea miles) and is narrower off Kathiawar (45 Sea miles). The study of the profiles (Fig. 22.2) indicates that the shelf break occurs at depths ranging from 85 m to 160 m, though more commonly between 120 to 140 metres. The profiles off Kori High indicates shelf break at shallower depth which may be due to prograding and piling of the sediments derived from the rivers, while the profiles off Kutch indicate shelf break at deeper depths which may be due to currents. However, in general the nature of shelf break varies considerably, it is abrupt and clearly recognisable off Kutch while it is gradual and unrecognisable off Kathiawar.

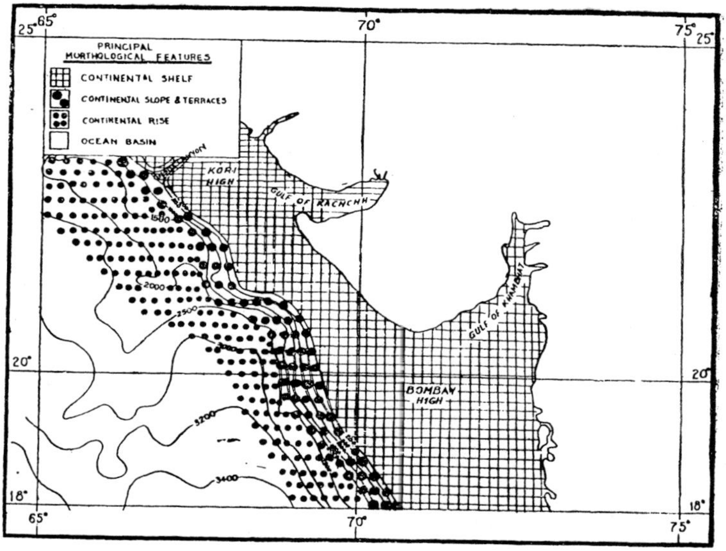

FIG. 22.1 : *Bathymetric Contoured Chart, North-East Arabian Sea Contours in Metres.*

An examination of the slopes (gradients) indicate that the area from Indus Canyon to the Gulf off Kutch has slopes of 1:8000 to 1:2,000 and it is covered by the thick sediments poured by the distribution of River Indus. The shelf of Kathiawar is narrow and the slopes are steeper in the north

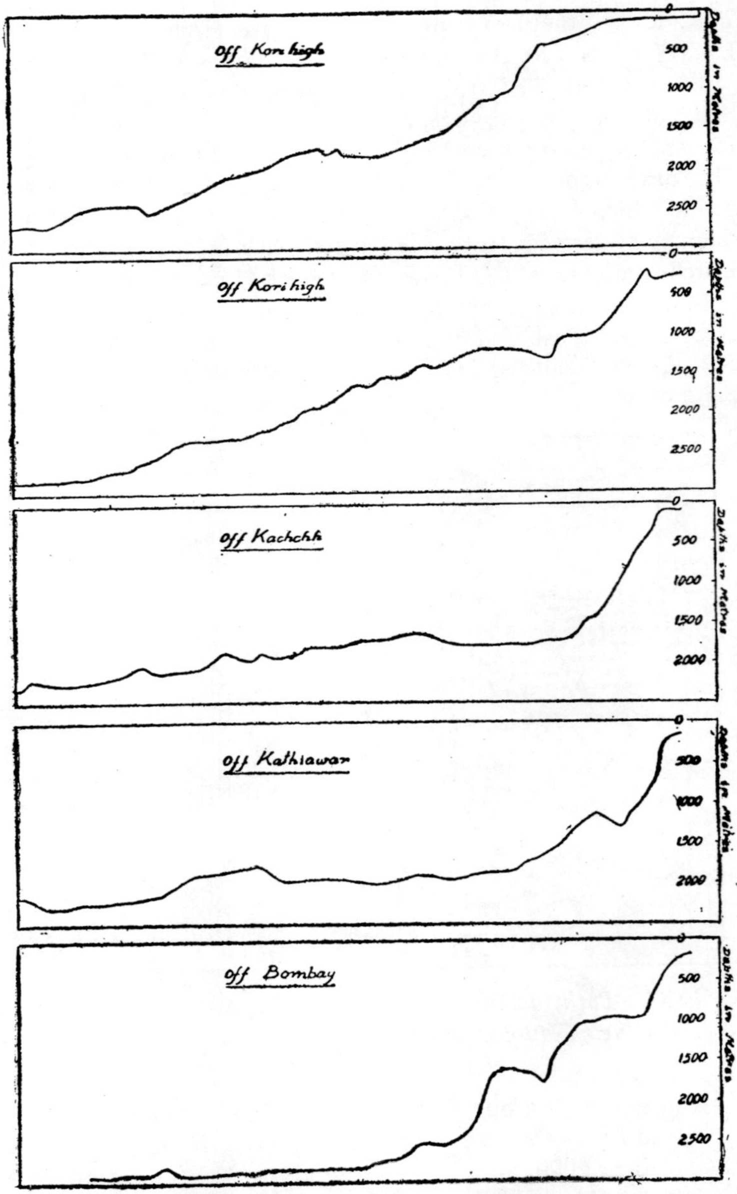

FIG. 22.2 : *Morphometric Profiles from Kori High to Bombay High.*

and gentlier (1:4000 to 1:8000) in the south. Shelf off Khambhat and Bombay is wide. At entrance to Khambhat which is covered by recent sediments the slopes are 1:2,000 but westwards the slopes are gentlier ranging from 1:8000 to 1:2000 where relict sediments are exposed.

The study of echograms throws light on the topography of the area. The areas of Gulf off Kutch and Khambhat, Lushington Shoal and the Bombay High have undulating topographic features where a number of scarps with the relative evaluation of 5 to 8 metres are seen—The undulations in the Gulfs may be related to the abrasion and deposition on the sea floor by the tide generated currents. However, the scarps off Kutch consists of coral and algal reefs while that of Kathiawar and Bombay High consists of oolitic limestones. The nearshore undulation off Bombay can be associated with Deccan trap exposures in the sea. Thus this topography appears to be the relicts of Holocene lowered sea level.

Continental Slope

The continental slope is defined to extend beyond the continental shelf break to the area where the gradient is less than 1:40. The edge of continental slope off Indus canyon lies at depth of 960 metres but southeastward it lies at greater depths and is observed at depth of 2900 metres off Bombay High which proves extensive deposits of sediments on the Kori High. The area covered by the continental slope is narrow off Indus and broadens gradually to the southeast, attaining maximum width off Bombay. The continental slope region is marked by benches, terraces, notches, hills, scarps and depressions, most remarkable of these are the hills and depressions ranging from less than 100 metres to 400 metres that lies south of Kathiawar. Remarkable terraces having width up to 10 km are seen in the area for southwest of Kathiawar. These terraces appear to be exposures of hard rocks and slopes and would have originated due to deposition of sediments in faulted zones (Fig. 22.3).

Continental Rise

The study of echograms and sounding profiles in the areas lying beyond the continental slope reveals that intensive

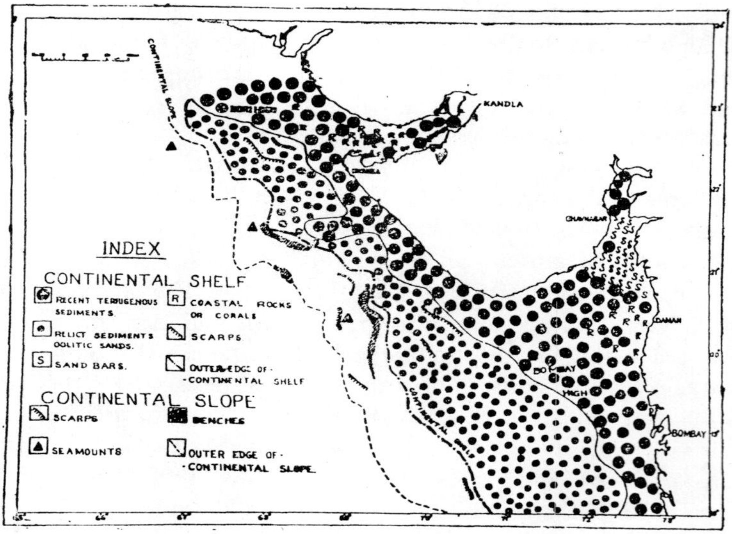

FIG. 22.3 : *Geomorphology of North-East Arabian Sea (Kori High to Bombay High).*

deposition of sediments has taken place in the offshore regions off the Indus cones and a number of depressions and rises having a relative difference over 100 metres, lies in this area which obviously represent the channels within the Indus cones. However, the presence of few abrupt elevations rising at places up to 400 metres, do support the theory that the Chagos-Lakshadweep aseismic ridge continues its northward trends and joins the Kori-Bombay High.

Seamounts

The interesting features of Seamounts have also been noticed during the evaluation of echograms and sounding profiles. The hill lying in position 20°30′N, 66°40′E off Indus cone has a depth of 400 metres, in the surrounding depths of 1500 metres (elevation 1100 metres) with a base of about 13 Sea miles having a summit of 5 sea miles. Another hill lying in position 21°30′N, 67°50′E has a least depth of 420 metres in the depths of 1900 metres (elevation 1480 metres) with a base of about 11 sea miles having a summit of 5 sea miles in extension. A similar submarine hill is seen in position 20°20′N, 69°00′E which has a least depth of 330 metres in depths of 1400 metres (elevation about 1070 metres) with a base of about 11 sea miles having a summit of 5 sea miles. A number of such lesser elevated hills, lie on the Indus cone which further supports the theory that the Chagos-Lakshadweep aseismic ridge extends further northward and merges with the western continental margins of India. These submarine hills mark a fracture zone and represent ancient volcanic activity. The most extensive volcanic activity on the adjacent peninsular India is found in the Deccan Trap period which took place from Early Palaeocene to Ecocene (65 M.Y to 34 M.Y.). Recent studies indicate that the Palaeocene volcanic activity on the Lakshadweep ridge may also be related to the Deccan traps.

CONCLUSION

The study of topography and geomorphology of the sea bed provides useful information about the geomorphic evolution of the area. These are in turn provide insight to the structure and geological history of the area. However, this needs to be substantiated by other evidences and surveys.

The continental slope marks the transition between continental and oceanic crusts and therefore its present features must have evolved over a relatively longer period of time then those of the shelf. The continental slope in this area can be broadly divided into two areas : an area of relatively simple topography i.e., the slope bordering the Gulf off Kutch and northern Kathiawar. Here presumably the topographic features have been marked by the subsequent deposition of sediments from the Indus cone and slumping from the shelf. This is evident from the profiles as also from the large pile of the sediments on the continental rise. On the other hand the continental slope off South Kathiwar has a complex topography marked by notches, benches, terraces, hills and depressions. The area apparently had a complex history and the primary features have not been covered by the Indus Cone sediments. The association of seamounts along with fractures indicates that both tectonic and volcanic activity in this area played a significant role in the evolution of the continental slope. The major tectonic and volcanic activity in the area was associated with the subsidence of the Lakshadweep Ridge in Palaeocene and erruption of Deccan basalts from Palaeocene to the Miocene. The continental margin probably started taking shape in this period along with the faulting and subsidence of the Lakshadweep ridge and the outpouring of the Deccan basalts. The terraces and benches on slopes are of either stratigraphic origin i.e., hard rock exposures or deposition of sediments in faulted areas.

The major features of the continental shelf are of comparatively recent origin. The uneven topography of the major part of the outer shelf particularly the wide shelf off Bombay was developed during the Holocene at about 10,000 years BP when the sea level was lower. During this period oolites were deposited on the shelf in shallow waters, coral and algae reefs thrived on the outer shelf off Bombay, Kathiawar and in the Gulf off Kutch. The undulating and rugged topography of the outer shelf represents the primary topographic features in carbonate sediments (oolite sands, oolitic limestones, coral ane algal reefs). With the rise in the sea level the coastline migrated to the east and largely terrigenous sediments were brought in by the numerous rivers drainage onto the western shelf. The distinction between the two different sedimentary environment is not only discernable in their sediment characteristics but also their topographic features i.e., the even topography of recent sediments of the inner shelf and the uneven topography of the relict sediments of the outer shelf. The transitional zone between the two sediments is marked by steep gradients. The terrigenous sediment supply throughout the period was insufficient to cover the wide continental shelf and possibly depositional environment. The primary unevenness of relict sediments in the transitional zone on the middle shelf is masked by a thin cover of recent sediments but is still seen in some echograms. In some of these areas particularly on the Kathiawar shelf and in the Gulf off Kutch because of the changes in drainage pattern and increased sediment supply, the coral reefs were probably swamped and killed by the turbid waters. The landward side of these coral reefs or oolitic limestone scarps formed the sediment dams behind which sediment was deposited. The Gulf off Kutch underwent a major change and the earlier transparent waters were replaced in the recent past by turbid waters and high tidal ranges generated swift currents causing an abrasion of the sea floor in the Gulf off Kutch. The continental shelf, therefore, represents two different stages in its development. The shelf surface was initially formed around 10,000 B.P. when the oolites were deposited and coral reefs thrived on the outer shelf. The inner shelf with the rise of the sea level was subsequently covered by the recent sediments, however, the present sediment supply is insufficient to cover the entire shelf and then the relict sediments and topographic features are still exposed on the outer shelf. The salient features of the above study may be pinpointed hereunder :

(a) The shelf break on the north western continental margin of India occurs between 85 to 160 metres but more commonly between 120-140 metres. The shelf may be divided into areas of even and undulating topography. The undulating topography on the outer shelf represents the relict oolitic sands and coral reefs of lowered sea levels while the one near the shore and in the Gulf off Kutch and Khambhat is associated with rock exposures, abrasion and deposition by tidal currents. Shoreward, the relict sediments are covered by the inner shelf clays and the topography is even. Some of the earlier coral reefs and oolitic limestones form scarps on the shelf and sediments have been deposited on their landward side.

(b) The edge of the continental slope occurs at 960 metres south of Indus Canyon and increases to 2900 m off Bombay indicating the extent and thickness of Indus sediments. The slope is marked by scarps, notches, benches, terraces, hills and depressions. The slope of Kathiawar has a complex topograph perhaps indicating its complex history and origin.

(c) A number of sea mounts occur on the continental slope and rise, some of these are marked by scarps on their landward side. These may indicate volcanoes along a fracture or continental fragments.

(d) It is inferred that the subsidence of the Chagos Lakshadweep ridge and erruption of Deccan Plateau basalts in Palaeocene marked the first stage in the shaping of the continental margin.

(e) The shelf surface was initially formed when oolites were deposited and coral reefs thrived and the outer shelf around 10,000 years B.P. The inner shelf was covered by terrigenous clays and with the rise of the sea level the shore line migrated to the east. The present terrigenous sediments supply is insufficient to cover the entire wide shelf and, therefore, on large areas on the outer shelf the relict sediments are exposed.

The information given in this paper has been collected from more than one hundred papers, charts, publications, etc. Only salient references are given below :

SUGGESTED READINGS

Agrawal, J.M. : A new concept-Submarine Terrain Evaluation.

Agrawal J.M. and Satya Prakash, (i) Cartographic Practices used in representation of submarine terrain, *National Cartographic Journals*. I.N.C.A

(ii) On the Cartographic Technique used in Oceanic Cartography, *National Cartographic Journal*. I.N.C.A.

Bezrukov P. L. and V. F. Kanaev, 1963 : Principal structural features of the bottom, of N.W. part of Indian Ocean. *Dokl. Akad Nauk* 153 (4) 926-929.

Belousoy, I.M., 1971 : Sea floor spreading in the south-west Indian Ocean, *J. Geophys. Res*. 76, pp. 6276-6282

Deep Sea Drilling Project 1973-74 : (DSDP) Reports.

Dietz, R.S. and Holden, J.C., 1970 : Re-construction of Pangea etc. *J. Geophys Res*. pp. 4939-4956.

Dietz, R.S., 1961 : Continent and Ocean basin evolution by sea floor spreading, *Nature*, 190, pp. 854-857.

Fairbridge, R.M., 1965 : The Indian Ocean and the status of Gondwana-land *Prog. Oceanogro*, Vol.3., pp. 83-136.

Gensser, A., 1966 : The Indian Ocean and the Himalayas, *Eclogae Geological Helvetiae*, 59, pp. 831-848.

Heezen, B.C. and M. Tharp, 1964 : Physiographic diagram of Indian Ocean Geol. Soc. *Ame. Boulder, Colo*.

Laughton, A.S., Mathews, D.H. and Fisher, R.L. 1971 : The structure of the Indian Ocean, *The see* 4 (2) Maxwell, chap 14.

McElhinney, M.W., 1970 : Formation of the Indian Ocean, *Nature* 228, pp. 977-979.

Mckenzie, D. and J.G. Sclater, 1971 : The evolution of the Indian Ocean since Late cretaceous, *Geophy. J. Roy. Asfrot Soc.* 25, pp. 437-528.

Naval Hydrographic Office Bathymetric/hydrographic Data and Charts held in archives of the Department.

Neprochnov, Yu. P., 1964 : Structure and thickness of the sedimentary layer of Arabian Sea, Bay of Bengal and Andaman Sea, *Trud. Inst. Okeanal* 64, pp. 214-226.

Satya Prakash, 1979 : *The evolution of the marine cartography of India* Pt. I, II and III-VIII, Conference on History of Cartography-Berlin

Udintsev, G.B., 1963 : Indian Ocean, Bathymetric Map, MOSCOW

Wiseman, J.D.H. and R.B Semour-Swell. The floor of Arabian Sea, *Geol. Mag*, 74, pp. 219-230.

TWENTY THREE

Ram Deo Gujar

SOME ASPECTS OF GEOMORPHOLOGY OF TONK DISTRICT (RAJASTHAN)

IN THE present paper some aspects of geomorphology of Tonk District have been studied. Geographically, Tonk District extends from 25° 42′ to 26° 33′ North latitude and from 75° 7′ to 76° 17′ East longitudes, with a maximum length and width of 94.25 and 116 kilometres from north to south and west to east respectively. The district under study sprawls over an area of about 7214.9 kilometres. This picturesque district appears roughly like an anvil with a top in the north and the base in the south. The district has a shape of the kite with its eastern and western sides bending somewhat inward and the southeastern portion protruding between Sawai Madhopur and Bundi districts to touch the northern tip of Kota district. (Rajasthan District Gazetteers, Tonk : *Gazetteer of India*, March 1970). The southwestern portion out also in Bhilwara district. The western and the northern boundaries have been formed by Ajmer and Jaipur districts respectively.

Climatically, the district falls within the semi-arid zone, and therefore has no marked distinctiveness from its adjacent districts. However, the amount of rainfall slightly increases towards the either sides of the district whereas the general trend in other districts as a whole is of decreasing rainfall from east and west.

Geomorphology is essentially a field science, hence the present study is based on author's field observations. Besides this, numerous morphometric techniques have been applied for the quantitative expression of the form of this study area.

Geology

Geological information met within this small area belong to the Archaean, the Purana and the Quarternary Super Groups. Most of the part is covered with alluvium of Recent and Sub-Recent periods (Fig. 23.1).

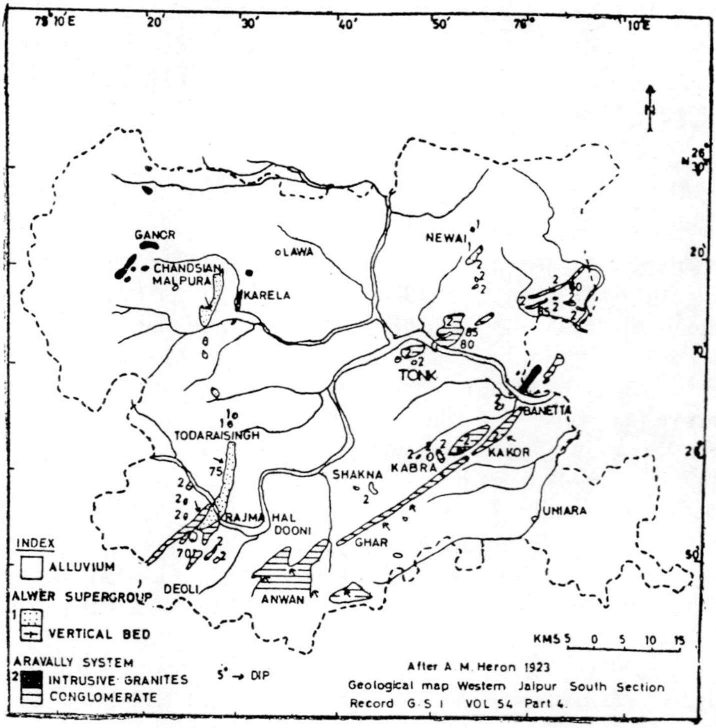

FIG. 23.1 : *Tonk District—Geology*

Physiography

In the study area there are generally three marked topographic units (Fig. 23.2).

The Hilly Area

Both the Aravalli and the Alwar rocks exhibit a hilly topography oscillating between 375 and 540 metres above mean sea level. These two arms come into conflict at the Rajmahal

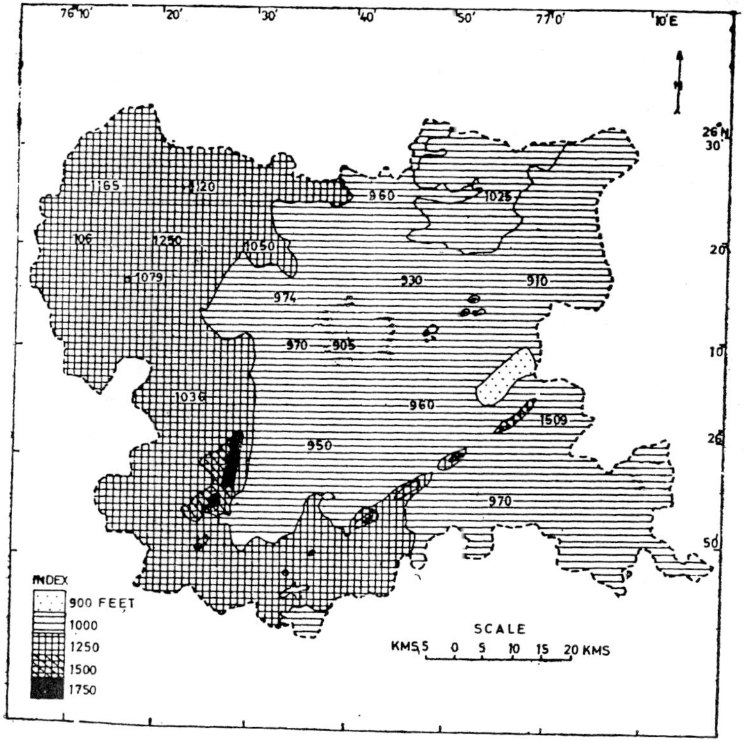

FIG. 23.2 : *Tonk District*—***Relief***

hill area and display the picturesque of two arms of a scissors. Besides these two marked hill continuations, the hilly pocket of Ganor and in the immediate south are worth mentioning. The northeastern region of the district is characterized by Tonk, Siris and Newai hills.

The Banas Plain

The another conspicuous feature in the relief is the alluvial plain of the Banas, generally below an elevation of 300 metres. The elevation decreases eastward and forms the level slope. Near Bachi, the plain represents the minimum elevation of about 262.5 metres above sea level.

The Piedmont Region

The third but no less impressive physiographic feature is represented by the area with an elevation between 300 and 375 metres, i.e., the transitional terrain between the plain and the hills. The alternation of the minor ups and downs in this transitional region catches the eyes of the observer. This unit encompasses about an area of 2936.5 square kilometres.

Drainage

Geomorphologically, drainage implies the stream course in relation to relief and structure. The Banas is only one major stream which intrudes into Tonk district near about four miles west of Deoli. In Tonk district the course of this river is rather complicated. Near the Rajmahal hills and farther downstream near the district headquarters (Tonk) meanders are the prominent features (Fig. 23.3).

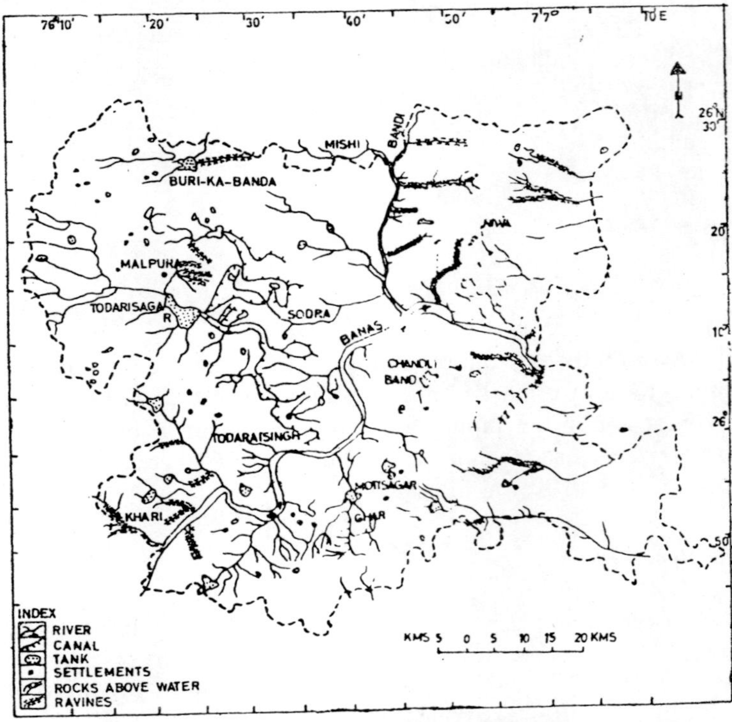

FIG. 23.3 : *Tonk District—Drainage*

Antecedent Stream Action

Dissection of the Banas near the Rajmahal hills is very interesting and significant from geomorphological point of view. Here the Banas at Rajmahal wandering through the extensive alluvial plain breaches isolated flat hills of Delhi quartzites in a gorge. One can infer from this very feature that Banas course was in existence before the Aravalli were uplifted out of the peneplain during the Cretaceous period and cut down vertically as the plain arched (Heron, 1938). Chatterjee and Bagchi (1957) have also studied this area and are inclined to believe that the Banas is a good example of the antecedent stream action at this point. Also the overwhelming incised meander further gives a clue to the understanding of antecedent drainage pattern (Fig. 23.3).

Inland Drainage

The district has also the characteristics of the inland drainage pattern. Conspicuously typical examples can be illustrated from Bilaspur area near Rajmahal and northwestern zone of Malpura. These streams are rather seasonal and lost themselves in the inland reservoirs.

Stream Frequency

Stream frequency is a helpful index to know the degree of dissection of any region. Here a careful attempt has been made to demarcate the district principally into three stream frequency regions :

I. The zone of moderately high frequency consisting of the Rajmahal, Sardarpura and Anwan, represents the hilly nature of the terrain which checks the infiltration of water and the slope boosts the flow of raining water.

II. The zone of medium frequency emerges from around Malpura, then tapers southward and afterwards girdles the Rajmahal-Anwan. The marked continuation can be traced up to Newai where it branches out into northwestern and northeastern arms. Thus the zone belongs to the transitional character of the land.

III. The rest of the area is characterized by the low frequency. The sandy areas along the Banas and its tributaries in Newai, Uniara, Deoli and Tonk tehsil are characterized by low frequency.

Table 23.1 pertaining to the stream frequency regions gives a percentage of the area confined to the respective frequency zones and their upper limits of stream frequency.

TABLE 23.1 : **The Stream Frequency Regions**

Zone	Frequency	Percentage of the area confined to the Total District area
The zone of low frequency	0 - 8	60
The zone of medium Frequency	9 - 16	37
The zone of moderately high frequency	17 - 20	3

Relative Relief

The Relative relief in this study area ranges from 0 to 24° metres and somewhere exceeds even 240 metres (Fig 23.4). A comprehensive study of the relative relief in Tonk District focuses on the recognition of three relative regions.

I. The region of maximum relative relief demarcated by the value exceeding 240 metres embraces the pockets around Malpura, Todaraising and the belt from Ghar to Kakor.

II. The region of minimum relative relief, with an upper limit of 30 metres encompasses the district bordering areas and the Banas plain.

III. The region of intermediate relative relief consists of the area between the hills and the Banas plain.

Analysis of Hypsometric Curve and Profiles

The relative percentage of area under elevations of varying orders has been obtained through the hypsometric curve of

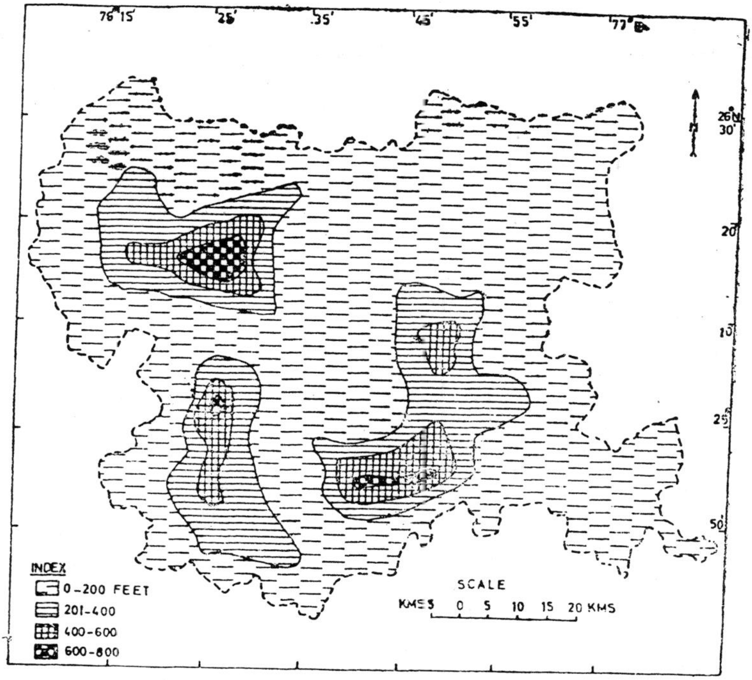

FIG. 23.4 : *Tonk District—Relative Relief*

the study area. It is quite apparent that the area between the elevation of 255-270 metres accounts for 60 per cent of the total district area. Approximately 37 per cent area consists of the land ranging from 300 to 375 metres whereas the remaining 3 per cent area is well above 375 metres.

The series of profiles have been drawn along the 26° 2′, 26° 9′, 26° 17′ and 26° 19′ 30″ north latitudes. The superimposed and serial profiles brings out the birds' eye view of the terrain (Fig. 23.5). The two surfaces are self evident.

The slope map of the area has been prepared following Wentworth's method. The slope swings between 0° to 7° 30′. In Tonk district three slope regions may be recognised (Fig. 23.6 and Table 23.2).

I. The moderate slope is restricted to the hilly areas of the Rajmahal, Rajkot, Sitarampura and Gothara, etc.

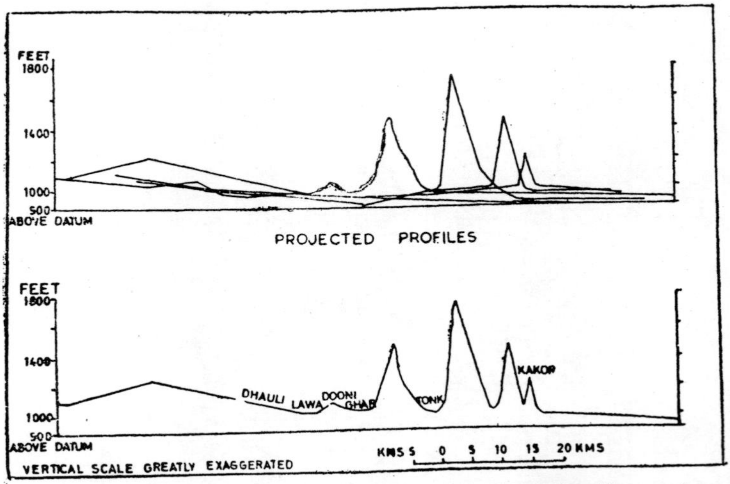

FIG. 23.5 : *Tonk District—Superimposed Profiles and Projected Profiles.*

II. The gentle slope runs along (1) the Anwan and (2) Gothara, Glan, Tonk and Newai hilly fragments of the Aravalli super groups.

III. The level slope below 2° pervades the region more than half of Malpura tehsil, south and east of Uniara, the Banas plain and more or less whole of Newai tehsil except a minor pocket of the gentle slope.

TABLE 23.2 : **Slope Regions**

Nomenclature	Code	Degree	Percentage of the confined area
Level slope	Ls	0°—2°	52
Gentle slope	Sg	2°—5°	45
Moderate slope	SM	5°—10°	3

Altimetric Analysis

Altimetric frequency denotes the frequent occurrence of summits. The general picture exhibited by the elevation between 285 and 330 metres accounts for the 50 per cent of the

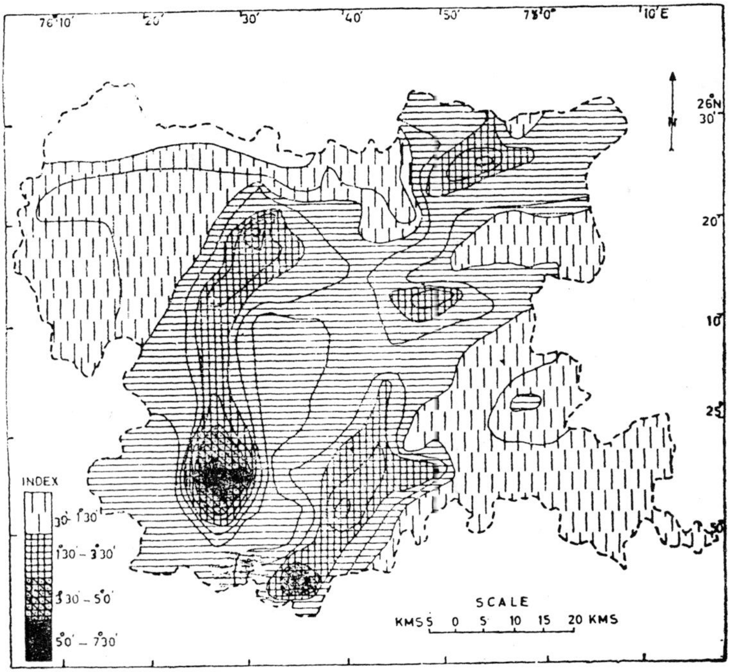

FIG. 23.6 : *Tonk District—Slope*

total district area followed by the elevations from 240 to 285 metres and 330 and 375 metres. The analysis of the altimetric frequency histogram indicates the presence of two erosion surfaces at the elevation of 375 metres and at below 330 metres. The elevation of hills coincides with the upper flat surface, hence these flat topped hills represent an erosion surface of earlier period. Heron (1938) has also considered this area as a part of Tertiary peneplain. So the present study upholds Heron's view quantitatively.

CONCLUSIONS

The above study of various geomorphic features of the area under investigation brings out the following facts :

1. Geologically the study area can be said to be a part of the stable Indian shield which has not been affected by any major organic movements since the Cretaceous period.
2. Being a part of the semi-arid climatic region the stream courses are mainly dry during the summers. The Banas is the only perennial stream.
3. The structural and geomorphic features of the Rajmahal area bring out the fact that the Banas is an antecedent stream whose course was established prior to the upliftment of the Aravalli in the Cretaceous period. The development of the valley at this part took place consequent with the arching of the Cretaceous peneplain.
4. The flat surfaces of the isolated hills and their quantitative expression through altimetric frequency histogram, give a clue of the presence of erosion surfaces.

SUGGESTED READINGS

Dury, G.R., 1963: *Face of the earth*, Pelican Books, p. 39.

Gupta, N.L., 1976: *Morphometric analysis of the terrain : A case study of Udaipur basin*, Udaipur District, Rajasthan.

Heron, A.M., 1923: The Geology of Western Jaipur—Southern Section, *Record of the Geological Survey of India*, Part IV, Vol. 54.

Heron, A.M. 1938: The physiography of Rajputana, *Proceedings of 25th Indian Science Congress*, Vol. 25., pp. 119-132.

Horton, R.E., Erosion developments of streams and their drainage basins, hydrological approach to quantitative morphology, *Bulletin Geographical Society of America*, Vol. 56, pp. 275-378.

Kesseli, I.F. : A neglected field of geomorphology, *Annals of the Association of American Geographers*, Vol. 39, p. 10.

Rao, G. Padmaja, 1976: Geomorphology of the Mej Basin, unpublished Ph.D. thesis, University of Rajasthan, Jaipur.

Robinson, G.A. : Consideration of the relations of geomorphology and geography, *Professional Geographer*, Vol. 12, No. 2, pp. 5-15.

Sharma, H.S., 1972, Erosion surfaces in the Lower Chambal Valley, *International Geography*, Canada, Vol. I, pp. 61-69.

Sharma, H.S., 1976 : Morphology of the ravines of the Morel basin, *International Geography*, Moscow, Vol. I, pp. 219-223.

Smith, G.H., 1933 : The relative relief of Ohio, *Geographical Review*, New York, Vol. 25, pp. 275-284.

Smith, G.H., 1938 : The Morphometry of landscape-Analysis, *Annals of the Association of American Geographers*, Vol. 28, pp. 63-64.

Wentworth, C.K., 1930: A simplified method of determining the average slope of land surfaces, *American Journal of Science*, Series 5, Vol. 20 New Haven, Conn.

INDEX